DATE

Nissan Altima Automotive Repair Manual

by Jeff Kibler
and John H Haynes
Member of the Guild of Motoring Writers

Models covered:
All Nissan Altima models
1993 through 2001

(10D5 - 72015)

ABCDE
FGHIJ
K

Haynes Publishing Group
Sparkford Nr Yeovil
Somerset BA22 7JJ England

Haynes North America, Inc
861 Lawrence Drive
Newbury Park
California 91320 USA

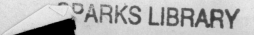

Acknowledgements

Technical writers who contributed to this project include Rob Maddox, Mike Stubblefield and Jay Storer.

© **Haynes North America, Inc. 1997, 2001**

With permission from J.H. Haynes & Co. Ltd.

A book in the Haynes Automotive Repair Manual Series

Printed in the U.S.A.

ISBN 1 56392 449 8

Library of Congress Control Number: 2001095960

While every attempt is made to ensure that the information in this manual is correct, no liability can be accepted by the authors or publishers for loss, damage or injury caused by any errors in, or omissions from, the information given.

Contents

Haynes photographer, mechanic and author with 1993 Nissan Altima

About this manual

Its purpose

The purpose of this manual is to help you get the best value from your vehicle. It can do so in several ways. It can help you decide what work must be done, even if you choose to have it done by a dealer service department or a repair shop; it provides information and procedures for routine maintenance and servicing; and it offers diagnostic and repair procedures to follow when trouble occurs.

We hope you use the manual to tackle the work yourself. For many simpler jobs, doing it yourself may be quicker than arranging an appointment to get the vehicle into a shop and making the trips to leave it and pick it up. More importantly, a lot of money can be saved by avoiding the expense the shop must pass on to you to cover its labor and overhead costs. An added benefit is the sense of satisfaction and accomplishment that you feel after doing the job yourself.

Using the manual

The manual is divided into Chapters. Each Chapter is divided into numbered Sections, which are headed in bold type between horizontal lines. Each Section consists of consecutively numbered paragraphs.

At the beginning of each numbered Section you will be referred to any illustrations which apply to the procedures in that Section. The reference numbers used in illustration captions pinpoint the pertinent Section and the Step within that Section. That is, illustration 3.2 means the illustration refers to Section 3 and Step (or paragraph) 2 within that Section.

Procedures, once described in the text, are not normally repeated. When it's necessary to refer to another Chapter, the reference will be given as Chapter and Section number. Cross references given without use of the word "Chapter" apply to Sections and/or paragraphs in the same Chapter. For example, "see Section 8" means in the same Chapter.

References to the left or right side of the vehicle assume you are sitting in the driver's seat, facing forward.

Even though we have prepared this manual with extreme care, neither the publisher nor the author can accept responsibility for any errors in, or omissions from, the information given.

NOTE

A **Note** provides information necessary to properly complete a procedure or information which will make the procedure easier to understand.

CAUTION

A **Caution** provides a special procedure or special steps which must be taken while completing the procedure where the Caution is found. Not heeding a Caution can result in damage to the assembly being worked on.

WARNING

A **Warning** provides a special procedure or special steps which must be taken while completing the procedure where the Warning is found. Not heeding a Warning can result in personal injury.

Introduction to the Nissan Altima

The Nissan Altima is a four-door sedan body style available in four models, the base model Altima XE, the Altima GXE, the Altima SE sports sedan and the fully equipped Altima GLE.

The transversely mounted 2.4-liter DOHC four-cylinder engine used in these models are equipped with a sequential multiport electronic fuel injection system.

The engine transmits power to the front wheels through either a five-speed manual transaxle or an electronically controlled four-speed automatic transaxle via independent driveaxles.

The Altima features an all steel uni-body and independent suspension with MacPherson strut/coil spring suspension and a stabilizer bar used on both the front and rear suspensions. The rack-and-pinion steering unit is mounted behind the engine with power-assist as standard equipment.

All models are equipped with power assisted front disc and rear drum brakes. A power assisted four wheel disc brake system is optional on SE models. Anti-lock Brake Systems (ABS) are optional on all models.

Vehicle identification numbers

The Vehicle Identification Number (VIN) is visible through the driver's side of the windshield

The Chassis Identification Number is stamped on the engine compartment firewall

Modifications are a continuing and unpublicized process in vehicle manufacturing. Since spare parts manuals and lists are compiled on a numerical basis, the individual vehicle numbers are essential to correctly identify the component required.

Vehicle Identification Number (VIN)

This very important identification number is located on a plate attached to the dashboard inside the windshield on the driver's side of the vehicle **(see illustration)**. The VIN also appears on the Vehicle Certificate of Title and Registration. It contains information such as where and when the vehicle was manufactured, the model year and the body style.

VIN engine and model year codes

Two particularly important pieces of information found in the VIN are the engine code and the model year code. Counting from the left, the engine code letter designation is the 4th digit and the model year code designation is the 10th digit.

On the models covered by this manual the engine codes are:

B........... A24DE 2.4L DOHC

On the models covered by this manual the model year codes are:

P........... 1993
R........... 1994
S........... 1995
T........... 1996
V........... 1997

Chassis Identification Number

The chassis identification number is stamped on the firewall in the engine compartment **(see illustration)**. Like the VIN it contains valuable information about the manufacturing of the vehicle such as the destination, model variations and transaxle information.

Manufacturer's Certification Regulation label

The manufacturer's Certification Regulation label is attached to the driver's side door end or post **(see illustration)**. The plate contains the name of the manufacturer, the month and year of production, the Gross

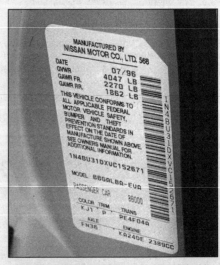

The Manufacturer's Certification label is affixed to the drivers side door end or post

The Engine Identification Number is stamped on the front side of the engine block, near the transaxle

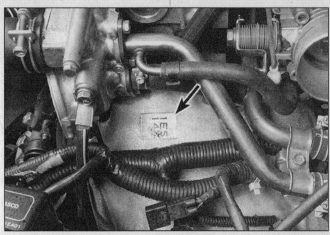

The manual transaxle ID number is located on top of the transaxle bellhousing

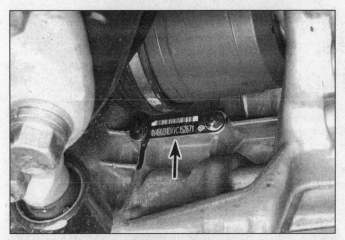

The automatic transaxle ID number (arrow) is stamped on a plate located by the inner end of the left driveaxle

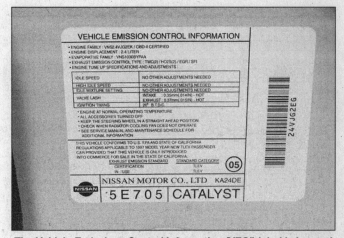

The Vehicle Emissions Control Information (VECI) label is located on the bottom side of the hood

Vehicle Weight Rating (GVWR), the Gross Axle Weight Rating (GAWR) and the certification statement.

Engine identification numbers

The engine code number can be found on a pad on the front (radiator) side of the cylinder block, near the transaxle (see illustration)

Transaxle identification numbers

The manual transaxle identification number is stamped on top of the bellhousing (see illustration). The automatic transaxle identification number is stamped on the rear side of the transaxle (see illustration).

Vehicle Emissions Control Information (VECI) label

The emissions control information label is found under the hood, normally on the radiator support or the bottom side of the hood. This label contains information on the emissions control equipment installed on the vehicle, as well as tune-up specifications (see illustration).

Buying parts

Replacement parts are available from many sources, which generally fall into one of two categories - authorized dealer parts departments and independent retail auto parts stores. Our advice concerning these parts is as follows:

Retail auto parts stores: Good auto parts stores will stock frequently needed components which wear out relatively fast, such as clutch components, exhaust systems, brake parts, tune-up parts, etc. These stores often supply new or reconditioned parts on an exchange basis, which can save a considerable amount of money. Discount auto parts stores are often very good places to buy materials and parts needed for general vehicle maintenance such as oil, grease, filters, spark plugs, belts, touch-up paint, bulbs, etc. They also usually sell tools and general accessories, have convenient hours, charge lower prices and can often be found not far from home.

Authorized dealer parts department: This is the best source for parts which are unique to the vehicle and not generally available elsewhere (such as major engine parts, transmission parts, trim pieces, etc.).

Warranty information: If the vehicle is still covered under warranty, be sure that any replacement parts purchased - regardless of the source - do not invalidate the warranty!

To be sure of obtaining the correct parts, have engine and chassis numbers available and, if possible, take the old parts along for positive identification.

Maintenance techniques, tools and working facilities

Maintenance techniques

There are a number of techniques involved in maintenance and repair that will be referred to throughout this manual. Application of these techniques will enable the home mechanic to be more efficient, better organized and capable of performing the various tasks properly, which will ensure that the repair job is thorough and complete.

Fasteners

Fasteners are nuts, bolts, studs and screws used to hold two or more parts together. There are a few things to keep in mind when working with fasteners. Almost all of them use a locking device of some type, either a lockwasher, locknut, locking tab or thread adhesive. All threaded fasteners should be clean and straight, with undamaged threads and undamaged corners on the hex head where the wrench fits. Develop the habit of replacing all damaged nuts and bolts with new ones. Special locknuts with nylon or fiber inserts can only be used once. If they are removed, they lose their locking ability and must be replaced with new ones.

Rusted nuts and bolts should be treated with a penetrating fluid to ease removal and prevent breakage. Some mechanics use turpentine in a spout-type oil can, which works quite well. After applying the rust penetrant, let it work for a few minutes before trying to loosen the nut or bolt. Badly rusted fasteners may have to be chiseled or sawed off or removed with a special nut breaker, available at tool stores.

If a bolt or stud breaks off in an assembly, it can be drilled and removed with a special tool commonly available for this purpose. Most automotive machine shops can perform this task, as well as other repair procedures, such as the repair of threaded holes that have been stripped out.

Flat washers and lockwashers, when removed from an assembly, should always be replaced exactly as removed. Replace any damaged washers with new ones. Never use a lockwasher on any soft metal surface (such as aluminum), thin sheet metal or plastic.

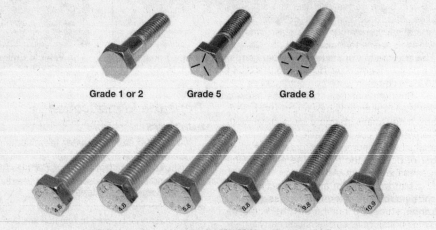

Grade 1 or 2   Grade 5   Grade 8

Bolt strength marking (standard/SAE/USS; bottom - metric)

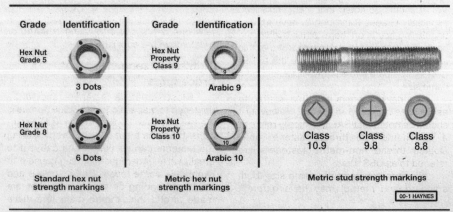

Grade	Identification	Grade	Identification
Hex Nut Grade 5	3 Dots	Hex Nut Property Class 9	Arabic 9
Hex Nut Grade 8	6 Dots	Hex Nut Property Class 10	Arabic 10

Standard hex nut strength markings

Metric hex nut strength markings

Class 10.9   Class 9.8   Class 8.8

Metric stud strength markings

00-1 HAYNES

Fastener sizes

For a number of reasons, automobile manufacturers are making wider and wider use of metric fasteners. Therefore, it is important to be able to tell the difference between standard (sometimes called U.S. or SAE) and metric hardware, since they cannot be interchanged.

All bolts, whether standard or metric, are sized according to diameter, thread pitch and length. For example, a standard 1/2 - 13 x 1 bolt is 1/2 inch in diameter, has 13 threads per inch and is 1 inch long. An M12 - 1.75 x 25 metric bolt is 12 mm in diameter, has a thread pitch of 1.75 mm (the distance between threads) and is 25 mm long. The two bolts are nearly identical, and easily confused, but they are not interchangeable.

In addition to the differences in diameter, thread pitch and length, metric and standard bolts can also be distinguished by examining the bolt heads. To begin with, the distance across the flats on a standard bolt head is measured in inches, while the same dimension on a metric bolt is sized in millimeters (the same is true for nuts). As a result, a standard wrench should not be used on a metric bolt and a metric wrench should not be used on a standard bolt. Also, most standard bolts have slashes radiating out from the center of the head to denote the grade or strength of the bolt, which is an indication of the amount of torque that can be applied to it. The greater the number of slashes, the greater the strength of the bolt. Grades 0 through 5 are commonly used on automobiles. Metric bolts have a property class (grade) number, rather than a slash, molded into their heads to indicate bolt strength. In this case, the higher the number, the stronger the bolt. Property class numbers 8.8, 9.8 and 10.9 are commonly used on automobiles.

Strength markings can also be used to distinguish standard hex nuts from metric hex nuts. Many standard nuts have dots stamped into one side, while metric nuts are marked with a number. The greater the number of dots, or the higher the number, the greater the strength of the nut.

Metric studs are also marked on their ends according to property class (grade). Larger studs are numbered (the same as metric bolts), while smaller studs carry a geometric code to denote grade.

It should be noted that many fasteners, especially Grades 0 through 2, have no distinguishing marks on them. When such is the case, the only way to determine whether it is standard or metric is to measure the thread pitch or compare it to a known fastener of the same size.

Standard fasteners are often referred to as SAE, as opposed to metric. However, it should be noted that SAE technically refers to a non-metric fine thread fastener only. Coarse thread non-metric fasteners are referred to as USS sizes.

Since fasteners of the same size (both standard and metric) may have different

Metric thread sizes	Ft-lbs	Nm
M-6	6 to 9	9 to 12
M-8	14 to 21	19 to 28
M-10	28 to 40	38 to 54
M-12	50 to 71	68 to 96
M-14	80 to 140	109 to 154

Pipe thread sizes		
1/8	5 to 8	7 to 10
1/4	12 to 18	17 to 24
3/8	22 to 33	30 to 44
1/2	25 to 35	34 to 47

U.S. thread sizes		
1/4 - 20	6 to 9	9 to 12
5/16 - 18	12 to 18	17 to 24
5/16 - 24	14 to 20	19 to 27
3/8 - 16	22 to 32	30 to 43
3/8 - 24	27 to 38	37 to 51
7/16 - 14	40 to 55	55 to 74
7/16 - 20	40 to 60	55 to 81
1/2 - 13	55 to 80	75 to 108

Standard (SAE and USS) bolt dimensions/grade marks

- G   Grade marks (bolt strength)
- L   Length (in inches)
- T   Thread pitch (number of threads per inch)
- D   Nominal diameter (in inches)

Metric bolt dimensions/grade marks

- P   Property class (bolt strength)
- L   Length (in millimeters)
- T   Thread pitch (distance between threads in millimeters)
- D   Diameter

strength ratings, be sure to reinstall any bolts, studs or nuts removed from your vehicle in their original locations. Also, when replacing a fastener with a new one, make sure that the new one has a strength rating equal to or greater than the original.

Tightening sequences and procedures

Most threaded fasteners should be tightened to a specific torque value (torque is the twisting force applied to a threaded component such as a nut or bolt). Overtightening the fastener can weaken it and cause it to break, while undertightening can cause it to eventually come loose. Bolts, screws and studs, depending on the material they are made of and their thread diameters, have

specific torque values, many of which are noted in the Specifications at the beginning of each Chapter. Be sure to follow the torque recommendations closely. For fasteners not assigned a specific torque, a general torque value chart is presented here as a guide. These torque values are for dry (unlubricated) fasteners threaded into steel or cast iron (not aluminum). As was previously mentioned, the size and grade of a fastener determine the amount of torque that can safely be applied to it. The figures listed here are approximate for Grade 2 and Grade 3 fasteners. Higher grades can tolerate higher torque values.

Fasteners laid out in a pattern, such as cylinder head bolts, oil pan bolts, differential cover bolts, etc., must be loosened or tightened in sequence to avoid warping the com-

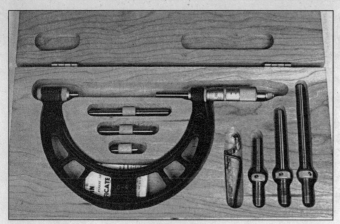

Micrometer set

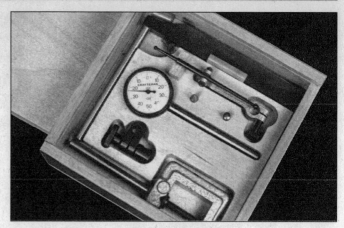

Dial indicator set

ponent. This sequence will normally be shown in the appropriate Chapter. If a specific pattern is not given, the following procedures can be used to prevent warping.

Initially, the bolts or nuts should be assembled finger-tight only. Next, they should be tightened one full turn each, in a criss-cross or diagonal pattern. After each one has been tightened one full turn, return to the first one and tighten them all one-half turn, following the same pattern. Finally, tighten each of them one-quarter turn at a time until each fastener has been tightened to the proper torque. To loosen and remove the fasteners, the procedure would be reversed.

Component disassembly

Component disassembly should be done with care and purpose to help ensure that the parts go back together properly. Always keep track of the sequence in which parts are removed. Make note of special characteristics or marks on parts that can be installed more than one way, such as a grooved thrust washer on a shaft. It is a good idea to lay the disassembled parts out on a clean surface in the order that they were removed. It may also be helpful to make sketches or take instant photos of components before removal.

When removing fasteners from a component, keep track of their locations. Sometimes threading a bolt back in a part, or putting the washers and nut back on a stud, can prevent mix-ups later. If nuts and bolts cannot be returned to their original locations, they should be kept in a compartmented box or a series of small boxes. A cupcake or muffin tin is ideal for this purpose, since each cavity can hold the bolts and nuts from a particular area (i.e. oil pan bolts, valve cover bolts, engine mount bolts, etc.). A pan of this type is especially helpful when working on assemblies with very small parts, such as the carburetor, alternator, valve train or interior dash and trim pieces. The cavities can be marked with paint or tape to identify the contents.

Whenever wiring looms, harnesses or connectors are separated, it is a good idea to identify the two halves with numbered pieces of masking tape so they can be easily reconnected.

Gasket sealing surfaces

Throughout any vehicle, gaskets are used to seal the mating surfaces between two parts and keep lubricants, fluids, vacuum or pressure contained in an assembly.

Many times these gaskets are coated with a liquid or paste-type gasket sealing compound before assembly. Age, heat and pressure can sometimes cause the two parts to stick together so tightly that they are very difficult to separate. Often, the assembly can be loosened by striking it with a soft-face hammer near the mating surfaces. A regular hammer can be used if a block of wood is placed between the hammer and the part. Do not hammer on cast parts or parts that could be easily damaged. With any particularly stubborn part, always recheck to make sure that every fastener has been removed.

Avoid using a screwdriver or bar to pry apart an assembly, as they can easily mar the gasket sealing surfaces of the parts, which must remain smooth. If prying is absolutely necessary, use an old broom handle, but keep in mind that extra clean up will be necessary if the wood splinters.

After the parts are separated, the old gasket must be carefully scraped off and the gasket surfaces cleaned. Stubborn gasket material can be soaked with rust penetrant or treated with a special chemical to soften it so it can be easily scraped off. A scraper can be fashioned from a piece of copper tubing by flattening and sharpening one end. Copper is recommended because it is usually softer than the surfaces to be scraped, which reduces the chance of gouging the part. Some gaskets can be removed with a wire brush, but regardless of the method used, the mating surfaces must be left clean and smooth. If for some reason the gasket surface is gouged, then a gasket sealer thick enough to fill scratches will have to be used during reassembly of the components. For most applications, a non-drying (or semi-drying) gasket sealer should be used.

Hose removal tips

Warning: *If the vehicle is equipped with air conditioning, do not disconnect any of the A/C hoses without first having the system depressurized by a dealer service department or a service station.*

Hose removal precautions closely parallel gasket removal precautions. Avoid scratching or gouging the surface that the hose mates against or the connection may leak. This is especially true for radiator hoses. Because of various chemical reactions, the rubber in hoses can bond itself to the metal spigot that the hose fits over. To remove a hose, first loosen the hose clamps that secure it to the spigot. Then, with slip-joint pliers, grab the hose at the clamp and rotate it around the spigot. Work it back and forth until it is completely free, then pull it off. Silicone or other lubricants will ease removal if they can be applied between the hose and the outside of the spigot. Apply the same lubricant to the inside of the hose and the outside of the spigot to simplify installation.

As a last resort (and if the hose is to be replaced with a new one anyway), the rubber can be slit with a knife and the hose peeled from the spigot. If this must be done, be careful that the metal connection is not damaged.

If a hose clamp is broken or damaged, do not reuse it. Wire-type clamps usually weaken with age, so it is a good idea to replace them with screw-type clamps whenever a hose is removed.

Tools

A selection of good tools is a basic requirement for anyone who plans to maintain and repair his or her own vehicle. For the owner who has few tools, the initial investment might seem high, but when compared to the spiraling costs of professional auto maintenance and repair, it is a wise one.

To help the owner decide which tools are needed to perform the tasks detailed in this manual, the following tool lists are offered: *Maintenance and minor repair, Repair/overhaul* and *Special.*

The newcomer to practical mechanics

Dial caliper

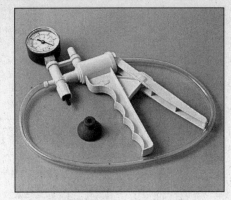

Hand-operated vacuum pump

Timing light

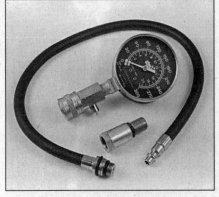

Compression gauge with spark plug
hole adapter

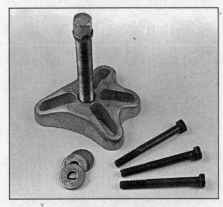

Damper/steering wheel puller

General purpose puller

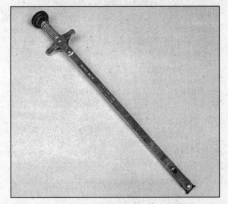

Hydraulic lifter removal tool

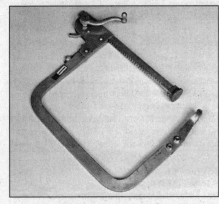

Valve spring compressor

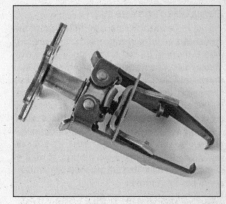

Valve spring compressor

Ridge reamer

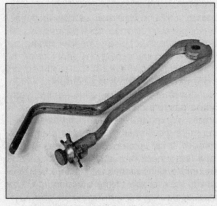

Piston ring groove cleaning tool

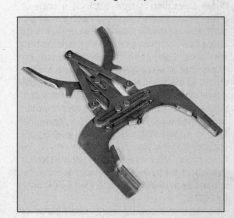

Ring removal/installation tool

Ring compressor

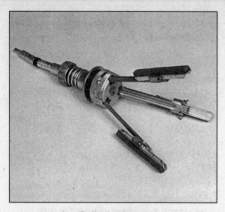

Cylinder hone

Brake hold-down spring tool

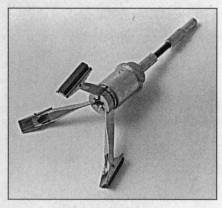

Brake cylinder hone

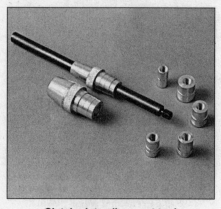

Clutch plate alignment tool

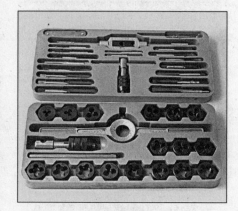

Tap and die set

should start off with the *maintenance and minor repair* tool kit, which is adequate for the simpler jobs performed on a vehicle. Then, as confidence and experience grow, the owner can tackle more difficult tasks, buying additional tools as they are needed. Eventually the basic kit will be expanded into the *repair and overhaul* tool set. Over a period of time, the experienced do-it-yourselfer will assemble a tool set complete enough for most repair and overhaul procedures and will add tools from the special category when it is felt that the expense is justified by the frequency of use.

Maintenance and minor repair tool kit

The tools in this list should be considered the minimum required for performance of routine maintenance, servicing and minor repair work. We recommend the purchase of combination wrenches (box-end and open-end combined in one wrench). While more expensive than open end wrenches, they offer the advantages of both types of wrench.

Combination wrench set (1/4-inch to 1 inch or 6 mm to 19 mm)
Adjustable wrench, 8 inch
Spark plug wrench with rubber insert
Spark plug gap adjusting tool
Feeler gauge set
Brake bleeder wrench
Standard screwdriver (5/16-inch x 6 inch)

Phillips screwdriver (No. 2 x 6 inch)
Combination pliers - 6 inch
Hacksaw and assortment of blades
Tire pressure gauge
Grease gun
Oil can
Fine emery cloth
Wire brush
Battery post and cable cleaning tool
Oil filter wrench
Funnel (medium size)
Safety goggles
Jackstands (2)
Drain pan

Note: *If basic tune-ups are going to be part of routine maintenance, it will be necessary to purchase a good quality stroboscopic timing light and combination tachometer/dwell meter. Although they are included in the list of special tools, it is mentioned here because they are absolutely necessary for tuning most vehicles properly.*

Repair and overhaul tool set

These tools are essential for anyone who plans to perform major repairs and are in addition to those in the maintenance and minor repair tool kit. Included is a comprehensive set of sockets which, though expensive, are invaluable because of their versatility, especially when various extensions and drives are available. We recommend the 1/2-inch drive over the 3/8-inch drive. Although the larger drive is bulky and more expensive,

it has the capacity of accepting a very wide range of large sockets. Ideally, however, the mechanic should have a 3/8-inch drive set and a 1/2-inch drive set.

Socket set(s)
Reversible ratchet
Extension - 10 inch
Universal joint
Torque wrench (same size drive as sockets)
Ball peen hammer - 8 ounce
Soft-face hammer (plastic/rubber)
Standard screwdriver (1/4-inch x 6 inch)
Standard screwdriver (stubby - 5/16-inch)
Phillips screwdriver (No. 3 x 8 inch)
Phillips screwdriver (stubby - No. 2)
Pliers - vise grip
Pliers - lineman's
Pliers - needle nose
Pliers - snap-ring (internal and external)
Cold chisel - 1/2-inch
Scribe
Scraper (made from flattened copper tubing)
Centerpunch
Pin punches (1/16, 1/8, 3/16-inch)
Steel rule/straightedge - 12 inch
Allen wrench set (1/8 to 3/8-inch or 4 mm to 10 mm)
A selection of files
Wire brush (large)
Jackstands (second set)
Jack (scissor or hydraulic type)

Note: *Another tool which is often useful is an electric drill with a chuck capacity of 3/8-inch and a set of good quality drill bits.*

Special tools

The tools in this list include those which are not used regularly, are expensive to buy, or which need to be used in accordance with their manufacturer's instructions. Unless these tools will be used frequently, it is not very economical to purchase many of them. A consideration would be to split the cost and use between yourself and a friend or friends. In addition, most of these tools can be obtained from a tool rental shop on a temporary basis.

This list primarily contains only those tools and instruments widely available to the public, and not those special tools produced by the vehicle manufacturer for distribution to dealer service departments. Occasionally, references to the manufacturer's special tools are included in the text of this manual. Generally, an alternative method of doing the job without the special tool is offered. However, sometimes there is no alternative to their use. Where this is the case, and the tool cannot be purchased or borrowed, the work should be turned over to the dealer service department or an automotive repair shop.

> *Valve spring compressor*
> *Piston ring groove cleaning tool*
> *Piston ring compressor*
> *Piston ring installation tool*
> *Cylinder compression gauge*
> *Cylinder ridge reamer*
> *Cylinder surfacing hone*
> *Cylinder bore gauge*
> *Micrometers and/or dial calipers*
> *Hydraulic lifter removal tool*
> *Balljoint separator*
> *Universal-type puller*
> *Impact screwdriver*
> *Dial indicator set*
> *Stroboscopic timing light (inductive pick-up)*
> *Hand operated vacuum/pressure pump*
> *Tachometer/dwell meter*
> *Universal electrical multimeter*
> *Cable hoist*
> *Brake spring removal and installation tools*
> *Floor jack*

Buying tools

For the do-it-yourselfer who is just starting to get involved in vehicle maintenance and repair, there are a number of options available when purchasing tools. If maintenance and minor repair is the extent of the work to be done, the purchase of individual tools is satisfactory. If, on the other hand, extensive work is planned, it would be a good idea to purchase a modest tool set from one of the large retail chain stores. A set can usually be bought at a substantial savings over the individual tool prices, and they often come with a tool box. As additional tools are needed, add-on sets, individual tools and a larger tool box can be purchased to expand the tool selection. Building a tool set gradually allows the cost of the tools to be spread over a longer period of time and gives the mechanic the freedom to choose only those tools that will actually be used.

Tool stores will often be the only source of some of the special tools that are needed, but regardless of where tools are bought, try to avoid cheap ones, especially when buying screwdrivers and sockets, because they won't last very long. The expense involved in replacing cheap tools will eventually be greater than the initial cost of quality tools.

Care and maintenance of tools

Good tools are expensive, so it makes sense to treat them with respect. Keep them clean and in usable condition and store them properly when not in use. Always wipe off any dirt, grease or metal chips before putting them away. Never leave tools lying around in the work area. Upon completion of a job, always check closely under the hood for tools that may have been left there so they won't get lost during a test drive.

Some tools, such as screwdrivers, pliers, wrenches and sockets, can be hung on a panel mounted on the garage or workshop wall, while others should be kept in a tool box or tray. Measuring instruments, gauges, meters, etc. must be carefully stored where they cannot be damaged by weather or impact from other tools.

When tools are used with care and stored properly, they will last a very long time. Even with the best of care, though, tools will wear out if used frequently. When a tool is damaged or worn out, replace it. Subsequent jobs will be safer and more enjoyable if you do.

How to repair damaged threads

Sometimes, the internal threads of a nut or bolt hole can become stripped, usually from overtightening. Stripping threads is an all-too-common occurrence, especially when working with aluminum parts, because aluminum is so soft that it easily strips out.

Usually, external or internal threads are only partially stripped. After they've been cleaned up with a tap or die, they'll still work. Sometimes, however, threads are badly damaged. When this happens, you've got three choices:

1) *Drill and tap the hole to the next suitable oversize and install a larger diameter bolt, screw or stud.*
2) *Drill and tap the hole to accept a threaded plug, then drill and tap the plug to the original screw size. You can also buy a plug already threaded to the original size. Then you simply drill a hole to the specified size, then run the threaded plug into the hole with a bolt and jam nut. Once the plug is fully seated, remove the jam nut and bolt.*
3) *The third method uses a patented thread repair kit like Heli-Coil or Slimsert. These easy-to-use kits are designed to repair damaged threads in straight-through holes and blind holes. Both are available as kits which can handle a variety of sizes and thread patterns. Drill the hole, then tap it with the special included tap. Install the Heli-Coil and the hole is back to its original diameter and thread pitch.*

Regardless of which method you use, be sure to proceed calmly and carefully. A little impatience or carelessness during one of these relatively simple procedures can ruin your whole day's work and cost you a bundle if you wreck an expensive part.

Working facilities

Not to be overlooked when discussing tools is the workshop. If anything more than routine maintenance is to be carried out, some sort of suitable work area is essential.

It is understood, and appreciated, that many home mechanics do not have a good workshop or garage available, and end up removing an engine or doing major repairs outside. It is recommended, however, that the overhaul or repair be completed under the cover of a roof.

A clean, flat workbench or table of comfortable working height is an absolute necessity. The workbench should be equipped with a vise that has a jaw opening of at least four inches.

As mentioned previously, some clean, dry storage space is also required for tools, as well as the lubricants, fluids, cleaning solvents, etc. which soon become necessary.

Sometimes waste oil and fluids, drained from the engine or cooling system during normal maintenance or repairs, present a disposal problem. To avoid pouring them on the ground or into a sewage system, pour the used fluids into large containers, seal them with caps and take them to an authorized disposal site or recycling center. Plastic jugs, such as old antifreeze containers, are ideal for this purpose.

Always keep a supply of old newspapers and clean rags available. Old towels are excellent for mopping up spills. Many mechanics use rolls of paper towels for most work because they are readily available and disposable. To help keep the area under the vehicle clean, a large cardboard box can be cut open and flattened to protect the garage or shop floor.

Whenever working over a painted surface, such as when leaning over a fender to service something under the hood, always cover it with an old blanket or bedspread to protect the finish. Vinyl covered pads, made especially for this purpose, are available at auto parts stores.

Booster battery (jump) starting

Observe these precautions when using a booster battery to start a vehicle:

a) *Before connecting the booster battery, make sure the ignition switch is in the Off position.*
b) *Turn off the lights, heater and other electrical loads.*
c) *Your eyes should be shielded. Safety goggles are a good idea.*
d) *Make sure the booster battery is the same voltage as the dead one in the vehicle.*
e) *The two vehicles MUST NOT TOUCH each other!*
f) *Make sure the transaxle is in Neutral (manual) or Park (automatic).*
g) *If the booster battery is not a maintenance-free type, remove the vent caps and lay a cloth over the vent holes.*

Connect the red jumper cable to the positive (+) terminals of each battery **(see illustration)**.

Connect one end of the black jumper cable to the negative (-) terminal of the booster battery. The other end of this cable should be connected to a good ground on the vehicle to be started, such as a bolt or bracket on the body.

Start the engine using the booster battery, then, with the engine running at idle speed, disconnect the jumper cables in the reverse order of connection.

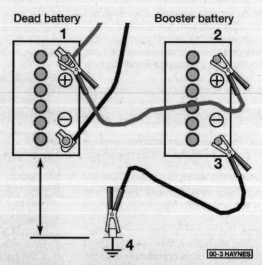

Make the booster battery cable connections in the numerical order shown (note that the negative cable of the booster battery is NOT attached to the negative terminal of the dead battery)

Jacking and towing

Jacking

Warning: *The jack supplied with the vehicle should only be used for changing a tire or placing jackstands under the frame. Never work under the vehicle or start the engine while this jack is being used as the only means of support.*

The vehicle should be on level ground. Place the shift lever in Park, if you have an automatic, or Reverse if you have a manual transaxle. Block the wheel diagonally opposite the wheel being changed. Set the parking brake.

Remove the spare tire and jack from stowage. Remove the wheel cover and trim ring (if so equipped) with the tapered end of the lug nut wrench by inserting and twisting the handle and then prying against the back of the wheel cover. Loosen the wheel lug nuts about 1/4-to-1/2 turn each.

Place the scissors-type jack under the side of the vehicle and adjust the jack height until it fits in the notch in the vertical rocker panel flange nearest the wheel to be changed. There is a front and rear jacking point on each side of the vehicle **(see illustration)**.

Turn the jack handle clockwise until the tire clears the ground. Remove the lug nuts and pull the wheel off. Replace it with the spare.

Install the lug nuts with the beveled edges facing in. Tighten them snugly. Don't attempt to tighten them completely until the vehicle is lowered or it could slip off the jack. Turn the jack handle counterclockwise to lower the vehicle. Remove the jack and tighten the lug nuts in a diagonal pattern.

Install the cover (and trim ring, if used) and be sure it's snapped into place all the way around.

Stow the tire, jack and wrench. Unblock the wheels.

Towing

As a general rule, the vehicle should be towed with the front (drive) wheels off the ground. If they can't be raised, place them on a dolly. The ignition key must be in the ACC position, since the steering lock mechanism isn't strong enough to hold the front wheels straight while towing.

Vehicles equipped with an automatic transaxle can be towed from the front only with all four wheels on the ground, provided that speeds don't exceed 30 mph and the distance is not over 40 miles. Before towing, check the transmission fluid level (see Chapter 1). If the level is below the HOT line on the dipstick, add fluid or use a towing dolly.

Caution: *Never tow a vehicle with an automatic transaxle from the rear with the front wheels on the ground.*

When towing a vehicle equipped with a manual transaxle with all four wheels on the ground, be sure to place the shift lever in neutral and release the parking brake.

Equipment specifically designed for towing should be used. It should be attached to the main structural members of the vehicle, not the bumpers or brackets.

Safety is a major consideration when towing and all applicable state and local laws must be obeyed. A safety chain system must be used at all times.

The jack fits over the rocker panel flange (there are two jacking points on each side of the vehicle, indicated by a notch in the rocker panel flange)

Automotive chemicals and lubricants

A number of automotive chemicals and lubricants are available for use during vehicle maintenance and repair. They include a wide variety of products ranging from cleaning solvents and degreasers to lubricants and protective sprays for rubber, plastic and vinyl.

Cleaners

Carburetor cleaner and choke cleaner is a strong solvent for gum, varnish and carbon. Most carburetor cleaners leave a dry-type lubricant film which will not harden or gum up. Because of this film it is not recommended for use on electrical components.

Brake system cleaner is used to remove grease and brake fluid from the brake system, where clean surfaces are absolutely necessary. It leaves no residue and often eliminates brake squeal caused by contaminants.

Electrical cleaner removes oxidation, corrosion and carbon deposits from electrical contacts, restoring full current flow. It can also be used to clean spark plugs, carburetor jets, voltage regulators and other parts where an oil-free surface is desired.

Demoisturants remove water and moisture from electrical components such as alternators, voltage regulators, electrical connectors and fuse blocks. They are non-conductive, non-corrosive and non-flammable.

Degreasers are heavy-duty solvents used to remove grease from the outside of the engine and from chassis components. They can be sprayed or brushed on and, depending on the type, are rinsed off either with water or solvent.

Lubricants

Motor oil is the lubricant formulated for use in engines. It normally contains a wide variety of additives to prevent corrosion and reduce foaming and wear. Motor oil comes in various weights (viscosity ratings) from 0 to 50. The recommended weight of the oil depends on the season, temperature and the demands on the engine. Light oil is used in cold climates and under light load conditions. Heavy oil is used in hot climates and where high loads are encountered. Multi-viscosity oils are designed to have characteristics of both light and heavy oils and are available in a number of weights from 5W-20 to 20W-50.

Gear oil is designed to be used in differentials, manual transmissions and other areas where high-temperature lubrication is required.

Chassis and wheel bearing grease is a heavy grease used where increased loads and friction are encountered, such as for wheel bearings, balljoints, tie-rod ends and universal joints.

High-temperature wheel bearing grease is designed to withstand the extreme temperatures encountered by wheel bearings in disc brake equipped vehicles. It usually contains molybdenum disulfide (moly), which is a dry-type lubricant.

White grease is a heavy grease for metal-to-metal applications where water is a problem. White grease stays soft under both low and high temperatures (usually from -100 to +190-degrees F), and will not wash off or dilute in the presence of water.

Assembly lube is a special extreme pressure lubricant, usually containing moly, used to lubricate high-load parts (such as main and rod bearings and cam lobes) for initial start-up of a new engine. The assembly lube lubricates the parts without being squeezed out or washed away until the engine oiling system begins to function.

Silicone lubricants are used to protect rubber, plastic, vinyl and nylon parts.

Graphite lubricants are used where oils cannot be used due to contamination problems, such as in locks. The dry graphite will lubricate metal parts while remaining uncontaminated by dirt, water, oil or acids. It is electrically conductive and will not foul electrical contacts in locks such as the ignition switch.

Moly penetrants loosen and lubricate frozen, rusted and corroded fasteners and prevent future rusting or freezing.

Heat-sink grease is a special electrically non-conductive grease that is used for mounting electronic ignition modules where it is essential that heat is transferred away from the module.

Sealants

RTV sealant is one of the most widely used gasket compounds. Made from silicone, RTV is air curing, it seals, bonds, waterproofs, fills surface irregularities, remains flexible, doesn't shrink, is relatively easy to remove, and is used as a supplementary sealer with almost all low and medium temperature gaskets.

Anaerobic sealant is much like RTV in that it can be used either to seal gaskets or to form gaskets by itself. It remains flexible, is solvent resistant and fills surface imperfections. The difference between an anaerobic sealant and an RTV-type sealant is in the curing. RTV cures when exposed to air, while an anaerobic sealant cures only in the absence of air. This means that an anaerobic sealant cures only after the assembly of parts, sealing them together.

Thread and pipe sealant is used for sealing hydraulic and pneumatic fittings and vacuum lines. It is usually made from a Teflon compound, and comes in a spray, a paint-on liquid and as a wrap-around tape.

Chemicals

Anti-seize compound prevents seizing, galling, cold welding, rust and corrosion in fasteners. High-temperature anti-seize, usually made with copper and graphite lubricants, is used for exhaust system and exhaust manifold bolts.

Anaerobic locking compounds are used to keep fasteners from vibrating or working loose and cure only after installation, in the absence of air. Medium strength locking compound is used for small nuts, bolts and screws that may be removed later. High-strength locking compound is for large nuts, bolts and studs which aren't removed on a regular basis.

Oil additives range from viscosity index improvers to chemical treatments that claim to reduce internal engine friction. It should be noted that most oil manufacturers caution against using additives with their oils.

Gas additives perform several functions, depending on their chemical makeup. They usually contain solvents that help dissolve gum and varnish that build up on carburetor, fuel injection and intake parts. They also serve to break down carbon deposits that form on the inside surfaces of the combustion chambers. Some additives contain upper cylinder lubricants for valves and piston rings, and others contain chemicals to remove condensation from the gas tank.

Miscellaneous

Brake fluid is specially formulated hydraulic fluid that can withstand the heat and pressure encountered in brake systems. Care must be taken so this fluid does not come in contact with painted surfaces or plastics. An opened container should always be resealed to prevent contamination by water or dirt.

Weatherstrip adhesive is used to bond weatherstripping around doors, windows and trunk lids. It is sometimes used to attach trim pieces.

Undercoating is a petroleum-based, tar-like substance that is designed to protect metal surfaces on the underside of the vehicle from corrosion. It also acts as a sound-deadening agent by insulating the bottom of the vehicle.

Waxes and polishes are used to help protect painted and plated surfaces from the weather. Different types of paint may require the use of different types of wax and polish. Some polishes utilize a chemical or abrasive cleaner to help remove the top layer of oxidized (dull) paint on older vehicles. In recent years many non-wax polishes that contain a wide variety of chemicals such as polymers and silicones have been introduced. These non-wax polishes are usually easier to apply and last longer than conventional waxes and polishes.

Conversion factors

Length (distance)

Inches (in)	X	25.4	= Millimetres (mm)	X 0.0394	= Inches (in)
Feet (ft)	X	0.305	= Metres (m)	X 3.281	= Feet (ft)
Miles	X	1.609	= Kilometres (km)	X 0.621	= Miles

Volume (capacity)

Cubic inches (cu in; in^3)	X	16.387	= Cubic centimetres (cc; cm^3)	X 0.061	= Cubic inches (cu in; in^3)
Imperial pints (Imp pt)	X	0.568	= Litres (l)	X 1.76	= Imperial pints (Imp pt)
Imperial quarts (Imp qt)	X	1.137	= Litres (l)	X 0.88	= Imperial quarts (Imp qt)
Imperial quarts (Imp qt)	X	1.201	= US quarts (US qt)	X 0.833	= Imperial quarts (Imp qt)
US quarts (US qt)	X	0.946	= Litres (l)	X 1.057	= US quarts (US qt)
Imperial gallons (Imp gal)	X	4.546	= Litres (l)	X 0.22	= Imperial gallons (Imp gal)
Imperial gallons (Imp gal)	X	1.201	= US gallons (US gal)	X 0.833	= Imperial gallons (Imp gal)
US gallons (US gal)	X	3.785	= Litres (l)	X 0.264	= US gallons (US gal)

Mass (weight)

Ounces (oz)	X	28.35	= Grams (g)	X 0.035	= Ounces (oz)
Pounds (lb)	X	0.454	= Kilograms (kg)	X 2.205	= Pounds (lb)

Force

Ounces-force (ozf; oz)	X	0.278	= Newtons (N)	X 3.6	= Ounces-force (ozf; oz)
Pounds-force (lbf; lb)	X	4.448	= Newtons (N)	X 0.225	= Pounds-force (lbf; lb)
Newtons (N)	X	0.1	= Kilograms-force (kgf; kg)	X 9.81	= Newtons (N)

Pressure

Pounds-force per square inch (psi; lbf/in^2; lb/in^2)	X	0.070	= Kilograms-force per square centimetre (kgf/cm^2; kg/cm^2)	X 14.223	= Pounds-force per square inch (psi; lbf/in^2; lb/in^2)
Pounds-force per square inch (psi; lbf/in^2; lb/in^2)	X	0.068	= Atmospheres (atm)	X 14.696	= Pounds-force per square inch (psi; lbf/in^2; lb/in^2)
Pounds-force per square inch (psi; lbf/in^2; lb/in^2)	X	0.069	= Bars	X 14.5	= Pounds-force per square inch (psi; lbf/in^2; lb/in^2)
Pounds-force per square inch (psi; lbf/in^2; lb/in^2)	X	6.895	= Kilopascals (kPa)	X 0.145	= Pounds-force per square inch (psi; lbf/in^2; lb/in^2)
Kilopascals (kPa)	X	0.01	= Kilograms-force per square centimetre (kgf/cm^2; kg/cm^2)	X 98.1	= Kilopascals (kPa)

Torque (moment of force)

Pounds-force inches (lbf in; lb in)	X	1.152	= Kilograms-force centimetre (kgf cm; kg cm)	X 0.868	= Pounds-force inches (lbf in; lb in)
Pounds-force inches (lbf in; lb in)	X	0.113	= Newton metres (Nm)	X 8.85	= Pounds-force inches (lbf in; lb in)
Pounds-force inches (lbf in; lb in)	X	0.083	= Pounds-force feet (lbf ft; lb ft)	X 12	= Pounds-force inches (lbf in; lb in)
Pounds-force feet (lbf ft; lb ft)	X	0.138	= Kilograms-force metres (kgf m; kg m)	X 7.233	= Pounds-force feet (lbf ft; lb ft)
Pounds-force feet (lbf ft; lb ft)	X	1.356	= Newton metres (Nm)	X 0.738	= Pounds-force feet (lbf ft; lb ft)
Newton metres (Nm)	X	0.102	= Kilograms-force metres (kgf m; kg m)	X 9.804	= Newton metres (Nm)

Vacuum

Inches mercury (in. Hg)	X	3.377	= Kilopascals (kPa)	X 0.2961	= Inches mercury
Inches mercury (in. Hg)	X	25.4	= Millimeters mercury (mm Hg)	X 0.0394	= Inches mercury

Power

Horsepower (hp)	X	745.7	= Watts (W)	X 0.0013	= Horsepower (hp)

Velocity (speed)

Miles per hour (miles/hr; mph)	X	1.609	= Kilometres per hour (km/hr; kph)	X 0.621	= Miles per hour (miles/hr; mph)

Fuel consumption*

Miles per gallon, Imperial (mpg)	X	0.354	= Kilometres per litre (km/l)	X 2.825	= Miles per gallon, Imperial (mpg)
Miles per gallon, US (mpg)	X	0.425	= Kilometres per litre (km/l)	X 2.352	= Miles per gallon, US (mpg)

Temperature

Degrees Fahrenheit = (°C x 1.8) + 32

Degrees Celsius (Degrees Centigrade; °C) = (°F - 32) x 0.56

*It is common practice to convert from miles per gallon (mpg) to litres/100 kilometres (l/100km), where mpg (Imperial) x l/100 km = 282 and mpg (US) x l/100 km = 235

Safety first!

Regardless of how enthusiastic you may be about getting on with the job at hand, take the time to ensure that your safety is not jeopardized. A moment's lack of attention can result in an accident, as can failure to observe certain simple safety precautions. The possibility of an accident will always exist, and the following points should not be considered a comprehensive list of all dangers. Rather, they are intended to make you aware of the risks and to encourage a safety conscious approach to all work you carry out on your vehicle.

Essential DOs and DON'Ts

DON'T rely on a jack when working under the vehicle. Always use approved jackstands to support the weight of the vehicle and place them under the recommended lift or support points.

DON'T attempt to loosen extremely tight fasteners (i.e. wheel lug nuts) while the vehicle is on a jack - it may fall.

DON'T start the engine without first making sure that the transmission is in Neutral (or Park where applicable) and the parking brake is set.

DON'T remove the radiator cap from a hot cooling system - let it cool or cover it with a cloth and release the pressure gradually.

DON'T attempt to drain the engine oil until you are sure it has cooled to the point that it will not burn you.

DON'T touch any part of the engine or exhaust system until it has cooled sufficiently to avoid burns.

DON'T siphon toxic liquids such as gasoline, antifreeze and brake fluid by mouth, or allow them to remain on your skin.

DON'T inhale brake lining dust - it is potentially hazardous (see *Asbestos* below).

DON'T allow spilled oil or grease to remain on the floor - wipe it up before someone slips on it.

DON'T use loose fitting wrenches or other tools which may slip and cause injury.

DON'T push on wrenches when loosening or tightening nuts or bolts. Always try to pull the wrench toward you. If the situation calls for pushing the wrench away, push with an open hand to avoid scraped knuckles if the wrench should slip.

DON'T attempt to lift a heavy component alone - get someone to help you.

DON'T rush or take unsafe shortcuts to finish a job.

DON'T allow children or animals in or around the vehicle while you are working on it.

DO wear eye protection when using power tools such as a drill, sander, bench grinder, etc. and when working under a vehicle.

DO keep loose clothing and long hair well out of the way of moving parts.

DO make sure that any hoist used has a safe working load rating adequate for the job.

DO get someone to check on you periodically when working alone on a vehicle.

DO carry out work in a logical sequence and make sure that everything is correctly assembled and tightened.

DO keep chemicals and fluids tightly capped and out of the reach of children and pets.

DO remember that your vehicle's safety affects that of yourself and others. If in doubt on any point, get professional advice.

Asbestos

Certain friction, insulating, sealing, and other products - such as brake linings, brake bands, clutch linings, torque converters, gaskets, etc. - may contain asbestos. Extreme care must be taken to avoid inhalation of dust from such products, since it is hazardous to health. If in doubt, assume that they do contain asbestos.

Fire

Remember at all times that gasoline is highly flammable. Never smoke or have any kind of open flame around when working on a vehicle. But the risk does not end there. A spark caused by an electrical short circuit, by two metal surfaces contacting each other, or even by static electricity built up in your body under certain conditions, can ignite gasoline vapors, which in a confined space are highly explosive. Do not, under any circumstances, use gasoline for cleaning parts. Use an approved safety solvent.

Always disconnect the battery ground (-) cable at the battery before working on any part of the fuel system or electrical system. Never risk spilling fuel on a hot engine or exhaust component. It is strongly recommended that a fire extinguisher suitable for use on fuel and electrical fires be kept handy in the garage or workshop at all times. Never try to extinguish a fuel or electrical fire with water.

Fumes

Certain fumes are highly toxic and can quickly cause unconsciousness and even death if inhaled to any extent. Gasoline vapor falls into this category, as do the vapors from some cleaning solvents. Any draining or pouring of such volatile fluids should be done in a well ventilated area.

When using cleaning fluids and solvents, read the instructions on the container carefully. Never use materials from unmarked containers.

Never run the engine in an enclosed space, such as a garage. Exhaust fumes contain carbon monoxide, which is extremely poisonous. If you need to run the engine, always do so in the open air, or at least have the rear of the vehicle outside the work area.

If you are fortunate enough to have the use of an inspection pit, never drain or pour gasoline and never run the engine while the vehicle is over the pit. The fumes, being heavier than air, will concentrate in the pit with possibly lethal results.

The battery

Never create a spark or allow a bare light bulb near a battery. They normally give off a certain amount of hydrogen gas, which is highly explosive.

Always disconnect the battery ground (-) cable at the battery before working on the fuel or electrical systems.

If possible, loosen the filler caps or cover when charging the battery from an external source (this does not apply to sealed or maintenance-free batteries). Do not charge at an excessive rate or the battery may burst.

Take care when adding water to a non maintenance-free battery and when carrying a battery. The electrolyte, even when diluted, is very corrosive and should not be allowed to contact clothing or skin.

Always wear eye protection when cleaning the battery to prevent the caustic deposits from entering your eyes.

Household current

When using an electric power tool, inspection light, etc., which operates on household current, always make sure that the tool is correctly connected to its plug and that, where necessary, it is properly grounded. Do not use such items in damp conditions and, again, do not create a spark or apply excessive heat in the vicinity of fuel or fuel vapor.

Secondary ignition system voltage

A severe electric shock can result from touching certain parts of the ignition system (such as the spark plug wires) when the engine is running or being cranked, particularly if components are damp or the insulation is defective. In the case of an electronic ignition system, the secondary system voltage is much higher and could prove fatal.

Troubleshooting

Contents

This section provides an easy reference guide to the more common problems which may occur during the operation of your vehicle. These problems and their possible causes are grouped under headings denoting various components or systems, such as Engine, Cooling system, etc. They also refer you to the chapter and/or section which deals with the problem.

Remember that successful troubleshooting is not a mysterious black art practiced only by professional mechanics. It is simply the result of the right knowledge combined with an intelligent, systematic approach to the problem. Always work by a process of elimination, starting with the simplest solution and working through to the most complex - and never overlook the obvious. Anyone can run the gas tank dry or leave the lights on overnight, so don't assume that you are exempt from such oversights.

Finally, always establish a clear idea of why a problem has occurred and take steps to ensure that it doesn't happen again. If the electrical system fails because of a poor connection, check the other connections in the system to make sure that they don't fail as well. If a particular fuse continues to blow, find out why - don't just replace one fuse after another. Remember, failure of a small component can often be indicative of potential failure or incorrect functioning of a more important component or system.

Engine

1 Engine will not rotate when attempting to start

1   Battery terminal connections loose or corroded (Chapter 1).
2   Battery discharged or faulty (Chapter 1).
3   Automatic transaxle not completely engaged in Park (Chapter 7) or clutch pedal not completely depressed (Chapter 8).
4   Broken, loose or disconnected wiring in the starting circuit (Chapters 5 and 12).
5   Starter motor pinion jammed in flywheel ring gear (Chapter 5).
6   Starter solenoid faulty (Chapter 5).
7   Starter motor faulty (Chapter 5).
8   Ignition switch faulty (Chapter 12).
9   Starter pinion or flywheel teeth worn or broken (Chapter 5).
10 Defective BATTERY fusible link (see Chapter 12)

2 Engine rotates but will not start

1   Fuel tank empty.
2   Battery discharged (engine rotates slowly) (Chapter 5).
3   Battery terminal connections loose or corroded (Chapter 1).
4   Leaking fuel injector(s), faulty fuel pump, pressure regulator, etc. (Chapter 4).
5   Broken or stripped timing chain (Chapter 2).
6   Ignition components damp or damaged (Chapter 5).
7   Worn, faulty or incorrectly gapped spark plugs (Chapter 1).
8   Broken, loose or disconnected wiring in the starting circuit (Chapter 5).
9   Loose distributor is changing ignition timing (Chapter 5).
10 Broken, loose or disconnected wires at the ignition coil or faulty coil (Chapter 5).
11 Defective MAF sensor (see Chapter 6).

3 Engine hard to start when cold

1   Battery discharged or low (Chapter 1).
2   Malfunctioning fuel system (Chapter 4).
3   Faulty coolant temperature sensor or intake air temperature sensor (Chapter 6).
4   Injector(s) leaking (Chapter 4B).
5   Faulty ignition system (Chapter 5).
6   Defective MAF sensor (see Chapter 12).

4 Engine hard to start when hot

1   Air filter clogged (Chapter 1).
2   Fuel not reaching the fuel injection system (Chapter 4).
3   Corroded battery connections, especially ground (Chapter 1).
4   Faulty coolant temperature sensor or intake air temperature sensor (Chapter 6).

5 Starter motor noisy or excessively rough in engagement

1   Pinion or flywheel gear teeth worn or broken (Chapter 5).
2   Starter motor mounting bolts loose or missing (Chapter 5).

6 Engine starts but stops immediately

1   Loose or faulty electrical connections at distributor, coil or alternator (Chapter 5).
2   Insufficient fuel reaching the fuel injector(s) (Chapters 1 and 4).
3   Vacuum leak at the gasket between the intake manifold/plenum and throttle body (Chapters 1 and 4).
4   Idle speed incorrect (Chapter 1).
5   Intake air leaks, broken vacuum lines (see Chapter 4)

7 Oil puddle under engine

1   Oil pan gasket and/or oil pan drain bolt washer leaking (Chapter 2).
2   Oil pressure sending unit leaking (Chapter 2).
3   Cylinder head covers leaking (Chapter 2).
4   Engine oil seals leaking (Chapter 2).
5   Oil pump housing leaking (Chapter 2).

8 Engine lopes while idling or idles erratically

1   Vacuum leakage (Chapters 2 and 4).
2   Leaking EGR valve (Chapter 6).
3   Air filter clogged (Chapter 1).
4   Fuel pump not delivering sufficient fuel to the fuel injection system (Chapter 4).
5   Leaking head gasket (Chapter 2).
6   Timing chain and/or sprockets worn (Chapter 2).
7   Camshaft lobes worn (Chapter 2).

9 Engine misses at idle speed

1   Spark plugs worn or not gapped properly (Chapter 1).
2   Faulty spark plug wires (Chapter 1).
3   Vacuum leaks (Chapters 2 and 4).
4   Incorrect ignition timing (Chapter 1).
5   Uneven or low compression (Chapter 2).
6   Problem with the fuel injection system (Chapter 4).

10 Engine misses throughout driving speed range

1   Fuel filter clogged and/or impurities in the fuel system (Chapter 1).
2   Low fuel output at the fuel injector(s) (Chapter 4).
3   Faulty or incorrectly gapped spark plugs (Chapter 1).
4   Incorrect ignition timing (Chapter 5).
5   Cracked distributor cap, disconnected distributor wires or damaged distributor components (Chapters 1 and 5).
6   Leaking spark plug wires (Chapters 1 or 5).
7   Faulty emission system components (Chapter 6).
8   Low or uneven cylinder compression pressures (Chapter 2).
9   Weak or faulty ignition system (Chapter 5).
10 Vacuum leak in fuel injection system, throttle body, intake manifold, IAC/AAC valve or vacuum hoses (Chapter 4).

11 Engine stumbles on acceleration

1   Spark plugs fouled (Chapter 1).
2   Problem with fuel injection system (Chapter 4).
3   Fuel filter clogged (Chapters 1 and 4).
4   Incorrect ignition timing (Chapter 5).
5   Intake manifold air leak (Chapters 2 and 4).
6   EGR system malfunction (Chapter 6).

12 Engine surges while holding accelerator steady

1   Intake air leak (Chapter 4).
2   Fuel pump or fuel pressure regulator faulty (Chapter 4).
3   Problem with fuel injection system (Chapter 4).
4   Problem with the emissions control system (Chapter 6).

13 Engine stalls

1   Idle speed incorrect (Chapter 1).
2   Fuel filter clogged and/or water and impurities in the fuel system (Chapters 1 and 4).
3   Distributor components damp or damaged (Chapter 5).
4   Faulty emissions system components (Chapter 6).
5   Faulty or incorrectly gapped spark plugs (Chapter 1).
6   Faulty spark plug wires (Chapter 1).
7   Vacuum leak in the fuel injection system, intake manifold or vacuum hoses (Chapters 2 and 4).
8   Valve clearances incorrectly set (Chapter 1).

14 Engine lacks power

1   Incorrect ignition timing (Chapter 5).
2   Excessive play in distributor shaft (Chapter 5).
3   Worn rotor, distributor cap, spark plug wires or faulty coil (Chapters 1 and 5).
4   Faulty or incorrectly gapped spark plugs (Chapter 1).
5   Problem with the fuel injection system (Chapter 4).
6   Plugged air filter (Chapter 1).
7   Brakes binding (Chapter 9).
8   Automatic transaxle fluid level incorrect (Chapter 1).
9   Clutch slipping (Chapter 8).
10 Fuel filter clogged and/or impurities in the fuel system (Chapters 1 and 4).
11 Emission control system not functioning properly (Chapter 6).
12 Low or uneven cylinder compression pressures (Chapter 2).
13 Obstructed exhaust system (Chapters 2 and 4).

15 Engine backfires

1   Emission control system not functioning properly (Chapter 6).
2   Ignition timing incorrect (Chapter 5).
3   Faulty secondary ignition system (cracked spark plug insulator, faulty plug wires, distributor cap and/or rotor) (Chapters 1 and 5).

4   Problem with the fuel injection system (Chapter 4).
5   Vacuum leak at fuel injector(s), intake manifold, air control valve or vacuum hoses (Chapters 2 and 4).
6   Valve clearances incorrectly set and/or valves sticking (Chapter 1).

16 Pinging or knocking engine sounds during acceleration or uphill

1   Incorrect grade of fuel.
2   Ignition timing incorrect (Chapter 5).
3   Fuel injection system faulty (Chapter 4).
4   Improper or damaged spark plugs or wires (Chapter 1).
5   Worn or damaged distributor components (Chapter 5).
6   EGR valve not functioning (Chapter 6).
7   Vacuum leak (Chapters 2 and 4).

17 Engine runs with oil pressure light on

1   Low oil level (Chapter 1).
2   Idle rpm below specification (Chapter 1).
3   Short in wiring circuit (Chapter 12).
4   Faulty oil pressure sender (Chapter 2).
5   Worn engine bearings and/or oil pump (Chapter 2).

18 Engine diesels (continues to run) after switching off

1   Idle speed too high (Chapter 1).
2   Excessive engine operating temperature (Chapter 3).
3   Ignition timing in need of adjustment (Chapter 5).
4   Excessive carbon deposits on valves and pistons (see Chapter 2)

Engine electrical system

19 Battery will not hold a charge

1   Alternator drivebelt defective or not adjusted properly (Chapter 1).
2   Battery electrolyte level low (Chapter 1).
3   Battery terminals loose or corroded (Chapter 1).
4   Alternator not charging properly (Chapter 5).
5   Loose, broken or faulty wiring in the charging circuit (Chapter 5).
6   Short in vehicle wiring (Chapter 12).
7   Internally defective battery (Chapters 1 and 5).

20 Alternator light fails to go out

1   Faulty alternator or charging circuit (Chapter 5).
2   Alternator drivebelt defective or out of adjustment (Chapter 1).
3   Alternator voltage regulator inoperative (Chapter 5).

21 Alternator light fails to come on when key is turned on

1   Warning light bulb defective (Chapter 12).
2   Fault in the printed circuit, dash wiring or bulb holder (Chapter 12).

Fuel system

22 Excessive fuel consumption

1   Dirty or clogged air filter element (Chapter 1).
2   Incorrectly set ignition timing (Chapter 5).
3   Emissions system not functioning properly (Chapter 6).
4   Fuel injection system not functioning properly (Chapter 4).
5   Low tire pressure or incorrect tire size (Chapter 1).

23 Fuel leakage and/or fuel odor

1   Leaking fuel feed or return line (Chapters 1 and 4).
2   Tank overfilled.
3   Charcoal canister filter clogged (Chapters 1 and 6).
4   Problem with fuel injection system (Chapter 4).

Cooling system

24 Overheating

1   Insufficient coolant in system (Chapter 1).
2   Water pump drivebelt defective or out of adjustment (Chapter 1).
3   Radiator core blocked or grille restricted (Chapter 3).
4   Thermostat faulty (Chapter 3).
5   Electric coolant fan inoperative or blades broken (Chapter 3).
6   Radiator cap not maintaining proper pressure (Chapter 3).
7   Ignition timing incorrect (Chapter 5).

25 Overcooling

1 Faulty thermostat (Chapter 3).
2 Inaccurate temperature gauge sending unit (Chapter 3)

26 External coolant leakage

1 Deteriorated/damaged hoses; loose clamps (Chapters 1 and 3).
2 Water pump defective (Chapter 3).
3 Leakage from radiator core or coolant reservoir bottle (Chapter 3).
4 Engine drain or water jacket core plugs leaking (Chapter 2).

27 Internal coolant leakage

1 Leaking cylinder head gasket (Chapter 2).
2 Cracked cylinder bore or cylinder head (Chapter 2).

28 Coolant loss

1 Too much coolant in system (Chapter 1).
2 Coolant boiling away because of overheating (Chapter 3).
3 Internal or external leakage (Chapter 3).
4 Faulty radiator cap (Chapter 3).

29 Poor coolant circulation

1 Inoperative water pump (Chapter 3).
2 Restriction in cooling system (Chapters 1 and 3).
3 Water pump drivebelt defective/out of adjustment (Chapter 1).
4 Thermostat sticking (Chapter 3).

Clutch

30 Pedal travels to floor - no pressure or very little resistance

1 Hydraulic release system leaking or air in the system (Chapter 8).
2 Broken release bearing or fork (Chapter 8).

31 Unable to select gears

1 Faulty transaxle (Chapter 7).
2 Faulty clutch disc or pressure plate (Chapter 8).
3 Faulty release lever or release bearing (Chapter 8).

4 Faulty shift lever assembly or rods (Chapter 8).

32 Clutch slips (engine speed increases with no increase in vehicle speed)

1 Clutch plate worn (Chapter 8).
2 Clutch plate is oil soaked by leaking rear main seal (Chapter 8).
3 Clutch plate not seated (Chapter 8).
4 Warped pressure plate or flywheel (Chapter 8).
5 Weak diaphragm springs in pressure plate (Chapter 8).
6 Clutch plate overheated. Allow to cool.
7 Piston stuck in bore of clutch release cylinder, preventing clutch from fully engaging (Chapter 8).

33 Grabbing (chattering) as clutch is engaged

1 Oil on clutch plate lining, burned or glazed facings (Chapter 8).
2 Worn or loose engine or transaxle mounts (Chapters 2 and 7).
3 Worn splines on clutch plate hub (Chapter 8).
4 Warped pressure plate or flywheel (Chapter 8).
5 Burned or smeared resin on flywheel or pressure plate (Chapter 8).

34 Transaxle rattling (clicking)

1 Release fork loose (Chapter 8).
2 Low engine idle speed (Chapter 1).

35 Noise in clutch area

Faulty bearing (Chapter 8).

36 Clutch pedal stays on floor

1 Broken release bearing or fork (Chapter 8).
2 Hydraulic release system leaking or air in the system (Chapter 8).

37 High pedal effort

1 Piston binding in bore of release cylinder (Chapter 8).
2 Pressure plate faulty (Chapter 8).
3 Incorrect size master or release cylinder (Chapter 8).

Manual transaxle

38 Knocking noise at low speeds

1 Worn driveaxle constant velocity (CV) joints (Chapter 8).
2 Worn side gear shaft counterbore in differential case (Chapter 7A).*

39 Noise most pronounced when turning

Differential gear noise (Chapter 7A).*

40 Clunk on acceleration or deceleration

1 Loose engine or transaxle mounts (Chapters 2 and 7A).
2 Worn differential pinion shaft in case.*
3 Worn side gear shaft counterbore in differential case (Chapter 7A).*
4 Worn or damaged driveaxle inboard CV joints (Chapter 8).

41 Clicking noise in turns

Worn or damaged outboard CV joint (Chapter 8).

42 Vibration

1 Rough wheel bearing (Chapters 1 and 10).
2 Damaged driveaxle (Chapter 8).
3 Out of round tires (Chapter 1).
4 Tire out of balance (Chapters 1 and 10).
5 Worn CV joint (Chapter 8).

43 Noisy in neutral with engine running

1 Damaged input gear bearing (Chapter 7A).*
2 Damaged clutch release bearing (Chapter 8).

44 Noisy in one particular gear

1 Damaged or worn constant mesh gears (Chapter 7A).*
2 Damaged or worn synchronizers (Chapter 7A).*
3 Bent reverse fork (Chapter 7A).*
4 Damaged fourth speed gear or output gear (Chapter 7A).*
5 Worn or damaged reverse idler gear or idler bushing (Chapter 7A).*

45 Noisy in all gears

1   Insufficient lubricant (Chapter 7A).
2   Damaged or worn bearings (Chapter 7A).*
3   Worn or damaged input gear shaft and/or output gear shaft (Chapter 7A).*

46 Slips out of gear

1   Worn or improperly adjusted linkage (Chapter 7A).
2   Transaxle loose on engine (Chapter 7A).
3   Shift linkage does not work freely, binds (Chapter 7A).
4   Input gear bearing retainer broken or loose (Chapter 7A).*
5   Dirt between clutch cover and engine housing (Chapter 7A).
6   Worn shift fork (Chapter 7A).*

47 Leaks lubricant

1   Side gear shaft seals worn (Chapter 7A.
2   Excessive amount of lubricant in transaxle (Chapters 1 and 7A).
3   Loose or broken input gear shaft bearing retainer (Chapter 7A).*
4   Input gear bearing retainer O-ring and/or lip seal damaged (Chapter 7A).*
5   Striking rod seal leaking (Chapter 7A).
6   Vehicle speed sensor O-ring leaking (Chapter 7A).

48 Hard to shift

Shift linkage loose or worn (Chapter 7A).
Although the corrective action necessary to remedy the symptoms described is beyond the scope of this manual, the above information should be helpful in isolating the cause of the condition so that the owner can communicate clearly with a professional mechanic.

Automatic transaxle
Note: *Due to the complexity of the automatic transaxle, it is difficult for the home mechanic to properly diagnose and service this component. For problems other than the following, the vehicle should be taken to a dealer or transaxle shop.*

49 Fluid leakage

1   Automatic transaxle fluid is a deep red color. Fluid leaks should not be confused with engine oil, which can easily be blown onto the transaxle by air flow.
2   To pinpoint a leak, first remove all built-up dirt and grime from the transaxle housing with degreasing agents and/or steam clean-

ing. Then drive the vehicle at low speeds so air flow will not blow the leak far from its source. Raise the vehicle and determine where the leak is coming from. Common areas of leakage are:
a) *Pan (Chapters 1 and 7)*
b) *Dipstick tube (Chapters 1 and 7)*
c) *Transaxle oil lines (Chapter 7)*
d) *Speed sensor (Chapter 7)*
e) *Driveaxle oil seals (Chapter 7).*

50 Transaxle fluid brown or has a burned smell

Transaxle fluid overheated (Chapter 1).

51 General shift mechanism problems

1   Chapter 7, Part B, deals with checking and adjusting the shift linkage on automatic transaxles. Common problems which may be attributed to poorly adjusted linkage are:
a) *Engine starting in gears other than Park or Neutral.*
b) *Indicator on shifter pointing to a gear other than the one actually being used.*
c) *Vehicle moves when in Park.*
2   Refer to Chapter 7B for the shift linkage adjustment procedure.

52 Transaxle will not downshift with accelerator pedal pressed to the floor

The transaxle is electronically controlled. This type of problem - which is caused by a malfunction in the control unit, a sensor or solenoid, or the circuit itself - is beyond the scope of this book. Take the vehicle to a dealer service department or a competent automatic transmission shop.

53 Engine will start in gears other than Park or Neutral

Neutral start switch out of adjustment or malfunctioning (Chapter 7B).

54 Transaxle slips, shifts roughly, is noisy or has no drive in forward or reverse gears

There are many probable causes for the above problems, but the home mechanic should be concerned with only one possibility - fluid level. Before taking the vehicle to a repair shop, check the level and condition of the fluid as described in Chapter 1. Correct

the fluid level as necessary or change the fluid and filter if needed. If the problem persists, have a professional diagnose the cause.

Driveaxles

55 Clicking noise in turns

Worn or damaged outboard CV joint (Chapter 8).

56 Shudder or vibration during acceleration

1   Excessive toe-in (Chapter 10).
2   Incorrect spring heights (Chapter 10).
3   Worn or damaged inboard or outboard CV joints (Chapter 8).
4   Sticking inboard CV joint assembly (Chapter 8).

57 Vibration at highway speeds

1   Out of balance front wheels and/or tires (Chapters 1 and 10).
2   Out of round front tires (Chapters 1 and 10).
3   Worn CV joint(s) (Chapter 8).

Brakes
Note: *Before assuming that a brake problem exists, make sure that:*
a) *The tires are in good condition and properly inflated (Chapter 1).*
b) *The front end alignment is correct (Chapter 10).*
c) *The vehicle is not loaded with weight in an unequal manner.*

58 Vehicle pulls to one side during braking

1   Incorrect tire pressures (Chapter 1).
2   Front end out of alignment (have the front end aligned).
3   Front, or rear, tire sizes not matched to one another.
4   Restricted brake lines or hoses (Chapter 9).
5   Malfunctioning drum brake or caliper assembly (Chapter 9).
6   Loose suspension parts (Chapter 10).
7   Loose calipers (Chapter 9).
8   Excessive wear of brake shoe or pad material or disc/drum on one side.

59 Noise (high-pitched squeal when the brakes are applied)

Front and/or rear disc brake pads worn out. The noise comes from the wear sensor rubbing against the disc (does not apply to all vehicles). Replace pads with new ones immediately (Chapter 9).

60 Brake roughness or chatter (pedal pulsates)

1 Excessive lateral runout (Chapter 9).
2 Uneven pad wear (Chapter 9).
3 Defective disc (Chapter 9).

61 Excessive brake pedal effort required to stop vehicle

1 Malfunctioning power brake booster (Chapter 9).
2 Partial system failure (Chapter 9).
3 Excessively worn pads or shoes (Chapter 9).
4 Piston in caliper or wheel cylinder stuck or sluggish (Chapter 9).
5 Brake pads or shoes contaminated with oil or grease (Chapter 9).
6 Brake disc grooved and/or glazed (Chapter 1).
7 New pads or shoes installed and not yet seated. It will take a while for the new material to seat against the disc or drum.

62 Excessive brake pedal travel

1 Partial brake system failure (Chapter 9).
2 Insufficient fluid in master cylinder (Chapters 1 and 9).
3 Air trapped in system (Chapters 1 and 9).

63 Dragging brakes

1 Incorrect adjustment of brake light switch (Chapter 9).
2 Master cylinder pistons not returning correctly (Chapter 9).
3 Restricted brakes lines or hoses (Chapters 1 and 9).
4 Incorrect parking brake adjustment (Chapter 9).

64 Grabbing or uneven braking action

1 Malfunction of proportioning valve (Chapter 9).
2 Malfunction of power brake booster unit (Chapter 9).

3 Binding brake pedal mechanism (Chapter 9).

65 Brake pedal feels spongy when depressed

1 Air in hydraulic lines (Chapter 9).
2 Master cylinder mounting bolts loose (Chapter 9).
3 Master cylinder defective (Chapter 9).

66 Brake pedal travels to the floor with little resistance

1 Little or no fluid in the master cylinder reservoir caused by leaking caliper piston(s) (Chapter 9).
2 Loose, damaged or disconnected brake lines (Chapter 9).

67 Parking brake does not hold

Parking brake linkage improperly adjusted (Chapters 1 and 9).

Suspension and steering systems

Note: *Before attempting to diagnose the suspension and steering systems, perform the following preliminary checks:*

a) *Tires for wrong pressure and uneven wear.*
b) *Steering universal joints from the column to the rack and pinion for loose connectors or wear.*
c) *Front and rear suspension and the rack-and-pinion assembly for loose or damaged parts.*
d) *Out-of-round or out-of-balance tires, bent rims and loose and/or rough wheel bearings.*

68 Vehicle pulls to one side

1 Mismatched or uneven tires (Chapter 10).
2 Broken or sagging springs (Chapter 10).
3 Wheel alignment out-of-specifications (Chapter 10).
4 Front brake dragging (Chapter 9).

69 Abnormal or excessive tire wear

1 Wheel alignment out-of-specifications (Chapter 10).
2 Sagging or broken springs (Chapter 10).
3 Tire out-of-balance (Chapter 10).
4 Worn strut damper (Chapter 10).
5 Overloaded vehicle.
6 Tires not rotated regularly.

70 Wheel makes a thumping noise

1 Blister or bump on tire (Chapter 10).
2 Improper strut damper action (Chapter 10).

71 Shimmy, shake or vibration

1 Tire or wheel out-of-balance or out-of-round (Chapter 10).
2 Loose or worn wheel bearings (Chapters 1, 8 and 10).
3 Worn tie-rod ends (Chapter 10).
4 Worn lower balljoints (Chapters 1 and 10).
5 Excessive wheel runout (Chapter 10).
6 Blister or bump on tire (Chapter 10).

72 Hard steering

1 Lack of lubrication at balljoints, tie-rod ends and rack and pinion assembly (Chapter 10).
2 Front wheel alignment out-of-specifications (Chapter 10).
3 Low tire pressure(s) (Chapters 1 and 10).

73 Poor returnability of steering to center

1 Lack of lubrication at balljoints and tie-rod ends (Chapter 10).
2 Binding in balljoints (Chapter 10).
3 Binding in steering column (Chapter 10).
4 Lack of lubricant in steering gear assembly (Chapter 10).
5 Front wheel alignment out-of-specifications (Chapter 10).

74 Abnormal noise at the front end

1 Lack of lubrication at balljoints and tie-rod ends (Chapters 1 and 10).
2 Damaged strut mounting (Chapter 10).
3 Worn control arm bushings or tie-rod ends (Chapter 10).
4 Loose stabilizer bar (Chapter 10).
5 Loose wheel nuts (Chapters 1 and 10).
6 Loose suspension bolts (Chapter 10)

75 Wander or poor steering stability

1 Mismatched or uneven tires (Chapter 10).
2 Lack of lubrication at balljoints and tie-rod ends (Chapters 1 and 10).
3 Worn strut assemblies (Chapter 10).
4 Loose stabilizer bar (Chapter 10).
5 Broken or sagging springs (Chapter 10).
6 Wheels out of alignment (Chapter 10).

76 Erratic steering when braking

1   Wheel bearings worn (Chapter 10).
2   Broken or sagging springs (Chapter 10).
3   Leaking wheel cylinder or caliper (Chapter 10).
4   Warped rotors or drums (Chapter 10).

77 Excessive pitching and/or rolling around corners or during braking

1   Loose stabilizer bar (Chapter 10).
2   Worn strut dampers or mountings (Chapter 10).
3   Broken or sagging springs (Chapter 10).
4   Overloaded vehicle.

78 Suspension bottoms

1   Overloaded vehicle.
2   Worn strut dampers (Chapter 10).
3   Incorrect, broken or sagging springs (Chapter 10).

79 Cupped tires

1   Front wheel or rear wheel alignment out-of-specifications (Chapter 10).
2   Worn strut dampers (Chapter 10).
3   Wheel bearings worn (Chapter 10).
4   Excessive tire or wheel runout (Chapter 10).
5   Worn balljoints (Chapter 10).

80 Excessive tire wear on outside edge

1   Inflation pressures incorrect (Chapter 1).
2   Excessive speed in turns.
3   Front end alignment incorrect (excessive toe-in). Have professionally aligned.
4   Suspension arm bent or twisted (Chapter 10).

81 Excessive tire wear on inside edge

1   Inflation pressures incorrect (Chapter 1).
2   Front end alignment incorrect (toe-out). Have professionally aligned.

3   Loose or damaged steering components (Chapter 10).

82 Tire tread worn in one place

1   Tires out-of-balance.
2   Damaged or buckled wheel. Inspect and replace if necessary.
3   Defective tire (Chapter 1).

83 Excessive play or looseness in steering system

1   Wheel bearing(s) worn (Chapter 10).
2   Tie-rod end loose (Chapter 10).
3   Steering gear loose (Chapter 10).
4   Worn or loose steering intermediate shaft (Chapter 10).

84 Rattling or clicking noise in steering gear

1   Steering gear loose (Chapter 10).
2   Steering gear defective.

Notes

Chapter 1
Tune-up and routine maintenance

Contents

1

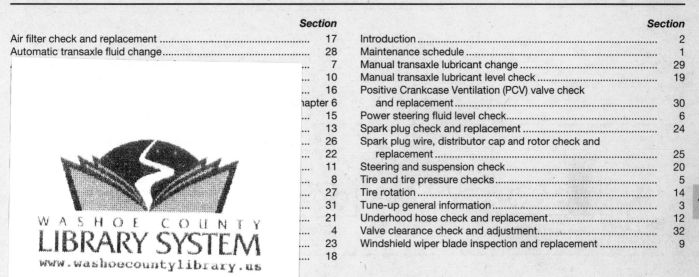

...time this manual was written. Manufacturers occasionally upgrade their fluid and ...re for current recommendations.

	API certified SG or SH energy conserving oil
	See accompanying chart
	Unleaded gasoline, 87 octane or higher
	DEXRON IIE or DEXRON III automatic transmission fluid
	API GL-4 75W-90 gear oil
	DOT 3 brake fluid or equivalent
	DEXRON IIE or DEXRON III automatic transmission fluid
	Nissan PSF II power steering fluid
	NLGI no. 2 lithium-base grease

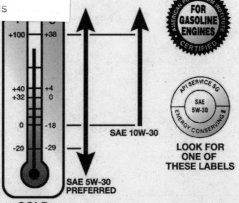

FOR GASOLINE ENGINES
AMERICAN PETROLEUM INSTITUTE
CERTIFIED

API SERVICE SG
SAE 5W-30
ENERGY CONSERVING II

+100 +38
+40 +4
+32 0
0 -18
-20 -29

SAE 10W-30

SAE 5W-30 PREFERRED

COLD WEATHER

LOOK FOR ONE OF THESE LABELS

Recommended engine oil viscosity

1-a3 HAYNES

Capacities*

Engine oil (including filter)	4.1 quarts
Coolant (including reservoir tank)	8.3 quarts
Transaxle	
Automatic (including torque convertor)	10 quarts
Manual	9.5 pints
Manual (with limited-slip differential)	9.1 pints

All capacities approximate. Add as necessary to bring up to appropriate level.

Ignition system

Spark plug type and gap	
Type	NGK BKR5E-11 or equivalent
Gap	0.043 inch
Spark plug wire resistance	10,000 to 25,000 ohms
Engine firing order	1-3-4-2

FRONT — **1994 and earlier**

FRONT — **1995 and later**

Cylinder location and distributor rotation

The blackened terminal shown on the distributor cap indicates the Number One spark plug wire position

Ignition timing

(See Chapter 5)

Idle speed adjustment

No adjustments necessary (See Chapter 4)

Valve clearance (engine hot)

Intake valve	0.012 to 0.015 inch (0.31 to 0.39 mm)
Exhaust valve	0.013 to 0.016 inch (0.33 to 0.41 mm)

Cooling system

Thermostat rating	
Starts to open	170-degrees F
Fully open	194-degrees F

Clutch pedal

1993 through 1994	
Freeplay	3/64 to 1/8 inch
Height	6-1/2 to 6-57/64 inches
1995	
Freeplay	23/64 to 5/8 inch
Height	6-39/64 to 7 inches
1996 on	
Freeplay	3/64 to 1/8 inch
Height	6-39/64 to 7 inches
Clutch interlock switch freeplay (all)	0.004 to 0.039 inch

Brakes

Disc brake pad lining thickness (minimum)	1/16 inch
Drum brake shoe lining thickness (minimum)	1/16 inch
Brake pedal	
Freeplay	1/32 to 7/64 inch
Free height	
Manual transaxle	6-21/32 to 7-3/64 inches
Automatic transaxle	6-31/32 to 7-23/64 inches
Depressed height (with engine running)	3-35/64
Parking brake adjustment	7 to 8 clicks

Suspension and steering

Steering wheel freeplay limit.. 1-3/8 inches
Balljoint allowable movement .... 0 inch

Torque specifications
Ft-lbs

Automatic transaxle drain plug... 22 to 29
Manual transaxle drain and filler plugs .... 14 to 22
Spark plugs... 14 to 22
Wheel lug nuts .... 72 to 87

Typical engine compartment components

1	Oil filler cap	8	Headlight adjustment level	15	Engine oil dipstick
2	EGR valve	9	Relay block	16	Upper radiator hose
3	Fuel filter	10	Automatic transaxle dipstick	17	Drivebelt
4	Brake fluid reservoir	11	Lower radiator hose	18	Coolant reservoir
5	Air cleaner housing	12	Distributor	19	Windshield washer fluid reservoir
6	Battery	13	Radiator cap	20	Power steering fluid reservoir
7	Fuse panel	14	Spark plugs		

Typical engine compartment underside components

1	Disc brake caliper	4	Engine oil pan drain plug	7	Exhaust system
2	Suspension strut and spring	5	Automatic transaxle drain plug	8	Steering gear boot
3	Air conditioning compressor	6	Driveaxle boot	9	Control arm bushing

Typical rear underside components

1	Fuel filler hose	3	Muffler	5	Gas tank
2	Stabilizer bar	4	Rear suspension control arms	6	Parking brake cable

1 Nissan Altima maintenance schedule

The maintenance intervals in this manual are provided with the assumption that you, not the dealer, will be doing the work. These are the minimum maintenance intervals recommended by the factory for vehicles that are driven daily. If you wish to keep your vehicle in peak condition at all times, you may wish to perform some of these procedures even more often. Because frequent maintenance enhances the efficiency, performance and resale value of your car, we encourage you to do so. If you drive in dusty areas, tow a trailer, idle or drive at low speeds for extended periods or drive for short distances (less than four miles) in below freezing temperatures, shorter intervals are also recommended.

When your vehicle is new, it should be serviced by a factory authorized dealer service department to protect the factory warranty. In many cases, the initial maintenance check is done at no cost to the owner.

Every 250 miles or weekly, whichever comes first

Check the engine oil level (Section 4)
Check the engine coolant level (Section 4)
Check the windshield washer fluid level (Section 4)
Check the battery electrolyte (Section 4)
Check the brake fluid level (Section 4)
Check the clutch fluid level (Section 4)
Check the tires and tire pressures (Section 5)

Every 3000 miles or 3 months, whichever comes first

All items listed above plus:
Check the power steering fluid level (Section 6)
Check the automatic transaxle fluid level (Section 7)
Change the engine oil and oil filter (Section 8)

Every 7500 miles or 6 months, whichever comes first

Inspect and replace if necessary the windshield wiper blades (Section 9)
Check and service the battery (Section 10)
Check and adjust if necessary the engine drivebelts (Section 11)
Inspect and replace if necessary all underhood hoses (Section 12)
Check the cooling system (Section 13)
Rotate the tires (Section 14)

Every 15,000 miles or 12 months, whichever comes first

All items listed above plus:
Check the clutch pedal and release lever for proper freeplay (Section 15)
Inspect the brake system (Section 16)*
Replace the air filter (Section 17)
Inspect the fuel system (Section 18)
Check the manual transaxle lubricant level (Section 19)
Inspect the suspension and steering components (Section 20)
Inspect the exhaust system (Section 21)
Check the driveaxle boots (Section 22)

Every 30,000 miles or 24 months, whichever comes first

All items listed above plus:
Replace the fuel filter (Section 23)
Check and replace if necessary the spark plugs (Section 24)
Inspect and replace if necessary the spark plug wires, distributor cap and rotor (Section 25)
Service the cooling system (drain, flush and refill) (Section 26)
Inspect the evaporative emissions control system (Section 27)
Change the automatic transaxle fluid (Section 28)**
Change the manual transaxle lubricant (Section 29)**
Check and replace if necessary the PCV valve (Section 30)
Check the operation of the Exhaust Gas Recirculation (EGR) system (Section 31)

Every 60,000 miles or 48 months, whichever comes first

Check and adjust if necessary, the valve clearance (Section 32)

* This item is affected by "severe" operating conditions as described below. If your vehicle is operated under "severe" conditions, perform all maintenance indicated with an asterisk (*) at 3000 mile/3 month intervals. Severe conditions are indicated if you mainly operate your vehicle under one or more of the following conditions:

Operating in dusty areas
Towing a trailer
Idling for extended periods and/or low speed operation
Operating when outside temperatures remain below freezing and when most trips are less than 4 miles

** If operated under one or more of the following conditions, change the, manual or automatic transaxle fluid and differential lubricant every 15,000 miles:

In heavy city traffic where the outside temperature regularly reaches 90-degrees F (32-degrees C) or higher
In hilly or mountainous terrain
Frequent trailer pulling

4.2 The engine oil dipstick (arrow) is located on the front side of the engine

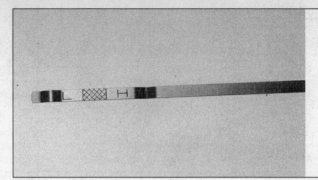

4.4 The oil level should be at or near the H mark - if it isn't, add enough oil to bring the level to near the H mark (it takes one full quart to raise the level from the L to the H mark)

2   Introduction

This Chapter is designed to help the home mechanic maintain the Nissan Altima for peak performance, economy, safety and long life.

On the following pages is a master maintenance schedule, followed by Sections dealing specifically with each item on the schedule. Visual checks, adjustments, component replacement and other helpful items are included. Refer to the accompanying illustrations of the engine compartment and the underside of the vehicle for the location of various components.

Servicing your Altima in accordance with the mileage/time maintenance schedule and the following Sections will provide it with a planned maintenance program that should result in a long and reliable service life. This is a comprehensive plan, so maintaining some items but not others at the specified service intervals will not produce the same results.

As you service your Altima, you will discover that many of the procedures can, and should, be grouped together because of the nature of the particular procedure you're performing or because of the close proximity of two otherwise unrelated components to one another.

For example, if the vehicle is raised for any reason, you should inspect the exhaust, suspension, steering and fuel systems while you're under the vehicle. When you're rotating the tires, it makes good sense to check the brakes and wheel bearings since the wheels are already removed.

Finally, let's suppose you have to borrow or rent a torque wrench. Even if you only need to tighten the spark plugs, you might as well check the torque of as many critical fasteners as time allows.

The first step of this maintenance program is to prepare yourself before the actual work begins. Read through all Sections pertinent to the procedures you're planning to do,

then make a list of and gather together all the parts and tools you will need to do the job. If it looks as if you might run into problems during a particular segment of some procedure, seek advice from your local parts man or dealer service department.

3   Tune-up general information

The term tune-up is used in this manual to represent a combination of individual operations rather than one specific procedure.

If, from the time the vehicle is new, the routine maintenance schedule is followed closely and frequent checks are made of fluid levels and high wear items, as suggested throughout this manual, the engine will be kept in relatively good running condition and the need for additional work will be minimized.

More likely than not, however, there will be times when the engine is running poorly due to lack of regular maintenance. This is even more likely if a used vehicle, which has not received regular and frequent maintenance checks, is purchased. In such cases, an engine tune-up will be needed outside of the regular routine maintenance intervals.

The first step in any tune-up or engine diagnosis to help correct a poor running engine would be a cylinder compression check. A check of the engine compression (Chapter 2 Part B) will give valuable information regarding the overall performance of many internal components and should be used as a basis for tune-up and repair procedures. If, for instance, a compression check indicates serious internal engine wear, a conventional tune-up will not help the running condition of the engine and would be a waste of time and money.

The following series of operations are those most often needed to bring a generally poor running engine back into a proper state of tune.

Minor tune-up

Check all engine related fluids (Section 4)
Clean, inspect and test the battery (Section 10)
Check and adjust the drivebelts (Section 11)
Check all underhood hoses (Section 12)
Check the cooling system (Section 13)
Check the air filter (Section 17)

Replace the spark plugs (Section 24)
Inspect the distributor cap and rotor (Section 25)
Inspect the spark plug and coil wires (Section 25)
Check the idle speed (Chapter 4)

Major tune-up

All items listed under Minor tune-up, plus . . .
Replace the air filter (Section 17)
Check the fuel system (Section 18)
Check the charging system (Chapter 5)
Replace the spark plug wires (Section 25)
Replace the distributor cap and rotor (Section 25)
Check the ignition system (Chapter 5)

4   Fluid level checks (every 250 miles or weekly)

1   Fluids are an essential part of the lubrication, cooling, brake, clutch and other systems. Because these fluids gradually become depleted and/or contaminated during normal operation of the vehicle, they must be periodically replenished. See *Recommended lubricants* and *fluids* and *capacities* at the beginning of this Chapter before adding fluid to any of the following components. **Note:** *The vehicle must be on level ground before fluid levels can be checked.*

Engine oil

Refer to illustrations 4.2, 4.4 and 4.6

2   The engine oil level is checked with a dipstick located at the front side of the engine **(see illustration)**. The dipstick extends through a metal tube from which it protrudes down into the engine oil pan.

3   The oil level should be checked before the vehicle has been driven, or about 15 minutes after the engine has been shut off. If the oil is checked immediately after driving the vehicle, some of the oil will remain in the upper engine components, producing an inaccurate reading on the dipstick.

4   Pull the dipstick from the tube and wipe all the oil from the end with a clean rag or paper towel. Insert the clean dipstick all the way back into its metal tube and pull it out again. Observe the oil at the end of the dipstick. At its highest point, the level should be between the L and H marks **(see illustration)**.

4.6 The threaded oil filler cap (arrow) is located on the valve cover - always make sure the area around the opening is clean before unscrewing the cap to prevent dirt from contaminating the engine

4.8 The coolant reservoir is located in the right front corner of the engine compartment - add coolant to bring the level near the MAX mark on the reservoir

5   It takes one quart of oil to raise the level from the L mark to the H mark on the dipstick. Do not allow the level to drop below the L mark or oil starvation may cause engine damage. Conversely, overfilling the engine (adding oil above the H mark) may cause oil fouled spark plugs, oil leaks or oil seal failures.

6   Remove the threaded cap from the valve cover to add oil (see illustration). Use a funnel to prevent spills. After adding the oil, install the filler cap hand tight. Start the engine and look carefully for any small leaks around the oil filter or drain plug. Stop the engine and check the oil level again after it has had sufficient time to drain from the upper block and cylinder head galleys.

7   Checking the oil level is an important preventive maintenance step. A continually dropping oil level indicates oil leakage through damaged seals, from loose connections, or past worn rings or valve guides. If the oil looks milky in color or has water droplets in it, a cylinder head gasket may be leaking. The cylinder head should be checked immediately. The condition of the oil should also be checked. Each time you check the oil level, slide your thumb and index finger up the dipstick before wiping off the oil. If you see small dirt or metal particles clinging to the dipstick, the oil should be changed (see Section 8).

Engine coolant

Refer to illustration 4.8

Warning: *Do not allow antifreeze to come in contact with your skin or painted surfaces of the vehicle. Flush contaminated areas immediately with plenty of water. Don't store new coolant or leave old coolant lying around where it's accessible to children or pets – they're attracted by its sweet smell. Ingestion of even a small amount of coolant can be fatal! Wipe up garage floor and drip pan spills immediately. Keep antifreeze containers covered and repair cooling system leaks as soon as they're noticed.*

8   All vehicles covered by this manual are equipped with a pressurized coolant recovery system. A white coolant reservoir located in the right front corner of the engine compartment is connected by a hose to the base of the coolant filler cap (see illustration). If the coolant gets too hot during engine operation, coolant can escape through a pressurized filler cap, then through a connecting hose into the reservoir. As the engine cools, the coolant is automatically drawn back into the cooling system to maintain the correct level.

9   The coolant level should be checked regularly. It must be between the Max and Min lines on the tank. The level will vary with the temperature of the engine. When the engine is cold, the coolant level should be at or slightly above the Min mark on the tank. Once the engine has warmed up, the level should be at or near the Max mark. If it isn't, allow the fluid in the tank to cool, then remove the cap from the reservoir and add coolant to bring the level up to the Max line. Use only ethylene/glycol type coolant and water in the mixture ratio recommended by your owner's manual. Do not use supplemental inhibitor additives. If only a small amount of coolant is required to bring the system up to the proper level, water can be used. However, repeated additions of water will dilute the recommended antifreeze and water solution. In order to maintain the proper ratio of antifreeze and water, it is advisable to top up the coolant level with the correct mixture. Refer to your owner's manual for the recommended ratio.

10   If the coolant level drops within a short time after replenishment, there may be a leak in the system. Inspect the radiator, hoses, engine coolant filler cap, drain plugs, air bleeder plugs and water pump. If no leak is evident, have the radiator cap pressure tested by your dealer. **Warning:** *Never remove the radiator cap or the coolant recovery reservoir cap when the engine is running or has just been shut down, because the cooling system is hot. Escaping steam and scalding liquid could cause serious injury.*

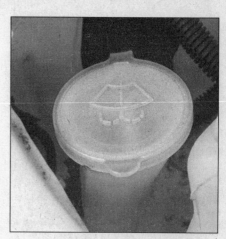

4.14 The windshield washer fluid reservoir is located on the right side of the engine compartment

11   If it is necessary to open the radiator cap, wait until the system has cooled completely, then wrap a thick cloth around the cap and turn it to the first stop. If any steam escapes, wait until the system has cooled further, then remove the cap.

12   When checking the coolant level, always note its condition. It should be relatively clear. If it is brown or rust colored, the system should be drained, flushed and refilled. Even if the coolant appears to be normal, the corrosion inhibitors wear out with use, so it must be replaced at the specified intervals.

13   Do not allow antifreeze to come in contact with your skin or painted surfaces of the vehicle. Flush contacted areas immediately with plenty of water.

Windshield washer fluid

Refer to illustration 4.14

14   Fluid for the windshield washer system is stored in a plastic reservoir which is located on the right (passenger) side of the engine compartment (see illustration). In milder climates, plain water can be used to top up the reservoir, but the reservoir should

4.15 Remove the cell caps to check the water level in the battery - if the level is low, add distilled water only

4.17a The brake fluid level should be kept between the MIN and MAX marks on the translucent plastic reservoir

4.17b The clutch fluid level should be kept between the MIN and MAX marks on the translucent plastic reservoir

be kept no more than two-thirds full to allow for expansion should the water freeze. In colder climates, the use of a specially designed windshield washer fluid, available at your dealer and any auto parts store, will help lower the freezing point of the fluid. Mix the solution with water in accordance with the manufacturer's directions on the container. Do not use regular antifreeze. It will damage the vehicle's paint.

Battery electrolyte

Refer to illustration 4.15

15   On models not equipped with a sealed battery, check the electrolyte level **(see illustration)** of all six battery cells. It must be between the upper and lower levels. If the level is low, remove the filler/vent cap and add distilled water. Install and securely re-tighten the cap. **Caution:** *Overfilling the cells may cause electrolyte to spill over during periods of heavy charging, causing corrosion or damage.*

Brake and clutch fluid

Refer to illustrations 4.17a and 4.17b

16   The brake master cylinder is mounted on the front of the power booster unit in the engine compartment. The hydraulic clutch master cylinder used on manual transaxle vehicles is located next to the brake master cylinder.

17   To check the fluid level of the brake and clutch master cylinders, simply look at the MAX and MIN marks on the reservoir **(see illustrations)**. The level should be within the specified distance from the maximum fill line.

18   If the level is low, wipe the top of the reservoir cover with a clean rag to prevent contamination of the brake system before lifting the cover.

19   Add only the specified brake fluid to the brake and clutch reservoirs (refer to *Recommended lubricants and fluids* at the front of this Chapter or to your owner's manual). Mixing different types of brake fluid can damage the system. Fill the brake master cylinder reservoir only to the MAX line. **Warning:** *Use*

caution when filling either reservoir - brake fluid can harm your eyes and damage painted surfaces. Do not use brake fluid that has been opened for more than one year or has been left open. Brake fluid absorbs moisture from the air. Excess moisture can cause a dangerous loss of braking.

20   While the reservoir cap is removed, inspect the master cylinder reservoir for contamination. If deposits, dirt particles or water droplets are present, the system should be drained and refilled.

21   After filling the reservoir to the proper level, make sure the lid is properly seated to prevent fluid leakage and/or system pressure loss.

22   The fluid in the brake master cylinder will drop slightly as the brake pads at each wheel wear down during normal operation. If either master cylinder requires repeated replenishing to keep it at the proper level, this is an indication of leakage in the brake or clutch system, which should be corrected immediately. If the brake system shows an indication of leakage check all brake lines and connections, along with the calipers, wheel cylinders and booster (see Section 16 for more information). If the hydraulic clutch system shows an indication of leakage check all clutch lines and connections, along with the clutch slave cylinder (see Chapter 8 for more information).

23   If, upon checking the brake or clutch master cylinder fluid level, you discover one or both reservoirs empty or nearly empty, the systems should be bled (see Chapter 9).

---

5   Tire and tire pressure checks (every 250 miles or weekly)

Refer to illustrations 5.2, 5.3, 5.4a, 5.4b and 5.8

1   Periodic inspection of the tires may spare you from the inconvenience of being stranded with a flat tire. It can also provide you with vital information regarding possible problems in the steering and suspension sys-

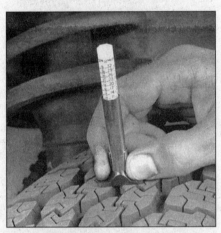

5.2 A tire tread depth indicator should be used to monitor tire wear - they are available at auto parts stores and service stations and cost very little

tems before major damage occurs.

2   Normal tread wear can be monitored with a simple, inexpensive device known as a tread depth indicator **(see illustration)**. When the tread depth reaches the specified minimum, replace the tire(s).

3   Note any abnormal tread wear **(see illustration)**. Tread pattern irregularities such as cupping, flat spots and more wear on one side than the other are indications of front end alignment and/or balance problems. If any of these conditions are noted, take the vehicle to a tire shop or service station to correct the problem.

4   Look closely for cuts, punctures and embedded nails or tacks. Sometimes a tire will hold its air pressure for a short time or leak down very slowly even after a nail has embedded itself into the tread. If a slow leak persists, check the valve stem core to make sure it is tight **(see illustration)**. Examine the tread for an object that may have embedded itself into the tire or for a "plug" that may have begun to leak (radial tire punctures are repaired with a plug that is installed in a puncture). If a punc-

UNDERINFLATION

CUPPING

Cupping may be caused by:
- **Underinflation and/or mechanical irreguarities such as out-of-balance condition of wheel and/or tire, and bent or damaged wheel.**
- **Loose or worn steering tie-rod or steering idler arm.**
- **Loose, damaged or worn front suspension parts.**

OVERINFLATION

INCORRECT TOE-IN OR EXTREME CAMBER

FEATHERING DUE TO MISALIGNMENT

5.3 This chart will help you determine the condition of your tires, the probable cause(s) of abnormal wear and the corrective action necessary

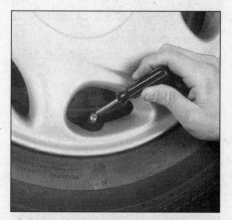

5.4a If a tire loses air on a steady basis, check the valve core first to make sure it's snug (special inexpensive wrenches are commonly available at auto parts stores)

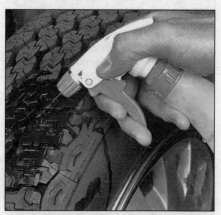

5.4b If the valve core is tight, raise the corner of the vehicle with the low tire and spray a soapy water solution onto the tread as the tire is turned slowly - slow leaks will cause small bubbles to appear

5.8 To extend the life of your tires, check the air pressure at least once a week with an accurate gauge (don't forget the spare!)

ture is suspected, it can be easily verified by spraying a solution of soapy water onto the puncture area **(see illustration)**. The soapy solution will bubble if there is a leak. Unless the puncture is inordinately large, a tire shop or gas station can usually repair the punctured tire.

5   Carefully inspect the inner sidewall of each tire for evidence of brake fluid leakage. If you see any, inspect the brakes immediately.

6   Correct tire air pressure adds miles to the lifespan of the tires, improves mileage and enhances overall ride quality. Tire pressure cannot be accurately estimated by look-

ing at a tire, particularly if it is a radial. A tire pressure gauge is therefore essential. Keep an accurate gauge in the glove box. The pressure gauges fitted to the nozzles of air hoses at gas stations are often inaccurate.

7   Always check tire pressure when the tires are cold. "Cold," in this case, means the vehicle has not been driven over a mile in the three hours preceding a tire pressure check. A pressure rise of four to eight pounds is not uncommon once the tires are warm.

8   Unscrew the valve cap protruding from the wheel or hubcap and push the gauge

firmly onto the valve **(see illustration)**. Note the reading on the gauge and compare this figure to the recommended tire pressure shown on the tire placard on the left door. Be sure to reinstall the valve cap to keep dirt and moisture out of the valve stem mechanism. Check all four tires and, if necessary, add enough air to bring them up to the recommended pressure levels.

9   Don't forget to keep the spare tire inflated to the specified pressure (consult your owner's manual). Note that the air pressure specified for the compact spare is significantly higher than the pressure of the regular tires.

6   Power steering fluid level check (every 3000 miles or 3 months)

Refer to illustration 6.6

1993 through 1999 models

1   Unlike manual steering, the power steering system relies on fluid which may, over a period of time, require replenishing.

2   The fluid reservoir for the power steering pump is located on the right inner fender panel near the shock tower.

3   For the check, the front wheels should be pointed straight ahead and the engine should be off.

4   Use a clean rag to wipe off the reservoir cap and the area around the cap. This will help prevent any foreign matter from entering the reservoir during the check.

5   Twist off the cap and check the temperature of the fluid at the end of the dipstick with your finger.

6   Wipe off the fluid with a clean rag, reinsert it, then withdraw it and read the fluid level **(see illustration)**. The level should be near the upper mark on the dipstick. On some models the dipstick is marked so the fluid can be checked either cold or hot. The level should be at the HOT mark if the fluid was hot to the touch. It should be at the COLD mark if the fluid was cool to the touch. At no time should the fluid level drop below the upper mark for each heat range.

7   If additional fluid is required, pour the specified type directly into the reservoir, using a funnel to prevent spills.

8   If the reservoir requires frequent fluid additions, all power steering hoses, hose connections, the power steering pump and the rack and pinion assembly should be carefully checked for leaks.

2000 and later models

9   To check the fluid level in the reservoir, simply look at the MIN and MAX on the side of the reservoir. The level should be within the specified distance from the maximum fill line. Use the HOT range at fluid temperatures of 122 to 176 degrees F (50 to 80 degrees C). Use the COLD range at fluid temperatures of 32 to 86 degrees F (0 to 30 degrees C). At no time should the fluid level drop below the minimum mark on the reservoir.

10   If additional fluid is required, use a clean rag to wipe off the reservoir cap and area around the cap. This will prevent any foreign matter from entering the reservoir during the check.

11   Unscrew the top cap and pour the specified type directly into the reservoir, using a funnel to prevent spills. Install the cap and tighten securely.

12   If the reservoir requires frequent fluid additions, all power steering hoses, hose connections, the power steering pump and the rack and pinion assembly should be carefully checked for leaks

6.6 The power steering fluid is checked with a dipstick which is part of the cap - the fluid level varies with temperature, so the fluid can be checked hot or cold

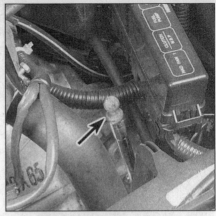

7.4a The automatic transaxle dipstick (arrow) is located in a tube which extends forward from the transaxle

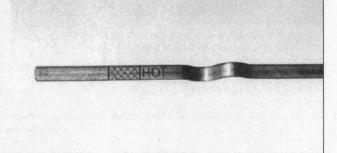

7.4b Check the automatic transaxle fluid with the engine idling at operating temperature and the gear selector in Park, then add fluid to bring the level to the upper portion of the cross-hatched area

7   Automatic transaxle fluid level check (every 3000 miles or 3 months)

Refer to illustrations 7.4a and 7.4b

1   The level of the automatic transaxle fluid should be carefully maintained. Low fluid level can lead to slipping or loss of drive, while overfilling can cause foaming, loss of fluid and transaxle damage.

2   The transaxle fluid level should only be checked when the transaxle is hot (at its normal operating temperature). If the vehicle has just been driven over 10 miles (15 miles in a frigid climate), and the fluid temperature is 160 to 175-degrees F, the transaxle is hot. **Caution:** *If the vehicle has just been driven for a long time at high speed or in city traffic in hot weather, or if it has been pulling a trailer, an accurate fluid level reading cannot be obtained. Allow the fluid to cool down for about 30 minutes.*

3   If the vehicle has not been driven, park the vehicle on level ground, set the parking brake, then start the engine and bring it to operating temperature. While the engine is idling, depress the brake pedal and move the selector lever through all the gear ranges, beginning and ending in Park.

4   With the engine still idling, remove the dipstick from its tube **(see illustration)**. Check the level of the fluid on the dipstick

(see illustration) and note its condition.

5   Wipe the fluid from the dipstick with a clean rag and reinsert it back into the filler tube until the cap seats.

6   Pull the dipstick out again and note the fluid level. If the transaxle is cold, the level should be in the COLD or COOL range on the dipstick. If it is hot, the fluid level should be in the HOT range. If the level is at the low side of either range, add the specified automatic transmission fluid through the dipstick tube with a funnel.

7   Add just enough of the recommended fluid to fill the transaxle to the proper level. It takes about one pint to raise the level from the low mark to the high mark when the fluid is hot, so add the fluid a little at a time and keep checking the level until it is correct.

8   The condition of the fluid should also be checked along with the level. If the fluid at the end of the dipstick is black or a dark reddish brown color, or if it emits a burned smell, the fluid should be changed (see Section 28). If you are in doubt about the condition of the fluid, purchase some new fluid and compare the two for color and smell.

8   Engine oil and oil filter change (every 3000 miles or 3 months)

Refer to illustrations 8.2, 8.7, 8.13 and 8.15

1   Frequent oil changes are the best pre-

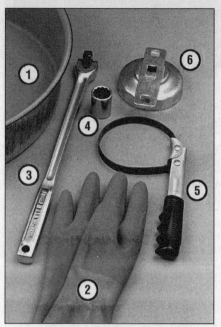

8.2 These tools are required when changing the engine oil and filter

1   *Drain pan - It should be fairly shallow in depth, but wide in order to prevent spills*
2   *Rubber gloves - When removing the drain plug and filter, it is inevitable that you will get oil on your hands (the gloves will prevent burns)*
3   *Breaker bar - Sometimes the oil drain plug is pretty tight and a long breaker bar is needed to loosen it*
4   *Socket – To be used with the breaker bar or a ratchet (must be the correct size to fit the drain plug)*
5   *Filter wrench - This is a metal band-type wrench, which requires clearance around the filter to be effective*
6   *Filter wrench - This type fits on the bottom of the filter and can be turned with a ratchet or breaker bar (different size wrenches are available for different types of filters)*

ventive maintenance the home mechanic can give the engine, because aging oil becomes diluted and contaminated, which leads to premature engine wear.

2   Make sure that you have all the necessary tools before you begin this procedure **(see illustration)**. You should also have plenty of rags or newspapers handy for mopping up any spills.

3   Access to the underside of the vehicle is greatly improved if the vehicle can be lifted on a hoist, driven onto ramps or supported by jackstands.

4   If this is your first oil change, get under the vehicle and familiarize yourself with the location of the oil drain plug. The engine and exhaust components will be warm during the actual work, so try to anticipate any potential problems before the engine and accessories are hot.

8.7 Use a proper size box-end wrench or socket to remove the oil drain plug and avoid rounding it off

5   Park the vehicle on a level spot. Start the engine and allow it to reach its normal operating temperature (the needle on the temperature gauge should be at least above the bottom mark). Warm oil and contaminates will flow out more easily. Turn off the engine when it's warmed up. Remove the filler cap in the valve cover.

6   Raise the vehicle and support it on jackstands. **Warning:** *To avoid personal injury, never get beneath the vehicle when it is supported by only by a jack. The jack provided with your vehicle is designed solely for raising the vehicle to remove and replace the wheels. Always use jackstands to support the vehicle when it becomes necessary to place your body underneath the vehicle.*

7   Being careful not to touch the hot exhaust components, place the drain pan under the drain plug in the bottom of the pan and remove the plug **(see illustration)**. You may want to wear gloves while unscrewing the plug the final few turns if the engine is really hot.

8   Allow the old oil to drain into the pan. It may be necessary to move the pan farther under the engine as the oil flow slows to a trickle. Inspect the old oil for the presence of metal shavings and chips.

9   After all the oil has drained, wipe off the drain plug with a clean rag. Even minute metal particles clinging to the plug would immediately contaminate the new oil.

10   Clean the area around the drain plug opening, reinstall the plug and tighten it securely, but do not strip the threads.

11   Move the drain pan into position under the oil filter.

12   Remove all tools, rags, etc. from under the vehicle, being careful not to spill the oil in the drain pan, then lower the vehicle.

13   Loosen the oil filter **(see illustration)** by turning it counterclockwise with the filter wrench. Any standard filter wrench should work. Once the filter is loose, use your hands

8.13 The oil filter is located on the back side of the engine facing the firewall - use an oil filter wrench for removal - DO NOT use the wrench to tighten the new filter

to unscrew it from the block. Just as the filter is detached from the block, immediately tilt the open end up to prevent the oil inside the filter from spilling out. **Warning:** *The engine exhaust manifold may still be hot, so be careful.*

14   With a clean rag, wipe off the mounting surface on the block. If a residue of old oil is allowed to remain, it will smoke when the block is heated up. It will also prevent the new filter from seating properly. Also make sure that the none of the old gasket remains stuck to the mounting surface. It can be removed with a scraper if necessary.

15   Compare the old filter with the new one to make sure they are the same type. Smear some engine oil on the rubber gasket of the new filter and screw it into place **(see illustration)**. Because over-tightening the filter will damage the gasket, do not use a filter wrench to tighten the filter. Tighten it by hand until the gasket contacts the seating surface. Then seat the filter by giving it an additional 3/4-turn.

8.15 Lubricate the oil filter gasket with clean engine oil before installing the filter on the engine

1

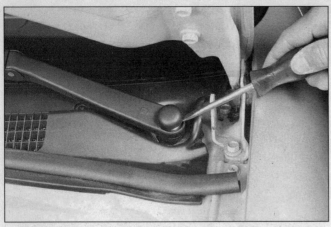

9.3 Gently pry off the trim cap and check the tightness of the wiper arm retaining nut

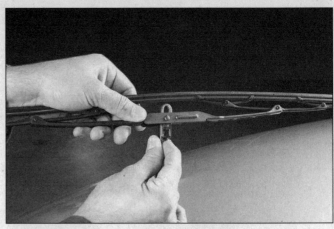

9.5 Press on the release tab and push the blade assembly down out of the hook in the arm

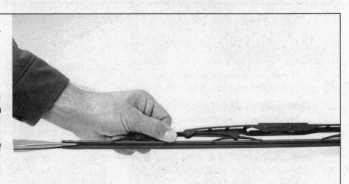

9.6 Use needle-nose pliers to compress the rubber element, then slide the element out - slide the new element in and lock the blade assembly fingers into the notches of the wiper element

16   Add new oil to the engine through the oil filler cap in the valve cover. Use a spout or funnel to prevent oil from spilling onto the top of the engine. Pour three quarts of fresh oil into the engine. Wait a few minutes to allow the oil to drain into the pan, then check the level on the oil dipstick (see Section 4 if necessary). If the oil level is at or near the H mark, install the filler cap hand tight, start the engine and allow the new oil to circulate.

17   Allow the engine to run for about a minute. While the engine is running, look under the vehicle and check for leaks at the oil pan drain plug and around the oil filter. If either is leaking, stop the engine and tighten the plug or filter slightly.

18   Wait a few minutes to allow the oil to trickle down into the pan, then recheck the level on the dipstick and, if necessary, add enough oil to bring the level to the H mark.

19   During the first few trips after an oil change, make it a point to check frequently for leaks and proper oil level.

20   The old oil drained from the engine cannot be reused in its present state and should be discarded. Oil reclamation centers, auto repair shops and gas stations will normally accept the oil, which can be recycled. After the oil has cooled, it can be drained into a suitable container (capped plastic jugs, topped bottles, milk cartons, etc.) for transport to one of these disposal sites.

9   Windshield wiper blade inspection and replacement (every 7500 miles or 6 months)

Refer to illustrations 9.3, 9.5 and 9.6

1   The windshield wiper and blade assembly should be inspected periodically for damage, loose components and cracked or worn blade elements.

2   Road film can build up on the wiper blades and affect their efficiency, so they should be washed regularly with a mild detergent solution.

3   The action of the wiping mechanism can loosen bolts, nuts and fasteners, so they should be checked and tightened, as necessary **(see illustration)**, at the same time the wiper blades are checked.

4   If the wiper blade elements are cracked, worn or warped, or no longer clean adequately, they should be replaced with new ones.

5   Lift the arm assembly away from the glass for clearance, press on the release lever, then slide the wiper blade assembly out of the hook in the end of the arm **(see illustration)**.

6   Use needle-nose pliers to compress the blade element, then slide the element out of the frame and discard it **(see illustration)**.

7   Installation is the reverse of removal.

10   Battery check, maintenance and charging (every 7500 miles or 6 months)

Refer to illustrations 10.1, 10.6a, 10.6b, 10.7a and 10.7b

Warning: *Certain precautions must be followed when checking and servicing the battery. Hydrogen gas, which is highly flammable, is always present in the battery cells, so keep lighted tobacco and all other open flames and sparks away from the battery. The electrolyte inside the battery is actually dilute sulfuric acid, which will cause injury if splashed on your skin or in your eyes. It will also ruin clothes and painted surfaces. When removing the battery cables, always detach the negative cable first and hook it up last!*

1   A routine preventive maintenance program for the battery in your vehicle is the only way to ensure quick and reliable starts. But before performing any battery maintenance, make sure that you have the proper equipment necessary to work safely around the battery **(see illustration)**.

2   There are also several precautions that should be taken whenever battery maintenance is performed. Before servicing the battery, always turn the engine and all accessories off and disconnect the cable from the negative terminal of the battery.

3   The battery produces hydrogen gas, which is both flammable and explosive. Never create a spark, smoke or light a match around the battery. Always charge the battery in a ventilated area.

4   Electrolyte contains poisonous and corrosive sulfuric acid. Do not allow it to get in your eyes, on your skin or on your clothes. Never ingest it. Wear protective safety glasses when working near the battery. Keep children away from the battery.

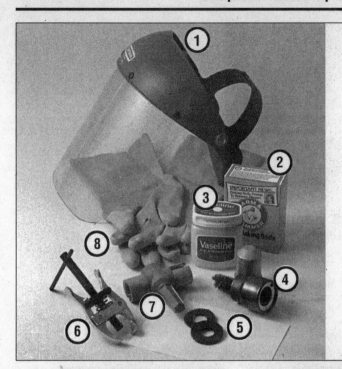

10.1 Tools and materials required for battery maintenance

1   *Face shield/safety goggles* - When removing corrosion with a brush, the acidic particles can easily fly up into your eyes

2   *Baking soda* - A solution of baking soda and water can be used to neutralize corrosion

3   *Petroleum jelly* - A layer of this on the battery posts will help prevent corrosion

4   *Battery post/cable cleaner* - This wire brush cleaning tool will remove all traces of corrosion from the battery posts and cable clamps

5   *Treated felt washers* - Placing one of these on each post, directly under the cable clamps, will help prevent corrosion

6   *Puller* - Sometimes the cable clamps are very difficult to pull off the posts, even after the nut/bolt has been completely loosened. This tool pulls the clamp straight up and off the post without damage

7   *Battery post/cable cleaner* - Here is another cleaning tool which is a slightly different version of number 4 above, but it does the same thing

8   *Rubber gloves* - Another safety item to consider when servicing the battery; remember that's acid inside the battery

1

10.6a Battery terminal corrosion usually appears as light, fluffy powder

10.6b Removing a cable from the battery post with a wrench - sometimes special battery pliers are required for this procedure if corrosion has caused deterioration of the nut hex (always remove the ground cable first and hook it up last!)

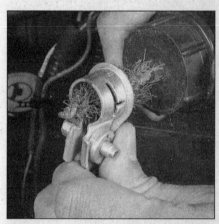

10.7a When cleaning the cable clamps, all corrosion must be removed (the inside of the clamp is tapered to match the taper on the post, so don't remove too much material)

5   Note the external condition of the battery. If the positive terminal and cable clamp on your vehicle's battery is equipped with a rubber protector, make sure it isn't torn or damaged. It should completely cover the terminal. Look for any corroded or loose connections, cracks in the case or cover or loose hold-down clamps. Also check the entire length of each cable for cracks and frayed conductors.

6   If corrosion, which looks like white, fluffy deposits **(see illustration)** is evident, particularly around the terminals, the battery should be removed for cleaning. Loosen the cable clamp bolts with a wrench, being careful to remove the ground cable first, and slide them off the terminals **(see illustration)**. Then disconnect the hold-down clamp bolt and nut, remove the clamp and lift the battery from the engine compartment.

7   Clean the cable clamps thoroughly with

a battery brush or a terminal cleaner and a solution of warm water and baking soda **(see illustration)**. Wash the terminals and the top of the battery case with the same solution but make sure that the solution doesn't get into the battery. When cleaning the cables, terminals and battery top, wear safety goggles and rubber gloves to prevent any solution from coming in contact with your eyes or hands. Wear old clothes too - even diluted, sulfuric acid splashed onto clothes will burn holes in them. If the terminals have been extensively corroded, clean them up with a terminal cleaner **(see illustration)**. Thoroughly wash all cleaned areas with plain water.

8   Make sure the battery tray is in good condition and the hold-down clamp bolt or

10.7b Regardless of the type of tool used to clean the battery posts, a clean, shiny surface should be the result

nut is tight. If the battery is removed from the tray, make sure no parts remain in the bottom of the tray when the battery is reinstalled. When reinstalling the hold-down clamp bolt or nut, do not over-tighten it.

9   Information on removing and installing the battery can be found in Chapter 5. Information on jump starting can be found at the front of this manual. For more detailed battery checking procedures, refer to the *Haynes Automotive Electrical Manual.*

Cleaning

10   Corrosion on the hold-down components, battery case and surrounding areas can be removed with a solution of water and baking soda. Thoroughly rinse all cleaned areas with plain water.

11   Any metal parts of the vehicle damaged by corrosion should be covered with a zinc-based primer, then painted.

Charging

Warning: *When batteries are being charged, hydrogen gas, which is very explosive and flammable, is produced. Do not smoke or allow open flames near a charging or a recently charged battery. Wear eye protection when near the battery during charging. Also, make sure the charger is unplugged before connecting or disconnecting the battery from the charger.*

12   Slow-rate charging is the best way to restore a battery that's discharged to the point where it will not start the engine. It's also a good way to maintain the battery charge in a vehicle that's only driven a few miles between starts. Maintaining the battery charge is particularly important in the winter when the battery must work harder to start the engine and electrical accessories that drain the battery are in greater use.

13   It's best to use a one or two-amp battery charger (sometimes called a "trickle" charger). They are the safest and put the least strain on the battery. They are also the least expensive. For a faster charge, you can use a higher amperage charger, but don't use one rated more than 1/10th the amp/hour rating of the

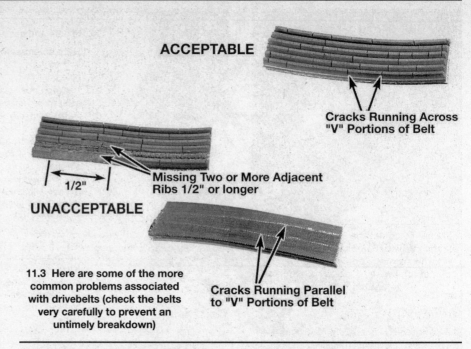

ACCEPTABLE

Cracks Running Across "V" Portions of Belt

1/2"

Missing Two or More Adjacent Ribs 1/2" or longer

UNACCEPTABLE

Cracks Running Parallel to "V" Portions of Belt

11.3 Here are some of the more common problems associated with drivebelts (check the belts very carefully to prevent an untimely breakdown)

battery. Rapid boost charges that claim to restore the power of the battery in one to two hours are hardest on the battery and can damage batteries not in good condition. This type of charging should only be used in emergency situations.

14   The average time necessary to charge a battery should be listed in the instructions that come with the charger. As a general rule, a trickle charger will charge a battery in 12 to 16 hours.

11   Drivebelt check, adjustment and replacement (every 7500 miles or 6 months)

Refer to illustrations 11.3, 11.4, 11.5a, 11.5b and 11.8

Check

1   The alternator, power steering pump

and air conditioning compressor drivebelts are located at the right end of the engine. The good condition and proper adjustment of the alternator belt is critical to the operation of the engine. Because of their composition and the high stresses to which they are subjected, drivebelts stretch and deteriorate as they get older. They must therefore be periodically inspected.

2   The number of belts used on a particular vehicle depends on the accessories installed. The main belt transmits power from the crankshaft to the water pump, alternator and the power steering pump. The second belt transmits power from the crankshaft to the air conditioning compressor.

3   With the engine off, open the hood and locate the drivebelts. With a flashlight, check each belt for separation of the adhesive rubber on both sides of the core, core separation from the belt side, a severed core, separation of the ribs from the adhesive rubber, cracking or separation of the ribs, and torn or worn ribs or cracks in the inner ridges of the ribs **(see illustration)**. Also check for fraying and glazing, which gives the belt a shiny appearance. Both sides of the belt should be inspected, which means you will have to twist the belt to check the underside. Use your fingers to feel the belt where you can't see it. If any of the above conditions are evident, replace the belt (go to Step 7).

4   Check the belt tension by pushing firmly on the belt with your thumb at a distance halfway between the pulleys and note how far the belt can be pushed (deflected). Measure this deflection with a ruler **(see illustration)**. The belt should deflect 1/4-inch if the distance from pulley center to pulley center is between 7 and 11 inches; the belt should deflect 1/2-inch if the distance from pulley center to pulley center is between 12 and 16 inches.

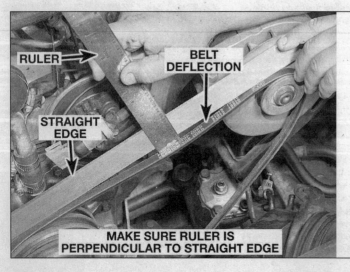

RULER

BELT DEFLECTION

STRAIGHT EDGE

11.4 Measuring drivebelt deflection with a straightedge and ruler

MAKE SURE RULER IS PERPENDICULAR TO STRAIGHT EDGE

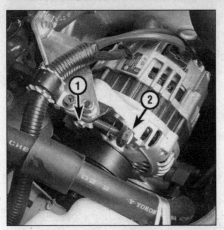

11.5a To adjust the alternator/power steering drivebelt, loosen the lock bolt (1) then turn the adjusting bolt (2) counterclockwise to loosen or clockwise to tighten the belt

11.5b To adjust the air conditioning compressor drivebelt, remove the plastic inner fender shield to access the lower idler pulley lock nut (1) and adjusting bolt (2)

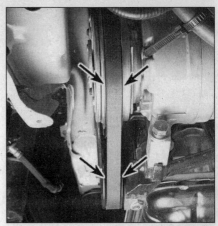

11.8 When installing ribbed V-belts, make sure the belt is centered between the pulley edges (arrows) - it must not overlap either edge of the pulley

Adjustment

5   Belt tension is adjusted by moving the idler pulley. To adjust the belt, loosen the locknut and turn the idler pulley adjusting bolt **(see illustrations)**. Measure the belt tension in accordance with the above method. Repeat this Step until the drivebelt is adjusted properly.

Replacement

6   To replace a belt, loosen the drivebelt adjustment bolt, slip the belt off the crankshaft pulley and remove it. If you are replacing the alternator/power steering pump belt, you'll have to remove the air conditioning compressor belt first because of the way they are arranged on the crankshaft pulley. Because of this and because belts tend to wear out more or less together, it is a good idea to replace both belts at the same time. Mark each belt and its appropriate pulley groove so the replacement belts can be installed in their proper positions.

7   Take the old belts to the parts store in order to make a direct comparison for length, width and design.

8   After replacing ribbed drivebelts, make sure that it fits properly in the ribbed grooves in the pulleys **(see illustration)**. It is essential that the belt be properly centered.

9   Adjust the belt(s) in accordance with the procedure outlined above.

---

12   Underhood hose check and replacement (every 7500 miles or 6 months)

Caution: *Replacement of air conditioning hoses must be left to a dealer service department or air conditioning shop that has the equipment to depressurize the system safely. Never remove air conditioning components or hoses until the system has been depressurized.*

General

1   High temperatures in the engine compartment can cause the deterioration of the rubber and plastic hoses used for engine, accessory and emission systems operation. Periodic inspection should be made for cracks, loose clamps, material hardening and leaks.

2   Information specific to the cooling system hoses can be found in Section 13.

3   Some, but not all, hoses are secured to the fittings with clamps. Where clamps are used, check to be sure they haven't lost their tension, allowing the hose to leak. If clamps aren't used, make sure the hose has not expanded and/or hardened where it slips over the fitting, allowing it to leak.

Vacuum hoses

4   It's quite common for vacuum hoses, especially those in the emissions system, to be color coded or identified by colored stripes molded into them. Various systems require hoses with different wall thickness, collapse resistance and temperature resistance. When replacing hoses, be sure the new ones are made of the same material.

5   Often the only effective way to check a hose is to remove it completely from the vehicle. If more than one hose is removed, be sure to label the hoses and fittings to ensure correct installation.

6   When checking vacuum hoses, be sure to include any plastic T-fittings in the check. Inspect the fittings for cracks and the hose where it fits over the fitting for distortion, which could cause leakage.

7   A small piece of vacuum hose (1/4-inch inside diameter) can be used as a stethoscope to detect vacuum leaks. Hold one end of the hose to your ear and probe around vacuum hoses and fittings, listening for the "hissing" sound characteristic of a vacuum leak. **Warning:** *When probing with the vacuum hose stethoscope, be very careful not to come into contact with moving engine components such as the drivebelts, cooling fan, etc.*

Fuel hose

Warning: *There are certain precautions which must be taken when inspecting or servicing fuel system components. Work in a well ventilated area and do not allow open flames (cigarettes, appliance pilot lights, etc.) or bare light bulbs near the work area. Mop up any spills immediately and do not store fuel-soaked rags where they could ignite.*

8   Check all rubber fuel lines for deterioration and chafing. Check especially for cracks in areas where the hose bends and just before fittings, such as where a hose attaches to the fuel filter.

9   High quality fuel line, meeting the manufacturer's original specifications, should be used for fuel line replacement. Never, under any circumstances, use unreinforced vacuum line, clear plastic tubing or water hose for fuel lines.

10   Spring-type clamps are commonly used on fuel lines. These clamps often lose their tension over a period of time, and can be "sprung" during removal. Replace all spring-type clamps with screw clamps whenever a hose is replaced.

Metal lines

11   Sections of metal line are often used for fuel line between the fuel pump and carburetor or fuel injection unit. Check carefully to be sure the line has not been bent or crimped and that cracks have not started in the line.

12   If a section of metal fuel line must be replaced, only seamless steel tubing should be used, since copper and aluminum tubing don't have the strength necessary to withstand normal engine vibration.

13   Check the metal brake lines where they enter the master cylinder and brake proportioning unit (if used) for cracks in the lines or loose fittings. Any sign of brake fluid leakage calls for an immediate thorough inspection of the brake system.

Check for a chafed area that could fail prematurely.

Check for a soft area indicating the hose has deteriorated inside.

Overtightening the clamp on a hardened hose will damage the hose and cause a leak.

Check each hose for swelling and oil-soaked ends. Cracks and breaks can be located by squeezing the hose

13.4 Hoses, like drivebelts, have a habit of failing at the worst possible time - to prevent the inconvenience of a blown radiator or heater hose, inspect them carefully as shown here

13   Cooling system check (every 7500 miles or 6 months)

Refer to illustration 13.4

1   Many major engine failures can be attributed to a faulty cooling system. If the vehicle is equipped with an automatic transaxle, the cooling system also cools the transaxle fluid and thus plays an important role in prolonging transaxle life.

2   The cooling system should be checked with the engine cold. Do this before the vehicle is driven for the day or after the engine has been shut off for at least three hours.

3   Remove the radiator cap by turning it to the left until it reaches a stop. If you hear a hissing sound (indicating there is still pressure in the system), wait until it stops. Now press down on the cap with the palm of your hand and continue turning to the left until the cap can be removed. Thoroughly clean the

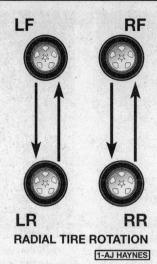

RADIAL TIRE ROTATION

1-AJ HAYNES

14.2 The recommended tire rotation pattern for these vehicles

cap, inside and out, with clean water. Also clean the filler neck on the radiator. All traces of corrosion should be removed. The coolant inside the radiator should be relatively transparent. If it's rust colored, the system should be drained and refilled (see Section 26). If the coolant level isn't up to the top, add additional antifreeze/coolant mixture (see Section 4).

4   Carefully check the large upper and lower radiator hoses along with the smaller diameter heater hoses which run from the engine to the firewall. Inspect each hose along its entire length, replacing any hose which is cracked, swollen or shows signs of deterioration. Cracks may become more apparent if the hose is squeezed **(see illustration)**. Regardless of condition, it's a good idea to replace hoses with new ones every two years.

5   Make sure that all hose connections are tight. A leak in the cooling system will usually show up as white or rust colored deposits on the areas adjoining the leak. If wire-type clamps are used at the ends of the hoses, it may be a good idea to replace them with more secure screw-type clamps.

6   Use compressed air or a soft brush to remove bugs, leaves, etc. from the front of the radiator or air conditioning condenser. Be careful not to damage the delicate cooling fins or cut yourself on them.

7   Every other inspection, or at the first indication of cooling system problems, have the cap and system pressure tested. If you don't have a pressure tester, most gas stations and repair shops will do this for a minimal charge.

14   Tire rotation (every 7500 miles or 6 months))

Refer to illustration 14.2

1   The tires should be rotated at the speci-

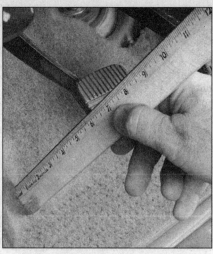

15.1 To check clutch pedal height, measure the distance between the pedal pad and the floor. To check clutch pedal freeplay, measure the distance between the natural resting place of the pedal and the point at which you encounter resistance

fied intervals and whenever uneven wear is noticed. Since the vehicle will be raised and the tires removed anyway, check the brakes (see Section 16) at this time.

2   Radial tires must be rotated in a specific pattern **(see illustration)**.

3   Refer to the information in *Jacking and towing* at the front of this manual for the proper procedures to follow when raising the vehicle and changing a tire. If the brakes are to be checked, do not apply the parking brake as stated. Make sure the tires are blocked to prevent the vehicle from rolling.

4   Preferably, the entire vehicle should be raised at the same time. This can be done on a hoist or by jacking up each corner and then lowering the vehicle onto jackstands placed under the frame rails. Always use four jackstands and make sure the vehicle is firmly supported.

5   After rotation, check and adjust the tire pressures as necessary and be sure to check the lug nut tightness.

6   For further information on the wheels and tires, refer to Chapter 10.

15   Clutch pedal height and freeplay check and adjustment (every 15,000 miles or 12 months)

Pedal height

Refer to illustrations 15.1 and 15.3

1   With the clutch pedal fully released, measure the distance from the top of the pad to the floor **(see illustration)**.

2   If the height is not as listed in the Specifications at the beginning of this Chapter it must be adjusted.

3   Loosen the locknut on the pedal stopper

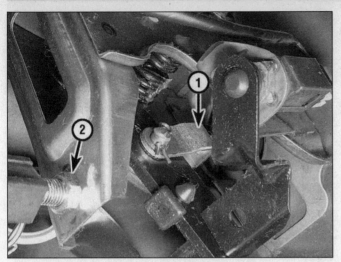

15.3 Master cylinder push rod (1) pedal stopper or cruise control switch (2)

16.6 You will find an inspection hole like this in each caliper - placing a steel ruler across the hole should enable you to determine the thickness of remaining pad material for both inner and outer pads

or cruise control (ASCD) switch **(see illustration)**.

4   Turn the pedal stopper or cruise control switch until the pedal height is correct.

5   Tighten the locknut and recheck the clutch pedal height.

Pedal freeplay

6   Press down lightly on the clutch pedal and, with a small steel ruler, measure the distance that it moves freely before the clutch resistance is felt **(see illustration 15.1)**. The freeplay should be within the specified limits listed at the beginning of this Chapter. If it isn't, it must be adjusted.

7   Loosen the locknut on the master cylinder push rod **(see illustration 15.3)** and turn the rod clockwise or counterclockwise to adjust the clutch pedal freeplay.

8   Tighten the locknut and recheck the clutch pedal freeplay.

---

16 Brake check (every 15,000 miles or 12 months)

Warning: The dust created by the brake system may contain asbestos, which is harmful to your health. Never blow it out with compressed air and don't inhale any of it. An approved filtering mask should be worn when working on the brakes. Do not, under any circumstances, use petroleum-based solvents to clean brake parts. Use brake system cleaner only! Try to use non-asbestos replacement parts whenever possible.

Note: For detailed photographs of the brake system, refer to Chapter 9.

1   In addition to the specified intervals, the brakes should be inspected every time the wheels are removed or whenever a defect is suspected. Any of the following symptoms could indicate a potential brake system defect: The vehicle pulls to one side when the brake pedal is depressed; the brakes make

squealing or dragging noises when applied; brake pedal travel is excessive; the pedal pulsates; brake fluid leaks, usually onto the inside of the tire or wheel.

2   The disc brake pads have built-in wear indicators which should make a high pitched squealing or scraping noise when they are worn to the replacement point. When you hear this noise, replace the pads immediately or expensive damage to the discs can result.

3   Loosen the wheel lug nuts.

4   Raise the vehicle and place it securely on jackstands.

5   Remove the wheels (see *Jacking and towing* at the front of this book, or your owner's manual, if necessary).

Disc brakes

Refer to illustrations 16.6 and 16.11

6   There are two pads (an outer and an inner) in each caliper. The pads are visible through inspection holes in each caliper **(see illustration)**.

7   Check the pad thickness by looking at each end of the caliper and through the inspection hole in the caliper body. If the lining material is less than the thickness listed in this Chapter's Specifications, replace the pads. **Note:** Keep in mind that the lining material is riveted or bonded to a metal backing plate and the metal portion is not included in this measurement.

8   If it is difficult to determine the exact thickness of the remaining pad material by the above method, or if you are at all concerned about the condition of the pads, remove the caliper(s), then remove the pads from the calipers for further inspection (refer to Chapter 9).

9   Once the pads are removed from the calipers, clean them with brake cleaner and re-measure them with a ruler or a vernier caliper.

10   Measure the disc thickness with a micrometer to make sure that it still has service life remaining. If any disc is thinner than

16.11 Check along the brake hoses and at each fitting (arrow) for deterioration and cracks

the specified minimum thickness, replace it (refer to Chapter 9). Even if the disc has service life remaining, check its condition. Look for scoring, gouging and burned spots. If these conditions exist, remove the disc and have it resurfaced (see Chapter 9).

11   Before installing the wheels, check all brake lines and hoses for damage, wear, deformation, cracks, corrosion, leakage, bends and twists, particularly in the vicinity of the rubber hoses at the calipers **(see illustration)**. Check the clamps for tightness and the connections for leakage. Make sure that all hoses and lines are clear of sharp edges, moving parts and the exhaust system. If any of the above conditions are noted, repair, reroute or replace the lines and/or fittings as necessary (see Chapter 9).

Rear drum brakes

Refer to illustrations 16.12, 16.14 and 16.16

12   On most models, it is possible to check the brake shoe lining thickness without removing the brake drums by removing the

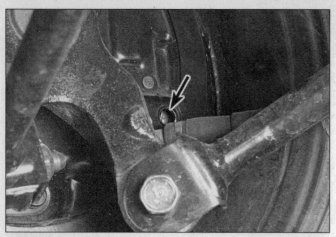

16.12 A quick check of the remaining drum brake shoe lining material can be made by removing the rubber plug in the backing plate and looking through the inspection hole (arrow)

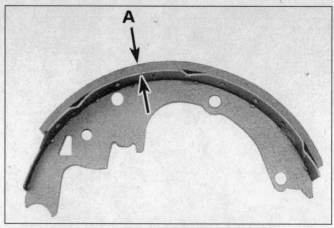

16.14 If the lining is bonded to the brake shoe, measure the lining thickness from the outer surface to the metal shoe, as shown here; if the lining is riveted to the shoe, measure from the lining outer surface to the rivet head

rubber plug from the backing plate and use a flashlight to inspect the linings **(see illustration)**. For a more thorough brake inspection, follow the procedure below.

13   Refer to Chapter 9 and remove the rear brake drums.

14   Note the thickness of the lining material on the rear brake shoes **(see illustration)** and look for signs of contamination by brake fluid and grease. If the lining material is within 1/16-inch of the recessed rivets or metal shoes, replace the brake shoes with new ones. The shoes should also be replaced if they are cracked, glazed (shiny lining surfaces) or contaminated with brake fluid or grease. See Chapter 9 for the replacement procedure.

15   Check the shoe return and hold-down springs and the adjusting mechanism to make sure they're installed correctly and in good condition. Deteriorated or distorted springs, if not replaced, could allow the linings to drag and wear prematurely.

16   Check the wheel cylinders for leakage by carefully peeling back the rubber boots **(see illustration)**. If brake fluid is noted behind the boots, the wheel cylinders must be replaced (see Chapter 9).

17   Check the drums for cracks, score marks, deep scratches and hard spots, which will appear as small discolored areas. If imperfections cannot be removed with emery cloth, the drums must be resurfaced by an automotive machine shop (see Chapter 9 for more detailed information).

18   Refer to Chapter 9 and install the brake drums.

19   Install the wheels and snug the wheel lug nuts finger tight.

20   Remove the jackstands and lower the vehicle.

21   Tighten the wheel lug nuts to the torque listed in this Chapter's Specifications.

Brake booster check

22   Sit in the driver's seat and perform the

16.16 Carefully peel back the wheel cylinder boots and check for leaking fluid indicating that the cylinder must be replaced or rebuilt

following sequence of tests.

23   With the brake fully depressed, start the engine - the pedal should move down a little when the engine starts.

24   With the engine running, depress the brake pedal several times - the travel distance should not change.

25   Depress the brake, stop the engine and hold the pedal in for about 30 seconds - the pedal should neither sink nor rise.

26   Restart the engine, run it for about a minute and turn it off. Then firmly depress the brake several times - the pedal travel should decrease with each application.

27   If your brakes do not operate as described above when the preceding tests are performed, the brake booster is either in need of repair or has failed. Refer to Chapter 9 for the removal procedure.

Parking brake

28   Slowly pull up on the parking brake and count the number of clicks you hear until the handle is up as far as it will go. The adjust-

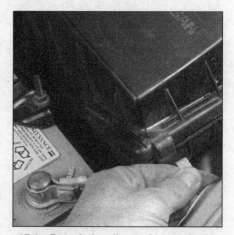

17.1a Detach the clips and separate the cover from the air cleaner housing

ment should be within the specified number of clicks listed in this Chapter's Specifications. If you hear more or fewer clicks, it's time to adjust the parking brake (refer to Chapter 9).

29   An alternative method of checking the parking brake is to park the vehicle on a steep hill with the parking brake set and the transaxle in Neutral (be sure to stay in the vehicle during this check!). If the parking brake cannot prevent the vehicle from rolling, it is in need of adjustment (see Chapter 9).

17   Air filter check and replacement (every 15,000 miles or 12 months)

Refer to illustrations 17.1a and 17.1b

1   The air filter is located inside a housing at the left (driver's) side of the engine compartment. To remove the air filter, release the four spring clips that secure the two halves of the air cleaner housing together, then lift the cover up and remove the air filter element **(see illustrations)**.

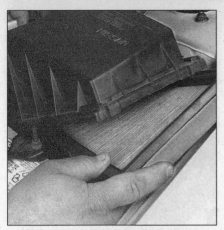

17.1b Lift the cover up and slide the element out of the housing

18.5 Inspect fuel filler hoses for cracks and make sure the clamps are tight (arrows)

18.6 Carefully inspect crimped fuel line clamps for loss of tension which can result in leaking fuel lines

2   Inspect the outer surface of the filter element. If it is dirty, replace it. If it is only moderately dusty, it can be reused by blowing it clean from the back to the front surface with compressed air. Because it is a pleated paper type filter, it cannot be washed or oiled. If it cannot be cleaned satisfactorily with compressed air, discard and replace it. While the cover is off, be careful not to drop anything down into the housing. **Caution:** *Never drive the vehicle with the air cleaner removed. Excessive engine wear could result and backfiring could even cause a fire under the hood.*

3   Wipe out the inside of the air cleaner housing.

4   Place the new filter into the air cleaner housing, making sure it seats properly.

5   Installation of the cover is the reverse of removal.

18   Fuel system check (every 15,000 miles or 12 months)

Refer to illustrations 18.5 and 18.6

Warning: *Gasoline is extremely flammable, so take extra precautions when you work on any part of the fuel system. Don't smoke or allow open flames or bare light bulbs near the work area, and don't work in a garage where a natural gas-type appliance (such as a water heater or clothes dryer) with a pilot light is present. Since gasoline is carcinogenic, wear latex gloves when there's a possibility of being exposed to fuel, and, if you spill any fuel on your skin, rinse it off immediately with soap and water. Mop up any spills immediately and do not store fuel-soaked rags where they could ignite. The fuel system is under constant pressure, so, if any fuel lines are to be disconnected, the fuel pressure in the system must be relieved first (see Chapter 4 for more information). When you perform any kind of work on the fuel system, wear safety glasses and have a Class B type fire extinguisher on hand.*

1   If you smell gasoline while driving or after the vehicle has been sitting in the sun,

inspect the fuel system immediately.

2   Remove the gas filler cap and inspect if for damage and corrosion. The gasket should have an unbroken sealing imprint. If the gasket is damaged or corroded, remove it and install a new one.

3   Inspect the fuel feed and return lines for cracks. Make sure the threaded flare nut type connectors which secure the metal fuel lines to fuel injection system and the clamps which secure the hoses to the in-line fuel filter are tight.

4   Since some components of the fuel system - the fuel tank and part of the fuel feed and return lines, for example - are underneath the vehicle, they can be inspected more easily with the vehicle raised on a hoist. If that's not possible, raise the vehicle and support it securely on jackstands.

5   With the vehicle raised and safely supported, inspect the gas tank and filler neck for punctures, cracks and other damage. The connection between the filler neck and the tank is particularly critical. Sometimes a rubber filler neck will leak because of loose clamps or deteriorated rubber **(see illustration)**. These are problems a home mechanic can usually rectify. **Warning:** *Do not, under any circumstances, try to repair a fuel tank (except rubber components). A welding torch or any open flame can easily cause fuel vapors inside the tank to explode.*

6   Carefully check all rubber hoses and metal lines leading away from the fuel tank **(see illustration)**. Check for loose connections, deteriorated hoses, crimped lines and other damage. Carefully inspect the lines from the tank to the fuel injection system. Repair or replace damaged sections as necessary (see Chapter 4).

19   Manual transaxle lubricant level check (every 15,000 miles or 12 months)

Refer to illustration 19.1

1   The manual transaxle does not have a

19.1 Remove the filler plug on the front side of the transaxle housing - place your finger in the filler plug hole to use as a dipstick to check the lubricant level

dipstick. To check the fluid level, raise the vehicle and support it securely on jackstands. On the lower front side of the transaxle housing, you will see a plug **(see illustration)**. Remove it. If the lubricant level is correct, it should be up to the lower edge of the hole.

2   If the transaxle needs more lubricant (if the level is not up to the hole), use a funnel or a gear oil pump to add more. Stop filling the transaxle when the lubricant begins to run out the hole.

3   Install the plug and tighten it securely. Drive the vehicle a short distance, then check for leaks.

20   Steering and suspension check (every 15,000 miles or 12 months)

Refer to illustrations 20.1, 20.6a and 20.6b

Note: *For detailed illustrations of the steering and suspension components, refer to Chapter 10.*

20.1 Steering wheel freeplay is the amount of travel between an initial steering input and the point at which the front wheels begin to turn (indicated by a slight resistance)

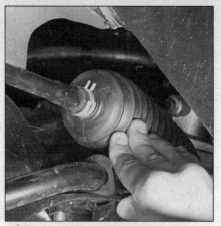

20.6a Check the steering gear dust boots for cracks and leaking steering fluid

20.6b Check the stabilizer bar bushings for deterioration at the front and the rear of the vehicle

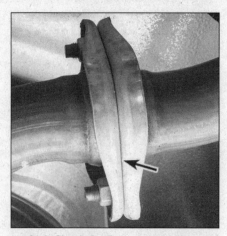

21.2 Check the flange connections (arrow) for exhaust leaks - also check that the retaining nuts are securely tightened

21.4 Check the exhaust system hangers (arrow) for damage and cracks

With the wheels on the ground

1   With the vehicle stopped and the front wheels pointed straight ahead, rock the steering wheel gently back and forth. If freeplay **(see illustration)** is excessive, a front wheel bearing, main shaft yoke, intermediate shaft yoke, lower arm balljoint or steering system joint is worn or the steering gear is out of adjustment or broken. Refer to Chapter 10 for the appropriate repair procedure.

2   Other symptoms, such as excessive vehicle body movement over rough roads, swaying (leaning) around corners and binding as the steering wheel is turned, may indicate faulty steering and/or suspension components.

3   Check the shock absorbers by pushing down and releasing the vehicle several times at each corner. If the vehicle does not come back to a level position within one or two bounces, the shocks/struts are worn and must be replaced. When bouncing the vehicle up and down, listen for squeaks and noises from the suspension components.

Under the vehicle

4   Raise the vehicle with a floor jack and support it securely on jackstands. See *jacking and towing* at the front of this book for proper jacking points.

5   Check the tires for irregular wear patterns and proper inflation. See Section 5 in this Chapter for information regarding tire wear.

6   Inspect the universal joint between the steering shaft and the steering gear housing. Check the steering gear housing for grease leakage. Make sure that the dust seals and boots are not damaged and that the boot clamps are not loose **(see illustration)**. Check the steering linkage for looseness or damage. Check the tie-rod ends for excessive play. Look for loose bolts, broken or disconnected parts and deteriorated rubber bushings on all suspension and steering

components **(see illustration)**. While an assistant turns the steering wheel from side to side, check the steering components for free movement, chafing and binding. If the steering components do not seem to be reacting with the movement of the steering wheel, try to determine where the slack is located.

7   Check the balljoints moving each lower arm up and down with a pry bar to ensure that its balljoint has no play. If any balljoint does have play, replace it. See Chapter 10 for the front balljoint replacement procedure.

8   Inspect the balljoint boots for damage and leaking grease. Replace the balljoints with new ones if they are damaged (see Chapter 10).

21   Exhaust system check (every 15,000 miles or 12 months)

Refer to illustrations 21.2 and 21.4

1   With the engine cold (at least three hours after the vehicle has been driven), check the complete exhaust system from its

starting point at the engine to the end of the tailpipe. This should be done on a hoist where unrestricted access is available.

2   Check the pipes and connections for evidence of leaks **(see illustration)**, severe corrosion or damage. Make sure that all brackets and hangers are in good condition and tight.

3   At the same time, inspect the underside of the body for holes, corrosion, open seams, etc. which may allow exhaust gases to enter the passenger compartment. Seal all body openings with silicone or body putty.

4   Rattles and other noises can often be traced to the exhaust system, especially the mounts and hangers **(see illustration)**. Try to move the pipes, muffler and catalytic converter. If the components can come in contact with the body or suspension parts, secure the exhaust system with new mounts.

5   Check the running condition of the engine by inspecting inside the end of the tailpipe. The exhaust deposits here are an indication of engine state-of-tune. If the pipe is black and sooty or coated with white deposits, the engine is in need of a tune-up, including a thorough fuel system inspection.

22.2 Check the driveaxle boot (arrow) for cracks and/or leaking grease

23.3 The fuel filter (arrow) is mounted in a clip to the right of the brake master cylinder

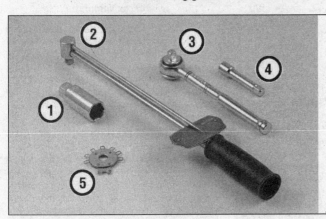

24.1 Tools required for changing spark plugs

1   *Spark plug socket - This will have special padding inside to protect the spark plug porcelain insulator*
2   *Torque wrench - Although not mandatory, use of this tool is the best way to ensure that the plugs are tightened properly*
3   *Ratchet - Standard hand tool to fit the plug socket*
4   *Extension - Depending on model and accessories, you may need special extensions and universal joints to reach one or more of the plugs*
5   *Spark plug gap gauge - This gauge for checking the gap comes in a variety of styles. Make sure the gap for your engine is included*

22 Driveaxle boot check (every 15,000 miles or 12 months)

Refer to illustration 22.2

1   The driveaxle boots are very important because they prevent dirt, water and foreign material from entering and damaging the constant velocity (CV) joints. Oil and grease can cause the boot material to deteriorate prematurely, so it's a good idea to wash the boots with soap and water.
2   Inspect the boots for tears and cracks as well as loose clamps **(see illustration)**. If there is any evidence of cracks or leaking lubricant, they must be replaced as described in Chapter 8.

23 Fuel filter replacement (every 30,000 miles or 24 months)

Refer to illustration 23.3

Warning: *Gasoline is extremely flammable, so take extra precautions when you work on any part of the fuel system. Don't smoke or allow open flames or bare light bulbs near the work area, and don't work in a garage where a natural gas-type appliance (such as a water heater or clothes dryer) with a pilot light is present. Since gasoline is carcinogenic, wear latex gloves when there's a possibility of being*

exposed to fuel, and, if you spill any fuel on your skin, rinse it off immediately with soap and water. Mop up any spills immediately and do not store fuel-soaked rags where they could ignite. The fuel system is under constant pressure, so, if any fuel lines are to be disconnected, the fuel pressure in the system must be relieved first (see Chapter 4 for more information). When you perform any kind of work on the fuel system, wear safety glasses and have a Class B type fire extinguisher on hand.

1   The canister filter is mounted in a clip on the firewall near the brake master cylinder.
2   Depressurize the fuel system (see Chapter 4), then disconnect the cable from the negative terminal of the battery.
3   Detach the filter from the bracket, loosen the screw clamps, then detach the hoses from the top and bottom of the fuel filter and remove it **(see illustration)**.
4   Note that the inlet and outlet pipes are clearly labeled on their respective ends of the filter. Make sure the new filter is installed so that it's facing the proper direction as noted above. When correctly installed, the filter should be installed so the outlet pipe faces up and the inlet pipe faces down.
5   Install the inlet and outlet fittings and tighten the screw clamps securely. Reconnect the battery cable, start the engine and check for leaks.

24 Spark plug check and replacement (every 30,000 miles or 24 months)

Refer to illustrations 24.1, 24.4a, 24.4b and 24.6

Note: *On 1999 and later models, the manufacturer suggests that checking and adjusting the spark plug gap is no longer necessary.*

1   Spark plug replacement requires a spark plug socket which fits onto a ratchet wrench. This socket is lined with a rubber grommet to protect the porcelain insulator of the spark plug and to hold the plug while you insert it into the spark plug hole. You will also need a wire-type feeler gauge to check and adjust the spark plug gap and a torque wrench to tighten the new plugs to the specified torque **(see illustration)**.
2   If you are replacing the plugs, purchase the new plugs, adjust them to the proper gap and then replace each plug one at a time.

Note: *When buying new spark plugs, it's essential that you obtain the correct plugs for your specific vehicle. This information can be found in the Specifications Section at the beginning of this Chapter, on the Vehicle Emissions Control Information (VECI) label located on the underside of the hood or in the owner's manual. If these sources specify different plugs, purchase the spark plug type*

24.4a Spark plug manufacturers recommend using a wire-type gauge when checking the gap - if the wire does not slide between the electrodes with a slight drag, adjustment is required

24.4b To change the gap, bend the side electrode only, as indicated by the arrows, and be very careful not to crack or chip the porcelain insulator surrounding the center electrode

24.6 When removing the spark plug wires, pull only on the boot and use a twisting/pulling motion

specified on the VECI label because that information is provided specifically for your engine.
3   Inspect each of the new plugs for defects. If there are any signs of cracks in the porcelain insulator of a plug, don't use it.
4   Check the electrode gaps of the new plugs. Check the gap by inserting the wire gauge of the proper thickness between the

24.8 Use a socket wrench with a long extension to unscrew the spark plugs

electrodes at the tip of the plug **(see illustration)**. The gap between the electrodes should be identical to that listed in this Chapter's Specifications or on the VECI label. If the gap is incorrect, use the notched adjuster on the feeler gauge body to bend the curved side electrode slightly **(see illustration)**.
5   If the side electrode is not exactly over the center electrode, use the notched adjuster to align them. **Caution:** *If the gap of a new plug must be adjusted, bend only the base of the ground electrode - do not touch the tip.*

Removal

Refer to illustration 24.8

6   To prevent the possibility of mixing up spark plug wires, work on one spark plug at a time. Remove the wire and boot from one spark plug. Grasp the boot - not the cable - as shown, give it a half twisting motion and pull straight up **(see illustration)**.
7   If compressed air is available, blow any dirt or foreign material away from the spark plug area before proceeding (a common bicycle pump will also work).
8   Remove the spark plug **(see illustration)**.

9   Whether you are replacing the plugs at this time or intend to reuse the old plugs, compare each old spark plug with the chart shown on the inside back cover of this manual to determine the overall running condition of the engine.

Installation

Refer to illustrations 24.10a and 24.10b

10   Prior to installation, apply a coat of anti-seize compound to the plug threads. It's often difficult to insert spark plugs into their holes without cross-threading them. To avoid this possibility, fit a short piece of 3/8-inch ID rubber hose over the end of the spark plug **(see illustrations)**. The flexible hose acts as a universal joint to help align the plug with the plug hole. Should the plug begin to cross-thread, the hose will slip on the spark plug, preventing thread damage. Tighten the plug to the torque listed in this Chapter's Specifications.
11   Attach the plug wire to the new spark plug, again using a twisting motion on the boot until it is firmly seated on the end of the spark plug.
12   Follow the above procedure for the remaining spark plugs, replacing them one at a time to prevent mixing up the spark plug wires.

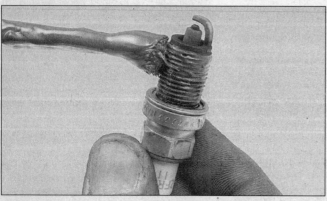

24.10a Apply a coat of anti-seize compound to the spark plug threads

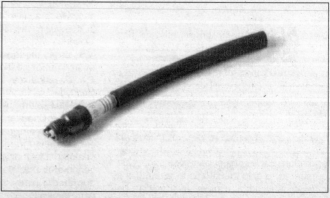

24.10b A length of 3/8-inch ID rubber hose will save time and prevent damaged threads when installing the spark plugs

25.8 Unsnap the rubber cover and pull it out of the way for access to the distributor cap screws

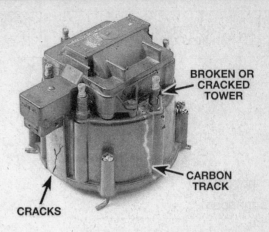

BROKEN OR CRACKED TOWER

CARBON TRACK

CRACKS

25.11a Remove the three distributor cap retaining screws - pull the cap out and up to access the rotor

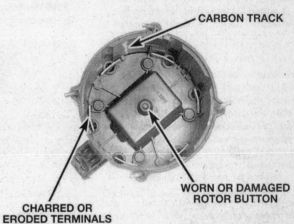

CARBON TRACK

WORN OR DAMAGED ROTOR BUTTON

CHARRED OR ERODED TERMINALS

25.11b Shown here are some of the common defects to look for when inspecting the distributor cap (if in doubt about its condition, install a new one)

25 Spark plug wire, distributor cap and rotor check and replacement (every 30,000 miles or 24 months)

Refer to illustrations 25.8, 25.11a, 25.11b, 25.12a and 25.12b

1 The spark plug wires should be checked whenever new spark plugs are installed.

2 Begin this procedure by making a visual check of the spark plug wires while the engine is running. In a darkened garage (make sure there is adequate ventilation) start the engine and observe each plug wire. Be careful not to come into contact with any moving engine parts. If there is a break in the wire, you will see arcing or a small spark at the damaged area. If arcing is noticed, make a note to obtain new wires, then allow the engine to cool and check the distributor cap and rotor.

3 The spark plug wires should be inspected one at a time to prevent mixing up the order, which is essential for proper engine operation. Each original plug wire should be numbered to help identify its location. If the number is illegible, a piece of tape can be marked with the correct number and wrapped around the plug wire.

4 Disconnect the plug wire from the spark plug. A removal tool can be used for this purpose or you can grasp the rubber boot, twist the boot half a turn and pull the boot free. Do not pull on the wire itself.

5 Check inside the boot for corrosion, which will look like a white crusty powder.

6 Push the wire and boot back onto the end of the spark plug. It should fit tightly onto the end of the plug. If it doesn't, remove the wire and use pliers to carefully crimp the metal connector inside the wire boot until the fit is snug.

7 Using a clean rag, wipe the entire length of the wire to remove built-up dirt and grease. Once the wire is clean, check for burns, cracks and other damage. Do not bend the wire sharply, because the conductor might break.

8 Remove the rubber boot (if equipped) and disconnect the wire from the distributor **(see illustration)**. Again, pull only on the rubber boot. Check for corrosion and a tight fit. Replace the wire in the distributor.

9 Inspect the remaining spark plug wires, making sure that each one is securely fastened at the distributor and spark plug when the check is complete.

10 If new spark plug wires are required, purchase a set for your specific engine model. Remove and replace the wires one at

a time to avoid mix-ups in the firing order.

11 Detach the distributor cap by loosening the three cap retaining screws. Look inside it for cracks, carbon tracks and worn, burned or loose contacts **(see illustrations)**.

12 Loosen the retaining screw and pull the rotor off the distributor shaft and examine it for cracks and carbon tracks **(see illustra-tions)**. Replace the cap and rotor if any dam-

25.12a Remove the rotor retaining screw then pull off the rotor and inspect it thoroughly

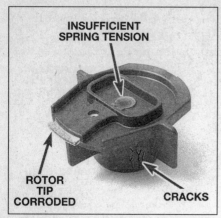

25.12b The ignition rotor should be checked for wear and corrosion as indicated here (if in doubt about its condition, buy a new one)

26.4a Push the radiator cap downward and rotate counterclockwise - never remove it when the engine is hot!

26.4b The cooling system air bleed plug is located below the distributor (arrow)

26.5 On most models you will have to remove a cover for access to the radiator drain fitting (arrow) located at the bottom of the radiator

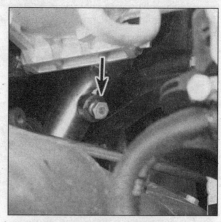

26.6 After draining the radiator, be sure to fully drain the cooling system by removing the drain plug located on the side of the lower water pipe (arrow)

age or defects are noted.

13   It is common practice to install a new cap and rotor whenever new spark plug wires are installed, but if you wish to continue using the old cap, check the resistance between the spark plug wires and the cap first. If the indicated resistance is more than the maximum value listed in this Chapter's Specifications, replace the cap and/or wires.

14   When installing a new cap, remove the wires from the old cap one at a time and attach them to the new cap in the exact same location **Note:** *If an accidental mix-up occurs, refer to the firing order Specifications at the beginning of this Chapter.*

26   Cooling system servicing (draining, flushing and refilling) (every 30,000 miles or 24 months)

Warning: *Do not allow engine coolant (antifreeze) to come in contact with your skin or painted surfaces of the vehicle. Rinse off spills immediately with plenty of water. Antifreeze is highly toxic if ingested. Never leave antifreeze laying around or in an open container or in puddles on the floor; children and pets are attracted by it's sweet smell and may drink it. Check with local authorities about disposing of used antifreeze. Many communities have collection centers which will see that antifreeze is disposed of safely.*

1   Periodically, the cooling system should be drained, flushed and refilled to replenish the antifreeze mixture and prevent formation of rust and corrosion, which can impair the performance of the cooling system and cause engine damage. When the cooling system is serviced, all hoses and the radiator cap should be checked and replaced if necessary.

Draining

Refer to illustrations 26.4a, 26.4b, 26.5 and 26.6

2   Apply the parking brake and block the

wheels. If the vehicle has just been driven, wait several hours to allow the engine to cool down before beginning this procedure.

3   On manually controlled air conditioning systems turn the ignition key to the ON position and move the temperature lever on the heater control panel to HOT - wait ten seconds and turn the ignition back to the OFF position. On automatically controlled air conditioning systems the computer must be put in the self diagnostic mode to allow coolant from the heater system to be drained (see Chapter 3 self diagnostic test 4)

4   Remove the radiator cap and the air bleed plug **(see illustrations).**

5   Move a large container under the radiator drain to catch the coolant. Then using a large screwdriver, open the radiator drain plug and direct the coolant into the container **(see illustration).**

6   After the coolant stops flowing out of the radiator, move the container under the lower water pipe drain plug **(see illustration).** Loosen the plug and allow the coolant in the block to drain.

7   While the coolant is draining, check the condition of the radiator hoses, heater hoses and clamps (refer to Section 12 if necessary).

8   Replace any damaged clamps or hoses (see Chapter 3).

Flushing

9   Once the system is completely drained, flush the radiator with fresh water from a garden hose until water runs clear at the drain. The flushing action of the water will remove sediments from the radiator but will not remove rust and scale from the engine and cooling tube surfaces.

10   These deposits can be removed by the chemical action of a cleaner. Follow the procedure outlined in the manufacturer's instructions. If the radiator is severely corroded, damaged or leaking, it should be removed (see Chapter 3) and taken to a radiator repair shop.

11   Remove the overflow hose from the coolant recovery reservoir. Drain the reservoir and flush it with clean water, then reconnect the hose.

Refilling

12   Close and tighten the radiator drain. Install and tighten the lower water pipe drain plug.

13   Make sure the heater temperature control is in the maximum heat position.

27.2 Check the charcoal canister for damage and the hose connections (arrows) for cracks and damage (1993 through 1997 shown)

28.7 On automatic transaxles, the drain plug is located on the bottom of the fluid pan

14   Slowly refill the radiator with a 50/50 mixture of water and antifreeze until coolant runs out the air bleed hole, then install the air bleed bolt and/or cap. Add coolant to the reservoir up to the lower mark.
15   Leave the radiator cap off and run the engine in a well-ventilated area until the thermostat opens (coolant will begin flowing through the radiator and the upper radiator hose will become hot). Race the engine two or three times under no load.
16   Turn the engine off and let it cool. Add more coolant mixture to bring the level back up to the lip on the radiator filler neck.
17   Squeeze the upper radiator hose to expel air, then add more coolant mixture if necessary. Replace the radiator cap.
18   Start the engine, allow it to reach normal operating temperature and check for leaks.

27 Evaporative emissions control system check (every 30,000 miles or 24 months)

Refer to illustration 27.2
1   The function of the evaporative emissions control system is to draw fuel vapors from the gas tank and fuel system, store them in a charcoal canister and then burn them during normal engine operation.
2   The most common symptom of a fault in the evaporative emissions system is a strong fuel odor in the engine compartment. If a fuel odor is detected, inspect the charcoal canister, located at the front of the engine compartment. Check the canister and all hoses for damage and deterioration **(see illustration)**.
3   The evaporative emissions control system is explained in more detail in Chapter 6.

28 Automatic transaxle fluid change (every 30,000 miles or 24 months)

Refer to illustration 28.7
1   At the specified time intervals, the automatic transaxle fluid should be drained and

replaced.
2   Before beginning work, purchase the specified transmission fluid (see *Recommended fluids and lubricants, and Capacities* at the beginning of this Chapter).
3   Other tools necessary for this job include jackstands to support the vehicle in a raised position, a wrench, a drain pan capable of holding at least four quarts, newspapers and clean rags.
4   The fluid should be drained after the vehicle has been driven and brought to operating temperature. Hot fluid is more effective than cold fluid at removing built up sediment. **Warning:** *Fluid temperature can exceed 350-degrees F in a hot transaxle. Wear protective gloves.*
5   Raise the vehicle and place it on jackstands.
6   Move the necessary equipment under the vehicle, being careful not to touch any of the hot exhaust components.
7   Place the drain pan under the drain plug in the transaxle housing or fluid pan and remove the drain plug **(see illustration)**. Be sure the drain pan is in position, as fluid will come out with some force. Once the fluid is drained, reinstall the drain plug securely.
8   Lower the vehicle.
9   With the engine off, add new fluid to the transaxle through the dipstick tube. Use a funnel to prevent spills. It is best to add a little fluid at a time, continually checking the level with the dipstick (see Section 7). Allow the fluid time to drain into the pan.
10   Start the engine and shift the selector into all positions from Park through Low then shift into Park and apply the parking brake.
11   With the engine idling, check the fluid level. Add fluid up to the Cool level on the dipstick.

29 Manual transaxle lubricant change (every 30,000 miles or 24 months)

Refer to illustration 29.4
1   At the specified time intervals, the man-

29.4 Use a box end wrench to remove the manual transaxle drain plug (arrow)

ual transaxle lubricant should be drained and replaced.
2   Before beginning work, purchase the specified transaxle lubricant (see *Recommended fluids and lubricants and Capacities* at the beginning of this Chapter).
3   Other tools necessary for this job include jackstands to support the vehicle in a raised position, 3/8-inch drive ratchet, a drain pan capable of holding at least four quarts, newspapers and clean rags.
4   Remove the drain plug **(see illustration)** and allow the old oil to drain into the pan.
5   Reinstall the drain plug securely.
6   Add new fluid until it begins to run out of the filler hole (see Section 19). Install the check/fill plug and tighten it securely.

30 Positive Crankcase Ventilation (PCV) valve check and replacement (every 30,000 miles or 24 months)

Refer to illustration 30.2
1   The PCV valve is located at the drivebelt end of the engine, under the intake manifold and attached to the breather/separator. Access is very difficult from above; access

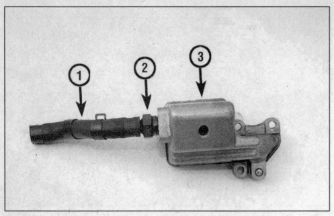

30.2 Details of the PCV valve

1 · Hose	3 · Breather/
2 · PCV valve	separator

31.2 Push up on the EGR valve diaphragm to make sure it moves easily

32.6a When the no. 1 piston is at TDC on the compression stroke, the valve clearance for the no. 1 and no. 3 cylinder exhaust valves and the no. 1 and no. 2 cylinder intake valves can be measured

32.6b Measure the clearance for each valve with a feeler gauge of the specified thickness - if the clearance is correct, you should feel a slight drag on the gauge as you pull it out

the valve from underneath the vehicle.

2   Unscrew the PCV valve from the breather/separator **(see illustration)**. Note: *It may be necessary to first unbolt the breather/separator, then unscrew the PCV valve.*

3   With the engine idling at normal operating temperature, place your finger over the end of the valve. If there's no vacuum at the valve, check for a plugged hose or valve. Replace any plugged or deteriorated hoses.

4   When purchasing a replacement PCV valve, make sure it's for your particular vehicle and engine size. Compare the old valve with the new one to make sure they're the same.

5   Installation is the reverse of removal.

31 Exhaust Gas Recirculation (EGR) valve check (every 30,000 miles or 24 months)

Refer to illustration 31.2

1   The EGR valve is located on the intake manifold, next to the firewall. The most com-

mon problem with this system is usually a stuck or corroded EGR valve.

2   With the engine cold to prevent burns, reach under the EGR valve and push up on the diaphragm. Using moderate pressure you should be able to press the diaphragm up-and-down within the housing **(see illustration)**.

3   If the diaphragm doesn't move, or moves only with much effort, replace the EGR valve with a new one. If in doubt about the condition of the valve, compare the free movement of your EGR valve with a new valve.

4   Refer to Chapter 6 for more information on the EGR system.

32 Valve clearance check and adjustment (every 60,000 miles or 48 months)

Refer to illustrations 32.6a, 32.6b, 32.7, 32.9a, 32.9b, 32.9c and 32.10
Note: *The manufacturer recommends adjusting the valve clearance at the specified inter-*

val only if the valve train is making excessive noise. The following procedure requires the use of special valve lifter tools. The tools are available from the dealer, specialty tool manufacturers and auto parts stores. It is impossible to perform this task without them.

1   Disconnect the cable from the negative terminal of the battery.

2   Remove the valve cover (see Chapter 2, Part A).

3   On manual transaxle vehicles set the parking brake and place the transaxle in the neutral position.

4   Remove the spark plugs (see Section 24).

5   Position the number 1 piston at TDC on the compression stroke and align the timing marks. Refer to the Top Dead Center Section in Chapter 2, Part A.

6   Measure the clearance of the indicated valves with a feeler gauge **(see illustrations)**. Record each measurement and compare your measurements with the desired valve clearance found in this Chapter's Specifications. Note which are out of specification, this data will be used later to determine the

32.7 When the no. 4 piston is at TDC on the compression stroke, the valve clearances for the no. 2 and no. 4 cylinder exhaust valves and the no. 3 and no. 4 cylinder intake valves can be measured

32.9a Install the valve lifter tool as shown - squeeze the handles together and rotate the tool away from the camshaft to depress the valve lifter

32.9b With the small tool wedged between the lifter and the camshaft pry the shim up with a small screwdriver at the notch . . .

32.9c . . . and remove the shim with a pair of tweezers or a magnet as shown

required replacement shims.

7   Turn the crankshaft one complete revolution and realign the timing marks. Measure and record the clearances of the remaining valves **(see illustration)**.

8   After the clearance of all the valves have been measured, rotate the crankshaft pulley until the camshaft lobe above the first valve which you intend to adjust is pointing up, away from the lifter.

9   Align the notch in the valve lifter with the notch in the cylinder head casting. Place the special valve lifter tool in position as shown, with the upper jaw over the camshaft, next to the lobe and the lower jaw on top of the shim **(see illustration)**. Depress the valve lifter by squeezing the handles of the valve lifter tool together and rotating the tool away from the camshaft. Insert the small tool between the edge of the lifter and the camshaft and release the lifter. Remove the adjusting shim with a small screwdriver and a magnet or a pair of tweezers **(see illustration)**.

10   Measure the thickness of the shim with a micrometer **(see illustration)**. To calculate the correct thickness of a replacement shim that will place the valve clearance within the specified value, use the following formula:

Intake side: $N = R + (M - 0.0138\text{-inch}$ [0.35 mm])

Exhaust side: $N = R + (M - 0.0146\text{-inch}$ [0.37 mm])

R = thickness of the old shim
M = valve clearance measured
N = thickness of the new shim

11   Select a shim with a thickness as close as possible to the valve clearance calculated. Shims, which are available in 37 sizes in increments of 0.0008-inch (0.020 mm). Available shims range in size from 0.0772 inch (1.96 mm) to 0.1055 inch (2.68 mm) (see accompanying chart below). **Note:** *Through careful analysis of the shim sizes needed to bring all the out-of-specification valve clearances within specification, it is often possible to simply move a shim that has to come out anyway to another valve lifter requiring a shim of that particular size, thereby reducing the number of new shims that must be purchased.*

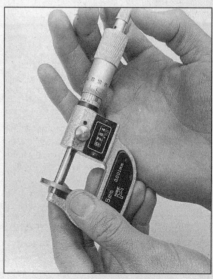

32.10 Measure the shim thickness with a micrometer or a dial caliper

Valve adjusting shim thickness chart

New valve shims are identified by a number stamped on the face of the shim which corresponds with the millimeter size. Example: the number 198 = 1.98 mm (0.0780 inch).

1.96 mm (0.0772 inch)	2.34 mm (0.0921 inch)
1.98 mm (0.0780 inch)	2.36 mm (0.0929 inch)
2.00 mm (0.0787 inch)	2.38 mm (0.0937 inch)
2.02 mm (0.0795 inch)	2.40 mm (0.0945 inch)
2.04 mm (0.0803 inch)	2.42 mm (0.0953 inch)
2.06 mm (0.0811 inch)	2.44 mm (0.0961 inch)
2.08 mm (0.0819 inch)	2.46 mm (0.0969 inch)
2.10 mm (0.0827 inch)	2.48 mm (0.0976 inch)
2.12 mm (0.0835 inch)	2.50 mm (0.0984 inch)
2.14 mm (0.0843 inch)	2.52 mm (0.0992 inch)
2.16 mm (0.0850 inch)	2.54 mm (0.1000 inch)
2.18 mm (0.0858 inch)	2.56 mm (0.1008 inch)
2.20 mm (0.0866 inch)	2.58 mm (0.1016 inch)
2.22 mm (0.0874 inch)	2.60 mm (0.1024 inch)
2.24 mm (0.0882 inch)	2.62 mm (0.1031 inch)
2.26 mm (0.0890 inch)	2.64 mm (0.1039 inch)
2.28 mm (0.0898 inch)	2.66 mm (0.1047 inch)
2.30 mm (0.0906 inch)	2.68 mm (0.1055 inch)
2.32 mm (0.0913 inch)	

12　Place the special valve lifter tools in position as shown in **illustration 32.9a**, depress the valve lifter and install the new adjusting shim. Measure the clearance with a feeler gauge to make sure that your calculations are correct.

13　Repeat this procedure until all the valves which are out of specification have been corrected.

14　The remainder of installation is the reverse of removal.

Chapter 2 Part A Engines

Contents

Specifications

General
Displacement .... 2.4L (146 cu. in.)
Cylinder numbers (drivebelt end-to-transaxle end) 1-2-3-4
Firing order 1-3-4-2

Warpage limits
Cylinder head-to-block surface .... 0.004 inch

Camshaft
Thrust clearance (endplay) .... 0.0028 to 0.0058 inch
Camshaft journal diameter
   1993 through 1996
      No. 1 .... 1.0998 to 1.1006 inches
      No. 2 through 5 .... 0.9423 to 0.9431 inches
   1997 and later .... 1.0998 to 1.1006 inches
Camshaft bearing inside diameter
   1993 through 1996
      No. 1 .... 1.1024 to 1.1033 inches
      No. 2 through 5 .... 0.9449 to 0.9459 inch
   1997 and later .... 1.1024 to 1.1033 inches

FRONT 1994 and earlier   FRONT 1995 and later

Cylinder location and distributor rotation

The blackened terminal shown on the distributor cap indicates the Number One spark plug wire position

Camshaft (continued)

Bearing oil clearance

Standard	0.0018 to 0.0035 inch
Service limit	0.0047 inch
Runout limit (all)	0.0016 inch maximum
Intake lobe height (all)	1.6699 to 1.6774 inches

Exhaust lobe height

1993 through 1995	1.6699 to 1.6931 inches

1996

Federal	1.6699 to 1.6931 inches
California	1.6699 to 1.6774 inches
1997 and later	1.6699 to 1.6774 inches

Oil pump

Body-to-outer rotor clearance	0.0045 to 0.0079 inch
Inner rotor-to-outer rotor tip clearance	0.0016 to 0.0071 inch
Inner rotor-to- cover clearance	0.0020 to 0.0035 inch
Outer rotor-to-cover clearance	0.0020 to 0.0043 inch
Inner rotor bearing clearance	0.0018 to 0.0036 inch

Torque specifications

Ft-lbs (unless otherwise indicated)

Intake manifold bolts/nuts	12 to 14
Exhaust manifold-to-block bolts/nuts	27 to 35
Crankshaft pulley-to-crankshaft bolt	105 to 112
Flywheel/driveplate bolts	105 to 112

Cylinder head bolts

Step 1	22
Step 2	59
Step 3	Loosen all bolts completely
Step 4	18 to 25
Step 5	Tighten an additional 90 degrees

Camshaft bearing cap bolts

Step 1	17 in-lbs
Step 2	80 to 104 in-lbs
Camshaft sprocket bolt	123 to 130
Idler sprocket bolt	48 to 61
Engine mount through-bolts	47 to 55
Engine mount to block bolts	22 to 29
Engine mount to crossmember bolts	47 to 55
Oil pump-to-front cover bolts	144 to 180 in-lbs
Oil pump-to-front cover screws	33 to 44 in-lbs
Oil pick-up to block bolts	144 to 168 in-lbs

Oil pan

Aluminum section-to-block

Large bolts	12 to 14
Small nuts	56 to 66 in-lbs
Steel pan-to-aluminum section	56 to 66 in-lbs
Drain plug	22 to 29

Valve cover nuts

Step 1	35 in-lbs
Step 2	69 to 95 in-lbs

1   General information

This Part of Chapter 2 is devoted to in-vehicle repair procedures for all engines. All information concerning engine removal and installation and engine block and cylinder head overhaul can be found in Part B of this Chapter.

The following repair procedures are based on the assumption that the engine is installed in the vehicle. If the engine has been removed from the vehicle and mounted on a stand, many of the steps outlined in this Part of Chapter 2 will not apply.

The Specifications included in this Part of Chapter 2 apply only to the procedures contained in this Part, basically those procedures that can be performed with the engine in the car. Part B of Chapter 2 contains the Specifications necessary for cylinder head and engine block rebuilding.

2   Repair operations possible with the engine in the vehicle

Warning: *The models covered by this manual are equipped with airbags. Always disconnect the negative battery cable, then the positive cable and wait 10 minutes before working in the vicinity of the impact sensors, steering column or instrument panel to avoid the possibility of accidental deployment of the airbag, which could cause personal injury (see Chapter 12). The airbag circuits are eas-ily identified by yellow insulation covering the entire wiring harness or just prior to the wire harness connectors. Do not use electrical test equipment on any of these wires or tamper with them in any way.*

Many major repair operations can be accomplished without removing the engine from the vehicle.

Clean the engine compartment and the exterior of the engine with some type of degreaser before any work is done. It will make the job easier and help keep dirt out of the internal areas of the engine.

Depending on the components involved, it may be helpful to remove the hood to improve access to the engine as repairs are performed (refer to Chapter 11 if necessary). Cover the fenders to prevent damage to the

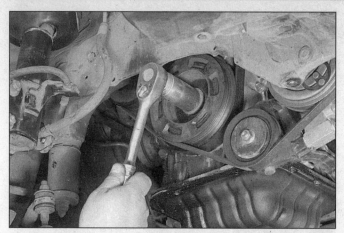

3.4 The crankshaft can be turned with a ratchet and socket on the crankshaft pulley bolt - it is necessary on some models to remove the plastic splash shield for access

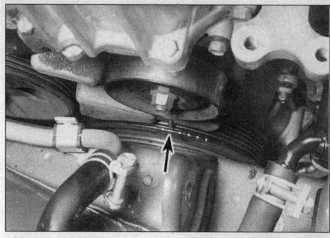

3.8 Align the red mark on the crankshaft pulley with the pointer

2A

paint. Special pads are available, but an old bedspread or blanket will also work.

If vacuum, exhaust, oil or coolant leaks develop, indicating a need for gasket or seal replacement, the repairs can generally be made with the engine in the vehicle. The intake and exhaust manifold gaskets, oil pan gasket, crankshaft oil seals and cylinder head gasket are all accessible with the engine in place.

Exterior engine components, such as the intake and exhaust manifolds, the oil pan, the oil pump, the water pump, the starter motor, the alternator, the distributor and the fuel system components can be removed for repair with the engine in place.

Since the cylinder head can be removed without pulling the engine, camshaft and valve component servicing can also be accomplished with the engine in the vehicle. Replacement of the timing chain and sprockets is also possible with the engine in the vehicle.

In extreme cases caused by a lack of necessary equipment, repair or replacement of piston rings, pistons, connecting rods and rod bearings is possible with the engine in the vehicle. However, this practice is not recommended because of the cleaning and preparation work that must be done to the components involved.

3 Top Dead Center (TDC) for number one piston - locating

Refer to illustrations 3.4 and 3.8
Note: *The following procedure is based on the assumption that the distributor is correctly installed. If you are trying to locate TDC to install the distributor correctly, piston position must be determined by feeling for compression at the number one spark plug hole, then aligning the ignition timing marks as described in step 8.*

1   Top Dead Center (TDC) is the highest point in the cylinder that each piston reaches as it travels up-and-down when the

crankshaft turns. Each piston reaches TDC on the compression stroke and again on the exhaust stroke, but TDC generally refers to piston position on the compression stroke.
2   Positioning the number one piston at TDC is an essential part of many procedures, such as camshaft, timing chain or distributor removal.
3   Before beginning this procedure, be sure to place the transmission in Neutral and apply the parking brake or block the rear wheels. Also, disable the ignition system by disconnecting the primary (low voltage) electrical connectors at the distributor) (see Chapter 5). Remove the spark plugs (see Chapter 1).
4   In order to bring any piston to TDC, the crankshaft must be turned using one of the methods outlined below. When looking at the front of the engine (timing chain end), normal crankshaft rotation is clockwise.

a)   *The preferred method is to turn the crankshaft with a socket and ratchet attached to the bolt threaded into the front of the crankshaft* **(see illustration)**.
b)   *A remote starter switch, which may save some time, can also be used. Follow the instructions included with the switch. Once the piston is close to TDC, use a socket and ratchet as described in the previous paragraph.*
c)   *If an assistant is available to turn the ignition switch to the Start position in short bursts, you can get the piston close to TDC without a remote starter switch. Make sure your assistant is out of the vehicle, away from the ignition switch, then use a socket and ratchet as described in Paragraph a) to complete the procedure.*

5   Note the position of the terminal for the number one spark plug wire on the distributor cap. If the terminal isn't marked, follow the plug wire from the number one cylinder spark plug to the cap.
6   Use a felt-tip pen or chalk to make a mark on the distributor body directly under the number one terminal.

7   Detach the cap from the distributor and set it aside (see Chapter 1 if necessary).
8   Turn the crankshaft (see Step 4) until the red notch in the crankshaft pulley is aligned with the pointer on the lower timing cover **(see illustration)**.
9   Look at the distributor rotor - it should be pointing directly at the mark you made on the distributor body.
10   If the rotor is 180-degrees off, the number one piston is at TDC on the exhaust stroke.
11   To get the piston to TDC on the compression stroke, turn the crankshaft one complete revolution (360-degrees) clockwise. The rotor should now be pointing at the mark on the distributor. When the rotor is pointing at the number one spark plug wire terminal in the distributor cap and the ignition timing marks are aligned, the number one piston is at TDC on the compression stroke. **Note:** *If it's impossible to align the ignition timing marks when the rotor is pointing at the mark on the distributor body, the timing chain may have jumped the teeth on the sprockets or may have been installed incorrectly.*
12   After the number one piston has been positioned at TDC on the compression stroke, TDC for any of the remaining pistons can be located by turning the crankshaft and following the firing order. Mark the remaining spark plug wire terminals on the distributor body just like you did for the number one terminal, then number the marks to correspond with the cylinder numbers. As you turn the crankshaft, the rotor will also turn. When it's pointing directly at one of the marks on the distributor, the piston for that particular cylinder is at TDC on the compression stroke.

4 Valve cover - removal and installation

Removal

Refer to illustrations 4.4a, 4.4b and 4.5
1   Disconnect the battery cable from the

4.4a Remove the PCV hoses, remove the throttle cable from its brackets and remove the retaining nuts (arrows)

4.4b Use a screwdriver to pry up the sealing washers around each valve cover stud

4.5 The spark plug tubes are sealed by a rubber strip (arrows) - replace if cracked or leaking

4.7 Apply a thin coat of RTV sealant where the rubber half-circles contact the front cover (arrow)

negative battery terminal.

2   Remove the spark plug wires from the spark plugs, remove the PCV hoses, and remove the throttle cable from its brackets and position it out of the way.

3   Detach the spark plug wires from the clips on the valve cover.

4   Remove the nuts around the cover's perimeter, then pry off the sealing washers on each stud **(see illustrations)**. If the cover is stuck to the cylinder head, bump the end with a wood block and a hammer to jar it loose. If that doesn't work, try to slip a flexible putty knife between the cylinder head and cover to break the seal. **Caution:** *Don't pry at the cover or housing-to-cylinder head joint or damage to the sealing surfaces may occur, leading to oil leaks after the cover is reinstalled.*

5   Remove the gasket from the valve cover. Check the condition of the spark plug tube seal **(see illustration)**, and replace it if necessary.

Installation

Refer to illustration 4.7

6   The mating surfaces of the valve cover and cylinder head must be clean when the cover is installed. Use a gasket scraper to

remove all traces of sealant from the half-circle areas at the front and rear of the camshafts, then clean the mating surfaces with lacquer thinner or acetone. If there's residue or oil on the mating surfaces when the cover is installed, oil leaks may develop. **Caution:** *Use care when scraping the soft aluminum of the cylinder head or valve cover. It is soft, and deep scratches may lead to oil leaks.*

7   Install the valve cover with a new gasket. Apply RTV sealant to the rubber half-circles **(see illustration)**. Also apply a thin coat of RTV sealant around the rear main cap of the exhaust camshaft.

8   Tighten the nuts or screws to the torque listed in this Chapter's Specifications, starting with the center fasteners and working out toward each end.

5   Intake manifold - removal and installation

Removal

Refer to illustrations 5.3, 5.5 and 5.6

1   Relieve the fuel system pressure (see Chapter 4). Disconnect the cable from the

negative battery terminal.

2   Drain the cooling system (see Chapter 1).

3   Label and detach all wire harnesses, control cables, coolant and vacuum hoses connected to the intake manifold **(see illustration)**.

5.3 The various wires and hoses should be marked to ensure correct reinstallation - this coolant hose (arrow) in front of the throttle body must be disconnected

5.5 A tubular brace (arrow) connected to the intake manifold must be unbolted from the power steering pump mounting bracket on the timing-chain end of the engine

5.6 Remove the two nuts and eight bolts that retain the intake manifold to the cylinder head (shown removed from the engine and the plenum removed for clarity)

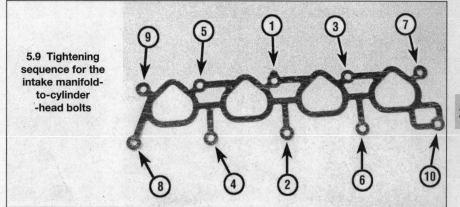

5.9 Tightening sequence for the intake manifold-to-cylinder-head bolts

4 Remove the fuel rail and injectors (see Chapter 4).
5 Unbolt the intake manifold from the two braces from underneath the vehicle. Some models may have only one brace, a tubular strut connecting the intake manifold to the power steering pump bracket (see illustration). There is also a bracket connected to the throttle body and bolted to the rear of the block (see Chapter 4, Section 13).
6 Remove the mounting nuts/bolts, then detach the manifold from the engine (see illustration). Note: *The intake manifold is a two-piece design, with a plenum located below the manifold. It is possible to remove the plenum with the intake manifold mounted to the engine, but the bolts are difficult to see and reach with a wrench. It is easier to remove the intake manifold and plenum as a unit.*

Installation

Refer to illustration 5.9
7 Use a scraper to remove all traces of old gasket material and sealant from the manifold

and cylinder head, then clean the mating surfaces with lacquer thinner or acetone.
8 Install a new gasket, then position the manifold on the cylinder head and install the nuts/bolts.
9 Tighten the nuts/bolts in three or four equal steps to the torque listed in this Chapter's Specifications. Follow the recommended tightening sequence (see illustration).
10 Install the remaining parts in the reverse order of removal. Refill the cooling system (see Chapter 1).

11 Before starting the engine, check the throttle linkage for smooth operation.
12 Run the engine and check for coolant and vacuum leaks.
13 Road test the vehicle and check for proper operation of all accessories.

6 Exhaust manifold - removal and installation

Warning: *The engine must be completely cool before beginning this procedure.*

Removal

Refer to illustrations 6.5, 6.6, 6.7a, 6.7b and 6.9
1 Disconnect the cable from the negative battery terminal.
2 Block the rear wheels and set the parking brake. Raise the front of the vehicle and support it securely on jackstands.
3 Remove the engine splash shields.
4 Disconnect the electrical connector from the oxygen sensor (see Chapter 6).
5 From underneath the vehicle, remove the lower heat shield from the exhaust pipe (see illustration).
6 Disconnect the exhaust pipe from the exhaust manifold (see illustration). Note: *Applying penetrating oil to the exhaust manifold fasteners may make removing the nuts/bolts easier.*

6.5 Remove the bolts retaining the lower heat shield to the exhaust manifold and remove the heat shield

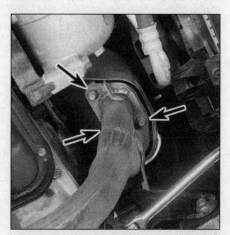

6.6 Remove the nuts (arrows) retaining the exhaust pipe to the exhaust manifold

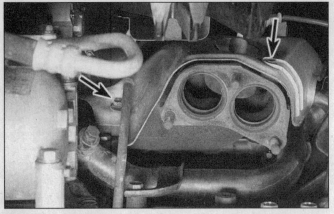

6.7a Two of the upper heat shield bolts (arrows) are best removed from underneath the vehicle - one retains the dipstick tube

6.7b Remove the upper heat shield bolts (arrows)

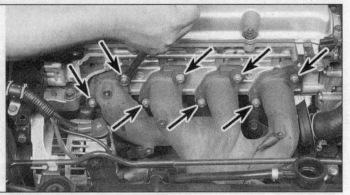

6.9 Remove the nuts and/or bolts retaining the exhaust manifold to the cylinder head and remove the manifold

7   Remove the upper heat shield from the exhaust manifold (see illustrations).

8   Refer to Chapter 6 and disconnect the EGR pipe from the exhaust manifold.

9   Remove the exhaust manifold-to-cylinder head nuts, working from the outside toward the middle, and detach the manifold and gaskets (see illustration).

Installation

10   Use a scraper to remove all traces of old gasket material and carbon deposits from the exhaust manifold and cylinder head mating surfaces.

11   Position the new exhaust manifold gaskets over the cylinder head studs.

12   Install the manifold and thread the mounting nuts into place.

13   Working from the center out, tighten the nuts to the torque listed in this Chapter's Specifications.

14   Reinstall the remaining parts in the reverse order of removal. Use anti-seize lubricant on the exhaust pipe studs and the EGR pipe nut.

15   Run the engine and check for exhaust leaks.

7   Crankshaft front oil seal - replacement

Refer to illustrations 7.3 and 7.5

1   Refer to Chapter 1 and remove the drivebelts. Remove the crankshaft pulley bolt

(see illustration 3.4). If necessary, have an assistant hold the crankshaft from turning by wedging a screwdriver in the flywheel/drive-plate teeth through the access hole (see Chapter 8).

2   The pulley can be removed by hand, without the need for a puller.

3   Use a seal-puller tool, or wrap the tip of a screwdriver with tape, to pry out the seal, being careful not to damage the seal bore or scratch the surface of the crankshaft snout (see illustration).

4   Clean the bore in the lower timing cover and coat the outer edge of the new seal with engine oil or multi-purpose grease. Also lubri-

7.3 Use a seal puller or screwdriver, with the tip wrapped in tape, to pry the seal (arrow) out of the front cover

cate the seal lips.

5   Using a socket with an outside diameter slightly smaller than the outside diameter of the seal, carefully drive the new seal into place with a hammer (see illustration). Make sure it's installed squarely and driven in to the same depth as the original. If a socket isn't available, a short section of large-diameter pipe will also work. Check the seal after installation to make sure the garter spring didn't pop out of place. **Caution:** *The oil-seal lip goes toward the engine, and the dust seal lip toward the pulley.*

6   Reinstall the crankshaft pulley and drivebelts.

7   Run the engine and check for oil leaks.

8   Timing chain and sprockets- removal, inspection and installation

Note 1: *The timing chain assembly on this engine consists of a lower chain connecting the crankshaft sprocket to an idler sprocket (attached to the front of the cylinder head), and an upper chain connecting the idler sprocket to the two camshaft sprockets. The timing chains should last for the life of the engine, and replacement with the engine in-*

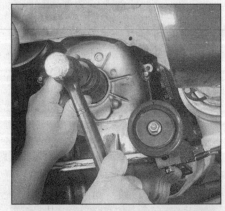

7.5 Drive the new seal squarely into the front cover with a large socket or section of pipe

8.10a Unbolt the engine brace from the cylinder head (arrow)

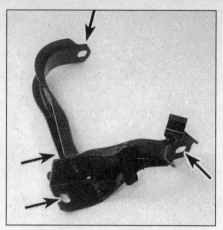

8.10b There are four bolts (arrows) retaining the brace to the engine

8.12 Remove the two bolts and the upper timing chain guide (arrow)

2A

vehicle is a difficult procedure, involving removing the oil pan (see Section 12), and the cylinder head (see Section 11). The procedure described below involves removing the intake and exhaust manifolds (see Sections 5 and 6) with the cylinder head. **Note 2:** *The following procedure is shown on a 1993 through 1997 model with a double row upper timing chain. Models since 1998 use a single row timing chain and the removal and installation procedures are identical.*

Removal

Refer to illustrations 8.10a, 8.10b, 8.12, 8.13, 8.14, 8.15, 8.16, 8.17a, 8.17b, 8.17c, 8.18a, 8.18b, 8.19a and 8.19b

1   Relieve the fuel system pressure (see Chapter 4).
2   Drain the cooling system (see Chapter 1).
3   Refer to Section 4 and remove the valve cover.
4   Remove the air duct connected to the intake manifold.
5   Refer to Chapter 5 and remove the alternator and bracket.
6   Remove the upper radiator hose (see Chapter 3).

7   Position the engine at TDC for number one cylinder (see Section 3).
8   Drain the engine oil and remove the oil pan and oil pump pick-up (see Section 12). **Note:** *This requires supporting the transmission with a jack, the engine with a crane from above and removing the longitudinal crossmember that retains the front and rear engine mounts.*
9   Lower the vehicle and support the engine with a jack and wood block placed under the front main bearing saddle so that the engine crane can be removed for cylinder head removal.* **Note:** *After the oil pan is removed, the crossmember and front and rear mounts can be temporarily installed again for support.*
10   Remove the engine mount at the front of the engine (see Section 16) and the brace around the front of the cylinder head **(see illustrations)**.
11   Remove the distributor (see Chapter 5).
12   Remove the upper timing chain guide **(see illustration)**.
13   Remove the engine mount bracket **(see illustration)**.
14   Remove the five bolts retaining the upper timing chain cover to the cylinder head and lower cover **(see illustration)**. Remove

the upper cover.
15   Remove the upper chain tensioner and the chain guide **(see illustration)**.
16   Use a large open end-wrench on the hex to hold the camshaft as you unbolt the camshaft sprocket bolts **(see illustration)**. Pull the sprockets from the camshafts and remove them with the upper timing chain.
17   Remove the crankshaft pulley (see Section 7). Unbolt the power steering pump upper bracket (it conceals one of the lower cover bolts) and remove the lower timing chain cover **(see illustrations)**.
18   Remove the lower timing chain tensioner, tension arm and the chain guide **(see illustrations)**.
19   Hold the idler sprocket with a screwdriver inserted through one of the holes in the sprocket while removing the idler sprocket bolt **(see illustration)**. Remove the idler sprocket and the lower timing chain **(see illustration)**. Remove the cylinder head (see Section 11). **Note:** *It is because of the cylinder head gasket design that the cylinder head must be removed for the timing chain installation procedure. Since the upper and lower timing chain covers must seal against the cylinder head gasket, the cylinder head gasket must be replaced.*

8.13 Remove the cast-iron engine mount bracket (arrow) from the front of the engine for access to the timing cover bolts

8.14 Remove the bolts (arrows) and the upper timing chain cover

8.15 Remove the two bolts and the upper chain tensioner (left arrows), then remove the chain guide (right arrows)

8.16 To remove either camshaft sprocket, hold the camshaft with a wrench as shown, while removing the sprocket bolt

8.17a Unbolt the power steering bracket (arrow) for access to . . .

8.17b . . . this front cover bolt

8.17c Remove these five bolts (arrows) from the lower cover - the aluminum section of the oil pan must be removed before removing the lower cover

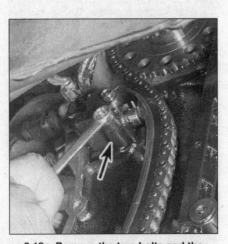

8.18a Remove the two bolts and the lower timing chain tensioner (arrow)

8.18b The tension arm (arrow) and the chain guide are retained by Torx fasteners

Inspection

20 Inspect the camshaft, idler and crankshaft sprockets for wear of the teeth and keyways. Inspect the chains for cracks or excessive wear of the rollers. Inspect the facing of the chain guides for excessive wear.

Installation

Refer to illustrations 8.21a, 8.21b, 8.22, 8.23a, 8.23b, 8.25, 8.26, and 8.28

21 Reinstall the lower timing chain guide and the tension arm. Install the lower chain with one of the silver-colored links aligned with the mark on the crankshaft sprocket **(see illustrations)**.

22 Retract the plunger in the lower chain tensioner, securing it in this position by flipping the latch over the pin **(see illustration)**, then install it on the block.

23 Apply a bead of RTV sealant to the timing cover sealing surface and install a new O-ring on the block **(see illustrations)**. Pull up on the lower timing chain to keep it aligned with the crankshaft sprocket, install the lower timing cover. **Note:** *Once the lower cover is*

8.19a Insert a screwdriver through one of the holes in the idler gear to keep it from turning while you remove the bolt

8.19b The idler gear and lower timing chain can be pulled up through the opening in the cylinder head gasket

bolted to the block, a cast-in projection in the cover will prevent the chain from disengaging the crankshaft sprocket, so you don't need to keep holding the other end of the chain.

24 Refer to Section 11 and install the cylin-

der head with a new gasket, install the bolts finger tight at this time. **Caution:** *Make sure the beveled edges of the cylinder head-bolt washers face the bolts, and the flat sides face the cylinder head.*

8.21a With the number one cylinder at TDC, align the silver link (arrow) with the dot on the crankshaft sprocket

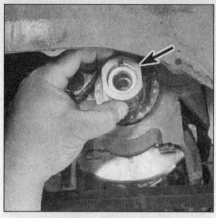

8.21b If the oil pump drive gear (arrow) had been removed before, reinstall it now

8.22 The plunger in the lower chain tensioner can be held in the retracted position by pushing this latch (arrow) over the pin

8.23a Install a new oil-passage O-ring (arrow) in the block before installing the lower timing cover

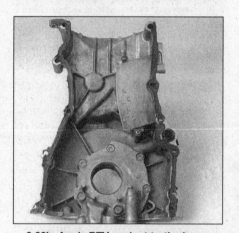

8.23b Apply RTV sealant to the lower cover-to-engine block sealing surface

8.25 Install the idler sprocket aligning the dot with the silver link in the chain (arrow)

8.26 Install the upper chain with its gold link (arrow) aligned with the dot on the idler sprocket, then install the upper cover

8.28 Install the camshaft sprockets with the upper chain's silver links (arrows) aligned with the dots on each camshaft sprocket

the chain can not slip off the idler sprocket.
27   Now tighten the cylinder head according to the procedure in Section 11 and reinstall the camshafts and lifters as described in Section 9.
28   Align the camshafts in their TDC positions. Slip the exhaust camshaft sprocket under the chain, aligning its dot with the silver link in the chain and bolt it to the exhaust camshaft. Repeat the procedure with the intake camshaft sprocket. Both camshaft sprockets should remain aligned with the silver links on the chain **(see illustration)**. Tighten the sprocket bolts to the torque listed in this Chapter's Specifications.
29   The remainder of the installation is the reverse of the disassembly sequence.

9   Camshaft and lifters- removal, inspection and installation

Removal

Refer to illustrations 9.3, 9.6, 9.8a and 9.8b
1   Detach the cable from the negative terminal of the battery.
2   Remove the valve cover (see Section 4).
3   Before removing the camshafts, use a

25   Position the idler sprocket on the chain, aligning the timing mark with the silver-colored link **(see illustration)**. Install the idler sprocket to the cylinder head and tighten the bolt to the torque listed in this Chapter's Specifications.
26   Install the upper timing chain on the idler

sprocket with its gold-colored link aligned with the dot on the front of the idler gear **(see illustration)**. With the chain held in this position, install the chain guide and upper tensioner, then install the upper timing chain cover with a bead of RTV sealant on the mating surface. Once the upper cover is in place,

9.3 With a dial indicator in place, pry the camshaft forward and back to check the camshaft endplay

9.6 The camshaft bearing caps are marked for direction of installation, intake (I) or exhaust (E) camshaft and position

9.8a Pull the lifters straight up, MAKE SURE you keep the shim with each lifter

dial indicator to check camshaft endplay **(see illustration)**. Mount the dial indicator so the gauge tip can be placed at the end of the camshaft. Move the camshaft all the way to the rear and zero the dial indicator. Next, use a screwdriver to pry it all the way forward. If the endplay (the total amount of movement) exceeds the limit listed in this Chapter's Specifications, replace the camshaft and/or

9.8b The lifters and shims can be stored in individually-marked plastic bags, or in a divided, marked box like this one

cylinder head.
4   Position the engine at TDC for number one cylinder (see Section 3) and remove the distributor (see Chapter 5).
5   Unbolt the camshaft sprockets and position the sprockets and upper timing chain out of the way (see Section 8). **Note:** *There is a cast-in protrusion inside the upper timing chain cover that will keep the upper*

timing chain from falling off the idler gear.
6   Loosen the camshaft bearing caps in two or three steps, in the reverse of the tightening sequence **(see illustration 9.15)**. They are numbered from 1 to 5, are stamped with an "I" or an "E" to indicate intake or exhaust, and have arrows to indicate which way faces the timing-chain end of the engine **(see illustration)**. **Caution:** *Keep the caps in order. They must go back in the same location they were removed from.*
7   Remove the bearing caps and lift the camshafts straight up and out.
8   Pull the lifters straight up, and store them in numbered plastic bags or a marked box, keeping the proper shim with each lifter **(see illustrations)**.

Inspection

Refer to illustrations 9.10a, 9.10b, 9.11a and 9.11b
9   Visually examine the camshaft lobes, journals, bearing caps and lifters. Check for score marks, pitting and evidence of overheating (blue, discolored areas). If wear is excessive or damage is evident, the component will have to be replaced.

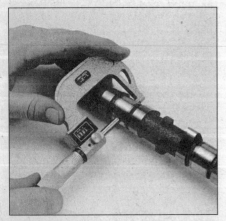

9.10a Measure each journal diameter with a micrometer (if any journal measures less than the specified limit, replace the camshaft)

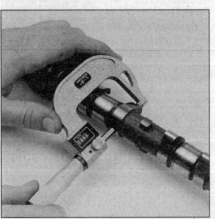

9.10b Measure the lobe heights - if any lobe height is less than the minimum listed in this Chapter's Specifications, replace the camshaft

9.11a Place a strip of Plastigage under each camshaft bearing cap and tighten the caps to Specifications

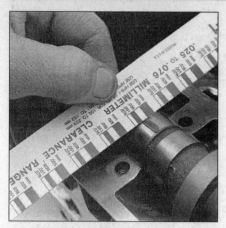

9.11b Compare the width of the crushed Plastigage to the scale on the envelope to determine the oil clearance

9.14 The distributor-drive notches (arrow) in the exhaust camshaft align with the cap/head split line when the camshaft is at TDC

12   Scrape off the Plastigage with your fingernail or the edge of a credit card - don't scratch or nick the journals or bearing caps.

Installation

Refer to illustrations 9.14 and 9.15

13   Apply moly-based engine assembly lubricant to the camshaft lobes and journals.

14   Install the camshafts in their original positions at TDC. **Note:** *The slotted end of the exhaust camshaft aligns with the cap as shown (see illustration).*

15   Install the bearing cap bolts and tighten them, in sequence **(see illustration)**, to the torque listed in this Chapter's Specifications.

16   Install the camshaft sprockets and timing chain (see Section 8).

17   The remainder of installation is the reverse of removal. If any part of the valve train was replaced, check and adjust the valve clearance (see Chapter 1).

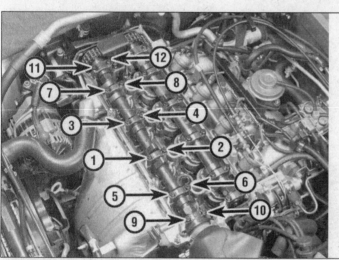

9.15 Camshaft bearing cap TIGHTENING sequence - the sequence is the same for both camshafts

2A

10 Valve springs, retainers and seals - replacement

Refer to illustrations 10.2a, 10.2b, 10.3, 10.4, 10.7, 10.8, 10.9, 10.10 and 10.11

Note: *Broken valve springs and leaking/worn valve stem seals can not be replaced without removing the cylinder head. The valve springs are located in deep, narrow pockets in the cylinder head that prevent common valve-spring tools from being used.*

1   Refer to Section 11 and remove the cylinder head.

2   With the cylinder head on a sturdy workbench, use a large, clamp-type valve spring compressor and adapter to compress each valve spring **(see illustrations)**. Make sure the bottom section of the tool is located under the valve which spring is being removed, and compress the spring just enough to remove the keepers with a magnet or small pliers.

10   Using a micrometer, measure camshaft journal diameter and lobe height **(see illustrations)**, and compare your measurements to this Chapter's Specifications. If the lobe height is less than the minimum allowable, the camshaft is worn and must be replaced.

11   Check the oil clearance for each camshaft journal as follows:

a) *Clean the bearing caps and the camshaft journals with lacquer thinner or acetone.*

b) *Carefully lay the camshafts in place in the cylinder head. DON'T use any lubrication.*

c) *Lay a strip of Plastigage on each journal.*

d) *Install the bearing caps with the arrows pointing toward the front (timing chain end) of the engine.*

e) *Tighten the bolts in sequence (see illustration 9.15) to the torque listed in this Chapter's Specifications in 1/4-turn increments.* **Caution:** *Don't turn the camshaft while the Plastigage is in place.*

f) *Remove the bolts, in the proper sequence, and detach the bearing caps*

g) *Compare the width of the crushed Plastigage (at it's widest point) to the scale on the Plastigage envelope (see illustrations).*

h) *If the clearance is greater than specified, replace the camshaft and/or cylinder head.*

10.2a Compress the valve spring and remove the keepers with needle-nose pliers or a magnet

10.2b Because of the tight quarters, a spring compressor must be used with an adapter (arrow) that has "windows" in the side to access the keepers

10.3 Remove the retainer and valve spring

10.4 Remove the valve stem seal (arrow) with needle-nose pliers

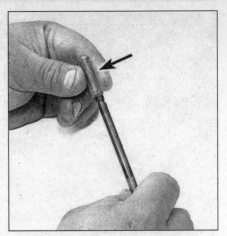

10.7 Slip a plastic sleeve (arrow) over the valve stem to protect the new seal during installation

3   Pull the valve spring out with its retainer **(see illustration)**.
4   Remove the old valve stem oil seal with pliers **(see illustration)**.
5   Remove the valve and inspect the valve stem for damage. Rotate the valve in the guide and check the end for eccentric movement, which would indicate that the valve is bent.
6   Move the valve up-and-down in the guide and make sure it doesn't bind. If the

valve stem binds, either the valve is bent or the guide is damaged. In either case, the cylinder head will require repair.
7   Lubricate the valve stem with engine oil and install it in the cylinder head. A plastic seal protector is usually provided with the seals that you slip over the valve stem **(see illustration)**. This protects the new seal from being torn as it passes over the keeper grooves in the valve. If you don't have the seal protector, apply a few wraps of cellulose

tape around the valve stem instead.
8   Lubricate the new seal with multi-purpose grease and push it down over the valve stem by hand **(see illustration)**. When it is down against the guide, remove the plastic protector or tape from the valve stem.
9   Use a deep socket of the appropriate size to lightly tap the new seal down against the top of the valve guide **(see illustration)**.
10   Install the spring in position over the valve, with the retainer in place **(see illustration)**.
11   Compress the valve spring and retainer and carefully position the keepers in the groove. Apply a small dab of grease to the inside of each keeper to retain it in place if necessary **(see illustrations)** and insert them with needle-nose pliers.
12   Remove the pressure from the spring tool and make sure the keepers are seated.
13   Refer to Section 9 and reinstall the lifters and camshafts.
14   The remainder of installation is the reverse of the removal procedure.
15   Start the engine, then check for oil leaks and unusual sounds coming from the valve cover area.

10.8 Lubricate the new seal and slip it onto the valve, past the protector

10.9 Tap the new seals down onto the valve guide with a deep socket (arrow)

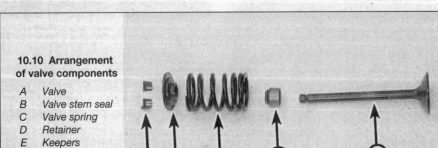

10.10 Arrangement of valve components

    A   Valve
    B   Valve stem seal
    C   Valve spring
    D   Retainer
    E   Keepers

10.11 Apply a small dab of grease to each keeper before installation to retain them in place on the valve stem until the spring is released

11.10 Unbolt the thermostat housing/water distribution block (arrow) from the left side of the cylinder head

11.11a Use a long hex bit to loosen the cylinder head bolts

11 Cylinder head - removal and installation

Caution: The engine must be completely cool before beginning this procedure.

Note: The cylinder head may be removed with both the intake and exhaust manifolds still attached, and the manifolds can be removed later on the workbench.

Removal

Refer to illustrations 11.10, 11.11a, 11.11b, 11.12 and 11.13

1   Relieve the fuel system pressure (see Chapter 4). Disconnect the cable from the negative battery terminal.

2   Drain the coolant from the engine block and radiator (see Chapter 1).

3   Drain the engine oil and remove the oil filter (see Chapter 1).

4   Remove the fuel rail and injectors (see Chapter 4).

5   Disconnect any hoses, cables or wires from the intake manifold (see Section 5).

6   Disconnect the EGR pipe, oxygen sensor electrical connector and exhaust pipe from the exhaust manifold (see Section 6).

7   Position cylinder number one at TDC (see Section 3). Remove the valve cover and camshafts (see Sections 4 and 9). Remove the upper timing chain and timing chain cover (see Section 8).

8   Remove the alternator (see Chapter 5) and the right engine mount (see Section 16).

9   Remove the power steering pump (see Chapter 10) and set it aside without disconnecting the hoses.

10   Label and remove any remaining items attached to the cylinder head, such as coolant fittings, tubes, cables, hoses or wiring harness. It is easier to unbolt the thermostat housing/water distribution block from the cylinder head than it is to disconnect every hose attached to it **(see illustration)**.

11   Using a breaker bar and the appropriate-sized hex bit, loosen the cylinder head bolts in 1/4-turn increments until they can be removed by hand **(see illustrations)**. Loosen the bolts in reverse of the tightening sequence **(see illustration 11.24)** to avoid warping or cracking the cylinder head. **Note:** *You will need a 10 mm hex bit at least 1-1/2 inches long to remove the cylinder head bolts under the camshafts.*

11.11b Don't overlook this small bolt (arrow) retaining the front of the cylinder head to the lower timing cover

12   Lift the cylinder head off the engine block. If it's stuck, very carefully pry up at the transaxle end, beyond the gasket surface, at a casting protrusion. There is also a pry spot at the right front corner of the cylinder head **(see illustration)**.

13   The cylinder head is easiest to remove

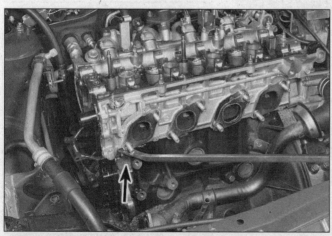

11.12 Pry under the cylinder head at the transaxle end and under this spot on the front - do not pry under the gasket surface

11.13 It's easier to remove the cylinder head with the intake manifold in place, but it is awkward to lift - be careful not to scratch body panels while removing it

11.16 Remove all traces of old gasket material - the cylinder head and block mating surfaces must be perfectly clean to ensure a good gasket seal

11.19 A die should be used to remove sealant and corrosion from the cylinder head bolt threads prior to installation

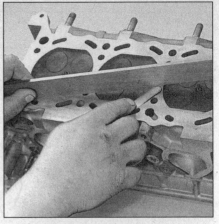

11.20 Check the cylinder head gasket surface for warpage by trying to insert a feeler gauge under the straightedge - see this Chapter's Specification for the maximum warpage allowed and use a feeler gauge of that thickness

with the intake manifold, and even the exhaust manifold in place **(see illustration).**
14   Remove all external components from the cylinder head to allow for thorough cleaning and inspection. **Note:** *See Chapter 2, Part B, for cylinder head servicing procedures.*

Installation

Refer to illustrations 11.16, 11.19, 11.20 and 11.24
15   The mating surfaces of the cylinder head and block must be perfectly clean when the cylinder head is installed.
16   Use a gasket scraper to remove all traces of carbon and old gasket material **(see illustration)**, then clean the mating surfaces with lacquer thinner or acetone. If there's oil on the mating surfaces when the cylinder head is installed, the gasket may not seal correctly and leaks could develop. When working on the block, stuff the cylinders with clean shop rags to keep out debris. Use a vacuum cleaner to remove material that falls into the cylinders.
17   Check the block and cylinder head mating surfaces for nicks, deep scratches and other damage. If damage is slight, it can be removed with a fine file; if it's excessive, machining may be the only alternative.
18   Use a tap of the correct size to chase

the threads in the cylinder head bolt holes, then clean the holes with compressed air - make sure that nothing remains in the holes. **Warning:** *Wear eye protection when using compressed air!*
19   Mount each bolt in a vise and run a die down the threads to remove corrosion and restore the threads **(see illustration)**. Dirt, corrosion, sealant and damaged threads will affect torque readings.
20   Once the cylinder head's gasket surface is clean, check the cylinder head for warpage **(see illustration)**. Check the cylinder head gasket, intake and exhaust manifold surfaces.
21   Install the components that were removed from the cylinder head, including the intake manifold (see Section 5).
22   Position the new cylinder head gasket over the dowel pins in the block and carefully set the cylinder head on the block without disturbing the gasket.
23   Before installing the cylinder head bolts, apply a small amount of clean engine oil to the threads and hardened washers. The chamfered side of the washers must face the bolt heads, and the flat side of the washers must face the cylinder head.
24   Install the bolts in their original locations and tighten them finger tight, then refer to Section 8 for installation of the upper timing

chain to the idler gear and installation of the upper timing chain cover. When the upper cover is bolted to the cylinder head, proceed to tighten all the cylinder head bolts, following the recommended sequence **(see illustration)**, to the torque listed in this Chapter's Specifications. Tighten the small bolt connecting the cylinder head to the lower timing cover to the specifications for timing cover bolts.
25   See Section 9 for installation of the lifters and camshafts. Refer to Chapter 1 and check and adjust the valve clearance. If any machine work was done to the cylinder head (a valve job), the valve clearances will have changed. See Section 8 for installation of the upper timing chain. The remaining installation steps are the reverse of removal.
26   Refill the cooling system, install a new oil filter and add oil to the engine (see Chapter 1).
27   Run the engine and check for leaks. Set the ignition timing (see Chapter 5) and road test the vehicle.

12   Oil pan - removal and installation

Removal

Refer to illustrations 12.2, 12.3a, 12.3b, 12.4, 12.5a, 12.5b, 12.6a, 12.6b, 12.6c and 12.7
1   Refer to Chapter 1 and drain the engine oil. Support the vehicle safely on jackstands.
2   The oil pan is a two-piece design. A steel pan is attached to an aluminum section which is bolted to the engine block. Remove the oil pan bolts **(see illustration)**, following the reverse of the recommended tightening sequence **(see illustration 12.13)**. Separate the steel pan by inserting a thin putty knife between the steel and aluminum sections. **Caution:** *Do not pry with a screwdriver between the steel pan and the aluminum flange or damage to the sealing surface may result.*

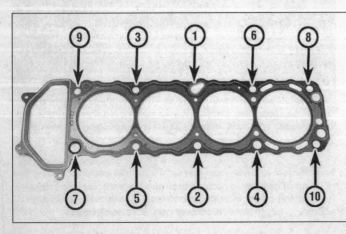

11.24 Cylinder head bolt TIGHTENING sequence

12.2 Remove the steel pan bolts and separate the pan with a putty knife - do not damage the sealing surface

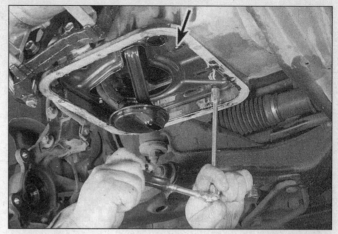

12.3a Remove the bolts and lower the oil baffle plate (arrow)

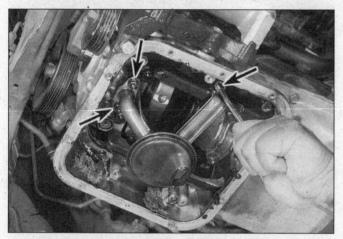

12.3b Remove the bolts (arrows) and the oil pump pickup

12.4 Remove the bolts retaining the exhaust heat-shield bracket and remove the bracket

2A

12.5a Remove this bolt (arrow) retaining part of the rear engine mount to the oil pan

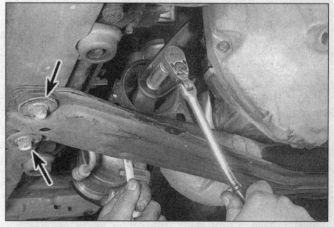

12.5b Support the transmission with a jack and remove the center crossmember - unbolt the engine mount throughbolts and remove the two bolts (arrows) at each end of the crossmember

3   Unbolt the oil baffle plate and the oil pump pickup tube (see illustrations).
4   Disconnect the exhaust pipe from the exhaust manifold and remove the heat shield (see Section 6), then remove the heat-shield bracket (see illustration).
5   Support the engine from above with an engine hoist and support the transaxle from below with a floor jack. Remove the center crossmember (see illustrations).
6   Remove the air conditioning compressor bracket-to-oil pan bolts and swing the compressor away, without disconnecting the refrigerant lines. Remove the flywheel inspection cover (see Chapter 8). Remove the bolts

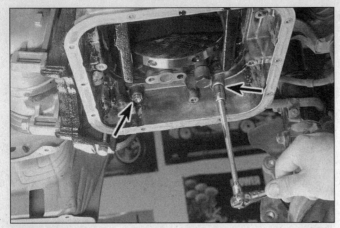

12.6a Remove the bolts from the inside pan area, including these two (arrows) connected to the lower timing cover

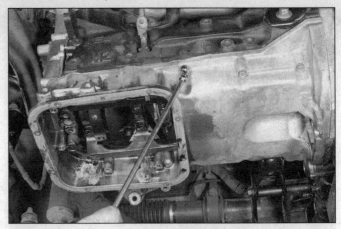

12.6b Remove the aluminum section-to-block bolts - start at the ends and work toward the middle

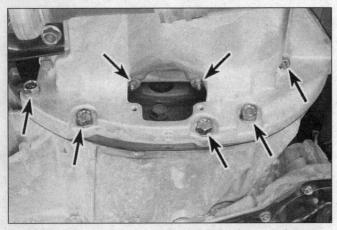

12.6c Remove the bolts retaining the aluminum section to the transaxle, including the two nuts inside the flywheel-inspection-cover area

12.7 After cutting the pan seal with a putty knife, pry at the rear corners near the transaxle - do not pry in the gasket area

attaching the aluminum section to the engine block and transaxle, working from the ends toward the center to prevent warpage **(see illustrations)**.

7   The sealant used to seal the aluminum section to the engine block can be very difficult to separate without damaging the aluminum section. Using a thin putty knife, work around the perimeter cutting the pan free before prying the pan down at the transaxle end **(see illustration)**

Installation

Refer to illustrations 12.11 and 12.13

8   Use a scraper to remove all traces of old gasket material and sealant from the block and oil pan. Clean the mating surfaces with lacquer thinner or acetone. **Caution:** *Be careful not to scratch or gouge the gasket surface of the block or oil pan. A leak could develop after the repairs have been completed.*

9   Make sure the threaded bolt holes in the block are clean.

10   Check the steel pan flange for distortion, particularly around the bolt holes. If necessary, place the pan on a wood block and use a hammer to flatten and restore the gasket surface.

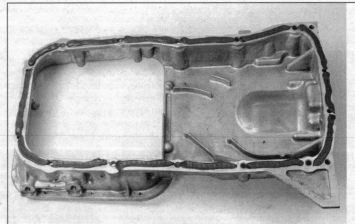

12.11 Apply a bead of RTV sealant around the perimeter of the aluminum section of the pan

11   Apply a 3/16-inch wide bead of RTV sealant around the perimeter of the aluminum section **(see illustration)**. **Note:** *The oil pan must be installed within 15 minutes once the sealant has been applied.*

12   Carefully position the aluminum section on the engine block and install the nuts and bolts. Working from the center out, tighten the oil pan-to-block fasteners in three or four steps to the torque listed in this Chapter's Specifications, then tighten the pan-to-transaxle bolts.

13   Apply a bead of RTV sealant around the perimeter of the steel pan and install it within 15 minutes of application. Tighten the bolts to the torque listed in this Chapter's Specifications in the recommended sequence **(see illustration)**.

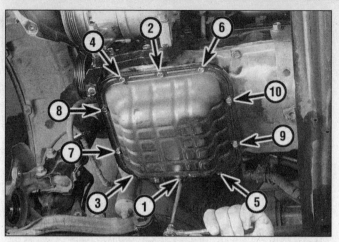

12.13 Tighten the oil pan bolts in the recommended sequence to the specified torque

13.3 The oil pump is located inside the front cover - remove the two bolts, four screws and the oil pump cover

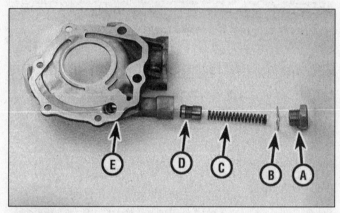

13.5 Remove the oil pump pressure regulator valve assembly for cleaning and inspection

| A | Cap | C | Spring | E | Oil pump |
| B | Washer | D | Relief valve | | cover |

13.6a Check the outer rotor-to-body clearance with a feeler gauge as shown

2A

14 The remainder of installation is the reverse of removal. Be sure to install a new oil filter and wait at least thirty minutes for the RTV to set-up before adding oil.

13 Oil pump - removal, inspection and installation

Removal

Refer to illustration 13.3

1 The oil pump is located inside the lower timing chain cover, and is driven from the crankshaft.
2 The front cover must be removed for access to the oil pump. Remove the oil pan (see Section 12), the cylinder head (see Section 11) and the front cover (see Section 8).
3 Remove the bolts/screws retaining the oil pump cover to the oil pump body (see illustration), disassemble the inner and outer rotors. Caution: *Be very careful with these components, the close tolerances are critical in creating the correct oil pressure. Any nicks or damage will require replacement of the complete pump/front cover assembly.*

Inspection

Refer to illustrations 13.5, 13.6a, 13.6b, 13.6c and 13.6d

4 Clean all the components, including the front cover and engine block gasket surfaces, with solvent, then inspect all surfaces for excessive wear and/or damage.
5 Disassemble the relief valve by removing the cap, washer, spring and regulator valve (see illustration). Check the oil pres-

sure regulator valve sliding surface and valve spring. The regulator, when clean and oiled, should slide easily in the valve bore. If either the spring or the valve is damaged, they must be replaced as a set. If no damage is found reassemble the relief valve parts, coating the parts with clean engine oil, and reinstall it in the oil pump cover.

6 Check the oil pump component clearance with a feeler gauge (see illustrations)

13.6b Check the clearance between the inner and outer rotor tips

13.6c With a precision straightedge placed over the pump body and rotors, check the clearance between the inner and outer rotors and the pump body cover

13.6d Measure the outer flanged surface of the inner rotor with a micrometer or precision calipers, then measure its bearing surface in the pump body (arrows) - the difference is the inner rotor bearing clearance

and compare the results to this Chapter's Specifications. If any of the Specifications are exceeded, replace both the front cover and the oil pump components.

Installation

7   Assemble oil pump components. Pour a generous amount of clean engine oil into the pump cavity and around the rotors. Install the cover to the pump body and tighten the fasteners to the torque listed in this Chapter's Specifications.

8   Refer to Section 8 to install the front cover, using a bead of RTV sealant on the cover-to-block surface. Note: Align the flats on the inner oil pump rotor with the crankshaft when installing the front cover. Refer to Sections 9 and 11 and install the cylinder head and other components.

9   Refer to Section 12 and install the oil pan. Install a new oil filter and add engine oil to the crankcase.

10   Start the engine and check for oil pressure and leaks.

11   Recheck the engine oil level.

14   Flywheel/driveplate - removal and installation

Removal

Refer to illustration 14.3

1   Raise the vehicle and support it securely on jackstands (see Chapter 1). Remove the transaxle (see Chapter 7).

2   If equipped with a manual transaxle, remove the pressure plate and clutch disc (see Chapter 8).

3   Use a center-punch, or paint, to make alignment marks on the flywheel/driveplate and crankshaft to ensure correct alignment during installation **(see illustration)**.

4   Remove the bolts that secure the flywheel/driveplate to the crankshaft.

5   Remove the flywheel/driveplate from the

14.3 Mark the flywheel/driveplate and the crankshaft so they can be reassembled in the same relative positions

crankshaft. **Warning:** *Since the flywheel is fairly heavy, be sure to support it while removing the last bolt. Flywheel teeth can be sharp, wear gloves or use rags to hold the flywheel.*

Installation

6   If equipped with a manual transaxle, clean the flywheel to remove grease and oil. Inspect the surface for cracks, rivet grooves, burned areas and score marks. Light scoring can be removed with emery cloth. Check for cracked and broken ring-gear teeth. Lay the flywheel on a flat surface and use a straightedge to check for warpage.

7   Clean and inspect the mating surfaces of the flywheel/driveplate and the crankshaft. If the crankshaft rear seal is leaking, replace it before reinstalling the flywheel/driveplate.

8   Position the flywheel/driveplate against the crankshaft. Be sure to install the spacer (if equipped) and align the marks made during removal. Some engines have an alignment dowel or staggered bolt holes to ensure cor-

rect installation. Before installing the bolts, apply thread-locking compound to the threads.

9   Wedge a screwdriver in the ring gear teeth to keep the flywheel/driveplate from turning as you tighten the bolts to the torque listed in this Chapter's Specifications. Follow a criss-cross pattern and work up to the final torque in three or four steps.

10   The remainder of installation is the reverse of the removal procedure.

15   Rear main oil seal - replacement

Refer to illustrations 15.3, 15.5 and 15.6

1   Remove the transaxle (see Chapter 7).

2   Remove the flywheel/driveplate (see Section 14).

3   Remove the rear oil seal retainer from the block **(see illustration)**.

4   Scrape any sealant or gasket material from the retainer, oil pan and the block.

5   Position the seal and retainer assembly between two wood blocks, to evenly support

15.3 Unbolt the rear oil seal retainer from the engine
block and oil pan

15.5 After removing the retainer from the block, support it
between two wood blocks and drive out the old seal
with a punch and hammer

15.6 Drive the new seal into the retainer with a wood block or a
section of pipe - make sure that you don't cock the
seal in the retainer bore

16.6 A long prybar can be used to check for relative movement
in the engine mounts

the aluminum housing, and drive the old seal out with a hammer and punch (see illustration). A screwdriver can also be used to pry the seal out, being careful not to gouge or nick the housing during seal removal.

6   Place the new seal squarely on the retainer and drive it into the retainer with a wood block (see illustration) or a section of pipe slightly smaller in diameter than the outside diameter of the seal.

7   Lubricate the crankshaft seal journal and the lip of the new seal with multi-purpose grease.

8   Apply a continuous 1/8-inch bead of RTV sealant around the perimeter of the seal retainer. Also apply sealant to the bottom of the retainer (oil pan mating surface).

9   Slowly and carefully push the seal onto the crankshaft. The seal lip is stiff, so work it onto the crankshaft with a smooth object such as the end of a socket extension as you push the retainer against the block.

10   Install and tighten the retainer bolts securely.

11   The remaining steps are the reverse of removal.

16   Engine mounts - check and replacement

1   Engine mounts seldom require attention, but broken or deteriorated mounts should be replaced immediately or the added strain placed on the driveline components may cause damage or wear.

Check

Refer to illustration 16.6

2   During the check, the engine must be raised slightly to remove the weight from the mounts.

3   Raise the vehicle and support it securely on jackstands and remove the splash shields.

4   Position a jack under the engine oil pan. Place a large wood block between the jack head and the oil pan, then carefully raise the

engine *just enough* to take the weight off the mounts. Do not place the wood block under the oil pan drain plug. **Warning:** *DO NOT place any part of your body under the engine when it's supported only by a jack!*

5   Check the mounts to see if the rubber is cracked, hardened or separated from the metal plates. Sometimes the rubber will split right down the center.

6   Check for relative movement between the mount plates and the engine or frame using a large screwdriver or prybar to attempt to move the mounts (see illustration). If movement is noted, lower the engine and tighten the mount fasteners.

7   Rubber preservative should be applied to the mounts to slow deterioration.

Replacement

Refer to illustrations 16.9 and 16.10

8   Disconnect the negative battery cable from the battery, then raise the vehicle and support it securely on jackstands. Support the engine as described in Step 4.

2A

16.9 With the engine supported from below, remove the throughbolt (larger arrow) and the three bolts (smaller arrows) on the engine side

16.10 The lower engine mounts are bolted (arrows) to the crossmember - remove the throughbolt and these two bolts to remove the mount

9   To remove the front engine mount, remove the engine mount through bolt at the chassis and the three bolts retaining the casting to the engine **(see illustration)**.

10   To remove either of the two lower engine mounts, remove the throughbolts and the bolts retaining the mount to the crossmember **(see illustration)**.

11   Remove the engine block-to-mount bolts and remove the mount.

12   Installation is the reverse of removal. Use thread-locking compound on the mount bolts/nuts and be sure to tighten them securely.

Chapter 2 Part B
General engine overhaul procedures

Contents

Specifications

General

Cylinder compression pressure	
Normal	178 psi
Minimum	149 psi
Compression variation between cylinders	14 psi
Displacement	2389 cc (146 cu. in.)
Oil pump pressure (at normal operating temperature)	
Idle	11 psi
3000 rpm	60 to 70 psi

Engine block

Deck surface warpage	0.004 inch
Bore diameter	
Grade 1	3.5039 to 3.5043 inches
Grade 2	3.5043 to 3.5047 inches
Grade 3	3.5047 to 3.5051 inches
Out-of-round limit	0.0006 inch
Taper limit	0.0004 inch

Crankshaft and connecting rods

Crankshaft runout	0.0016 inch maximum
Endplay	0.012 inch maximum
Main bearing journal	
Diameter	
Grade 0	2.3609 to 2.3612 inches
Grade 1	2.3606 to 2.3609 inches
Grade 2	2.3603 to 2.3606 inches
Taper	0.0001 inch maximum
Out-of-round	0.0002 inch maximum
Main bearing oil clearance	
Standard	0.0008 to 0.0019 inch
Limit	0.004 inch

2B

Crankshaft and connecting rods (continued)

Connecting rod journal diameter
    Grade 0.. 1.9672 to 1.9675 inches
    Grade 1.. 1.9670 to 1.9672 inches
    Grade 2.. 1.9668 to 1.9670 inches
Connecting rod oil clearance
    Standard.. 0.0004 to 0.0014 inch
    Limit .... 0.0035 inch
Connecting rod side clearance (endplay)
    Standard.. 0.008 to 0.016 inch
    Limit .... 0.024 inch

Piston and rings

Piston diameter
    Grade 1.. 3.5027 to 3.5031 inches
    Grade 2.. 3.5031 to 3.5035 inches
    Grade 3.. 3.5035 to 3.5039 inches
Piston-to-bore clearance.. 0.0008 to 0.0016 inch
Piston ring end gap
    Top compression ring .... 0.0110 to 0.0205 inch
    Second compression ring .... 0.0177 to 0.0272 inch
    Oil ring .... 0.0079 to 0.0272 inch
    Service limit (all).. 0.039 inch
Piston ring side clearance
    Top compression ring .... 0.0016 to 0.0031 inch
    Second compression ring .... 0.0012 to 0.0028 inch
    Service limit (all).. 0.004 inch

Valves and related components

Valve face angle.. 45 degrees
Valve spring free length .... 1.8028 inches
Valve spring out of square limit .... 0.079 inch
Valve margin width
    Intake.. 0.0374 to 0.0492 inch
    Exhaust;... 0.0453 to 0.0571 inch
Valve stem diameter
    Intake.. 0.2742 to 0.2748 inch
    Exhaust... 0.2734 to 0.2740 inch
Valve stem-to-guide clearance
    Intake.. 0.0008 to 0.0021 inch
        Limit .... 0.0031 inch
    Exhaust... 0.0016 to 0.0029 inch
        Limit .... 0.0004 inch
Valve lifter-to-guide clearance.. 0.0010 to 0.0024 inch

Torque specifications*   **Ft-lbs** (unless otherwise indicated)

Main bearing cap bolts .... 34 to 41
Connecting rod cap nuts
    Step 1.. 120 to 144 in-lbs
    Step 2.. Tighten an additional 60 to 65 degrees

* **Note:** *Refer to Part A for additional torque specifications.*

1   General information

Included in this portion of Chapter 2 are the general overhaul procedures for the cylinder head and internal engine components.

The information ranges from advice concerning preparation for an overhaul and the purchase of replacement parts to detailed, step-by-step procedures covering removal and installation of internal engine components and the inspection of parts.

The following Sections have been written based on the assumption that the engine has been removed from the vehicle. For information concerning in-vehicle engine repair, as well as removal and installation of the external components necessary for the overhaul, see Section 7 and Part A of this Chapter.

The Specifications included in this Part are only those necessary for the inspection and overhaul procedures which follow. Refer to Part A for additional Specifications.

2   Engine overhaul - general information

Refer to illustration 2.4

It's not always easy to determine when, or if, an engine should be completely overhauled, as a number of factors must be considered.

High mileage is not necessarily an indication that an overhaul is needed, while low mileage doesn't preclude the need for an overhaul. Frequency of servicing is probably the most important consideration. An engine that's had regular and frequent oil and filter changes, as well as other required maintenance, will most likely give many thousands of miles of reliable service. Conversely, a neglected engine may require an overhaul very early in its life.

Excessive oil consumption is an indication that piston rings, valve seals and/or valve

2.4 Oil pressure sending unit location (arrow) -next to the oil filter (view from below)

guides are in need of attention. Make sure that oil leaks aren't responsible before deciding that the rings and/or guides are bad. Perform a cylinder compression check to determine the extent of the work required (see Section 4).

Check the oil pressure with a gauge installed in place of the oil pressure sending unit and compare it to the Specifications **(see illustration)**. If it's extremely low, the bearings and/or oil pump are probably worn out.

Loss of power, rough running, knocking or metallic engine noises, excessive valve train noise and high fuel consumption rates may also point to the need for an overhaul, especially if they're all present at the same time. If a complete tune-up doesn't remedy the situation, major mechanical work is the only solution.

An engine overhaul involves restoring the internal parts to the specifications of a new engine. During an overhaul, the piston rings are replaced and the cylinder walls are reconditioned (rebored and/or honed). If a rebore is done by an automotive machine shop, new oversize pistons will also be installed. The main bearings and connecting rod bearings are generally replaced with new ones and, if necessary, the crankshaft may be reground to restore the journals. The timing chain and its sprockets are replaced with a new set. Generally, the valves are serviced as well, since they're usually in less-than-perfect condition at this point. The end result should be a like-new engine that will give many trouble free miles. **Note:** *Critical cooling system components such as the hoses, drivebelts, thermostat and water pump MUST be replaced with new parts when an engine is overhauled. The radiator should be checked carefully to ensure that it isn't clogged or leaking (see Chapter 3). If you purchase a rebuilt engine or short block, some rebuilders will not warranty their engines unless the radiator has been professionally flushed.*

Before beginning the engine overhaul, read through the entire procedure to familiarize yourself with the scope and requirements

of the job. Overhauling an engine isn't difficult, but it is time consuming. Plan on the vehicle being tied up for a minimum of two weeks, especially if parts must be taken to an automotive machine shop for repair or reconditioning. Check on availability of parts and make sure that any necessary special tools and equipment are obtained in advance. Most work can be done with typical hand tools, although a number of precision measuring tools are required for inspecting parts to determine if they must be replaced. Often an automotive machine shop will handle the inspection of parts and offer advice concerning reconditioning and replacement. **Note:** *Always wait until the engine has been completely disassembled and all components, especially the engine block, have been inspected before deciding what service and repair operations must be performed by an automotive machine shop. Since the block's condition will be the major factor to consider when determining whether to overhaul the original engine or buy a rebuilt one, never purchase parts or have machine work done on other components until the block has been thoroughly inspected. As a general rule, time is the primary cost of an overhaul, so it doesn't pay to install worn or substandard parts.*

As a final note, to ensure maximum life and minimum trouble from a rebuilt engine, everything must be assembled with care in a spotlessly-clean environment.

3 Vacuum gauge diagnostic checks

Refer to illustration 3.4

A vacuum gauge provides valuable information about what is going on in the engine at a low-cost. You can check for worn rings or cylinder walls, leaking cylinder head or intake manifold gaskets, incorrect carburetor adjustments, restricted exhaust, stuck or burned valves, weak valve springs, improper ignition or valve timing and ignition problems.

Unfortunately, vacuum gauge readings are easy to misinterpret, so they should be used in conjunction with other tests to confirm the diagnosis.

Both the absolute readings and the rate of needle movement are important for accurate interpretation. Most gauges measure vacuum in inches of mercury (in-Hg). As a point of reference, normal atmospheric pressure at sea level is about 30 in-Hg. As vacuum increases (or atmospheric pressure decreases), the reading will decrease. Also, for every 1,000 foot increase in elevation above sea level; the gauge readings will decrease about one inch of mercury.

Connect the vacuum gauge directly to intake manifold vacuum, not to ported vacuum **(see illustration)**. Be sure no hoses are left disconnected during the test or false readings will result.

Before you begin the test, allow the engine to warm up completely. Block the

3.4 A simple vacuum gauge, properly interpreted, can tell you a lot about an engine's operating conditions - pull the vacuum hose from the fuel pressure regulator and attach the gauge to the metal vacuum line leading to it (arrow)

wheels and set the parking brake. With the transmission in Park (automatic) or Neutral (manual), start the engine and allow it to run at normal idle speed. **Warning:** *Carefully inspect the fan blades for cracks or damage before starting the engine. Keep your hands and the vacuum tester clear of the fan and do not stand in front of the vehicle or in line with the fan when the engine is running.*

Read the vacuum gauge; an average, healthy engine should normally produce between 17 and 22 inches of vacuum with a fairly steady needle.

Refer to the following vacuum gauge readings and what they indicate about the engines condition:

1   A low steady reading usually indicates a leaking gasket between the intake manifold and carburetor or throttle body, a leaky vacuum hose, late ignition timing or incorrect camshaft timing. Check ignition timing with a timing light and eliminate all other possible causes, utilizing the tests provided in this Chapter before you remove the timing chain cover to check the timing marks.

2   If the reading is three to eight inches below normal and it fluctuates at that low reading, suspect an intake manifold gasket leak at an intake port or a faulty injector.

3   If the needle has regular drops of about two to four inches at a steady rate the valves are probably leaking. Perform a compression or leak-down test to confirm this.

4   An irregular drop or down-flick of the needle can be caused by a sticking valve or an ignition misfire. Perform a compression or leak-down test and read the spark plugs.

5   A rapid fluctuation of about four in-Hg at idle combined with exhaust smoke indicates worn valve guides. Perform a leak-down test to confirm this. If the rapid fluctuation occurs with an increase in engine speed, check for a leaking intake manifold gasket or cylinder head gasket, weak valve springs, burned valves or ignition misfire.

2B

4.6 A compression gauge with a threaded fitting for the spark plug hole is preferred over the type that requires hand pressure to maintain the seal - be sure to open the throttle valve as far as possible during the compression check

6   A slight fluctuation, say one inch up and down, may mean ignition problems. Check all the usual tune-up items and, if necessary, run the engine on an ignition analyzer.

7   If there is a large fluctuation, perform a compression or leak-down test to look for a weak or dead cylinder or a blown cylinder head gasket.

8   If the needle moves slowly through a wide range, check for a clogged PCV system, incorrect idle fuel mixture, throttle body or intake manifold gasket leaks.

9   Check for a slow return after revving the engine by quickly snapping the throttle open until the engine reaches about 2,500 rpm and let it shut. Normally the reading should drop to near zero, rise above normal idle reading (about 5 in-Hg over) and then return to the previous idle reading. If the vacuum returns slowly and doesn't peak when the throttle is snapped shut, the rings may be worn. If there is a long delay, look for a restricted exhaust system (often the muffler or catalytic converter). An easy way to check this is to temporarily disconnect the exhaust ahead of the suspected part and redo the test.

4   Compression check

Refer to illustration 4.6

1   A compression check can indicate the condition of the upper end (pistons, rings, valves, cylinder head gasket, etc.). Specifically, it can tell you if the compression is down due to leakage caused by worn piston rings, defective valves and seats or a blown cylinder head gasket. **Note:** *The engine must be at normal operating temperature and the battery must be fully charged for this check.*

2   Begin by cleaning the area around the spark plugs before you remove them (compressed air should be used, if available, otherwise a small brush or even a bicycle tire pump will work). The idea is to prevent dirt

from getting into the cylinders as the compression check is being done.

3   Remove all of the spark plugs from the engine (see Chapter 1).

4   Block the throttle wide open.

5   Detach the high-tension coil wire from the distributor cap and ground it to the engine block (see Chapter 5). Disable the fuel pump circuit and relieve the fuel pressure (see Chapter 4).

6   Install the compression gauge in the spark plug hole **(see illustration)**.

7   Crank the engine over at least five compression strokes and watch the gauge. The compression should build up quickly in a healthy engine. Low compression on the first stroke, followed by gradually increasing pressure on successive strokes, indicates worn piston rings. A low compression reading on the first stroke, which doesn't build up during successive strokes, indicates leaking valves or a blown cylinder head gasket (a cracked cylinder head could also be the cause). Deposits on the undersides of the valve heads can also cause low compression. Record the highest gauge reading obtained.

8   Repeat the procedure for the remaining cylinders and compare the results to the Specifications.

9   Add some engine oil (about three squirts from a plunger-type oil can) to each cylinder, through the spark plug holes, and repeat the test.

10   If the compression increases after the oil is added, the piston rings are definitely worn. If the compression doesn't increase significantly, the leakage is occurring at the valves or cylinder head gasket. Leakage past the valves may be caused by burned valve seats and/or faces or warped, cracked or bent valves.

11   If two adjacent cylinders have equally low compression, there's a strong possibility that the cylinder head gasket between them is blown. The appearance of coolant in the combustion chambers or the crankcase would verify this condition.

12   If one cylinder is 20-percent lower than the others, and the engine has a slightly rough idle, a worn exhaust lobe on the camshaft could be the cause.

13   If the compression is unusually high, the combustion chambers are probably coated with carbon deposits. If that's the case, the cylinder head should be removed and decarbonized.

14   If compression is very low or varies greatly between cylinders, it would be a good idea to have a leak-down test performed by an automotive repair shop. This test will pinpoint exactly where the leakage is occurring and how severe it is.

5   Engine removal - methods and precautions

If you've decided that an engine must be removed for overhaul or major repair work, several preliminary steps should be taken.

Locating a suitable place to work is extremely important. Adequate work space, along with storage space for the vehicle, will be needed. If a shop or garage isn't available, at the very least a flat, level, clean work surface made of concrete or asphalt is required.

Cleaning the engine compartment and engine before beginning the removal procedure will help keep tools clean and organized.

An engine hoist or A-frame will also be necessary. Make sure the equipment is rated in excess of the combined weight of the engine and transaxle. Safety is of primary importance, considering the potential hazards involved in lifting the engine out of the vehicle.

If the engine is being removed by a novice, a helper should be available. Advice and aid from someone more experienced would also be helpful. There are many instances when one person cannot simultaneously perform all of the operations required when lifting the engine out of the vehicle.

Plan the operation ahead of time. Arrange for or obtain all of the tools and equipment you'll need prior to beginning the job. Some of the equipment necessary to perform engine removal and installation safely and with relative ease are (in addition to an engine hoist) a heavy duty floor jack, complete sets of wrenches and sockets as described in the front of this manual, wooden blocks and plenty of rags and cleaning solvent for mopping up spilled oil, coolant and gasoline. If the hoist must be rented, make sure that you arrange for it in advance and perform all of the operations possible without it beforehand. This will save you money and time.

Plan for the vehicle to be out of use for quite a while. A machine shop will be required to perform some of the work which the do-it-yourselfer can't accomplish without special equipment. These shops often have a busy schedule, so it would be a good idea to consult them before removing the engine in order to accurately estimate the amount of time required to rebuild or repair components that may need work.

Always be extremely careful when removing and installing the engine. Serious injury can result from careless actions. Plan ahead, take your time and a job of this nature, although major, can be accomplished successfully.

6   Engine - removal and installation

Refer to illustrations 6.7, 6.19 and 6.23
Warning 1: *Gasoline is extremely flammable, so take extra precautions when disconnecting any part of the fuel system. Don't smoke or allow open flames or bare light bulbs in or near the work area and don't work in a garage where a natural gas appliance (such as a clothes dryer or water heater) is installed. If you spill gasoline on your skin, rinse it off immediately. Have a fire extinguisher rated for gasoline fires handy and know how to use it.*

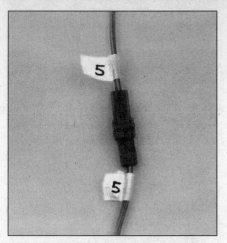

6.7 Label each hose and wire connector before disassembly

Warning 2: *The models covered by this manual are equipped with airbags. The airbag is armed and can deploy (inflate) any time the battery is connected. To prevent accidental deployment (and possible injury), disconnect the negative battery cable, then the positive battery cable, whenever working near airbag components. After the battery is disconnected, wait at least ten minutes before beginning work (the system has a back-up capacitor that must fully discharge). See Chapter 12 for more information.*

Note: *Read through the entire Section before beginning this procedure. The engine and transaxle may be removed as a unit, to be separated afterward on the garage floor, or either unit may be supported with jacks and/or chain hoists while the other is removed separately. The engine should be removed from the top when taken out separately, while the transaxle should be lowered out the bottom. To remove both engine and transaxle as a unit, lift out the unit from the top. For more on transaxle removal, see Chapter 7.*

Removal

1 Relieve the fuel system pressure (see Chapter 4). Disconnect the negative battery cable, then the positive battery cable.

2 Place protective covers on the fenders and cowl and remove the hood (see Chapter 11).

3 Remove the battery and battery tray (see Chapter 5).

4 Remove the air cleaner assembly (see Chapter 4).

5 Remove the cruise control actuator cable, if equipped.

6 Raise the vehicle and support it securely on jackstands. Drain the cooling system and engine oil and remove the drivebelts (see Chapter 1). **Note:** *Don't raise the vehicle any higher than necessary.*

7 Clearly label and disconnect all vacuum lines, coolant and emissions hoses, electrical connectors, ground straps and fuel lines (for fuel line removal see Chapter 4). Masking tape and/or a touch up paint applicator work well for marking items **(see illustration)**. Take instant photos or sketch the locations of components and brackets, if necessary.

8 Remove the cooling fan(s), shroud and radiator (see Chapter 3).

9 Release the residual fuel pressure in the tank by removing the gas cap, then disconnect the fuel lines between the engine and chassis (see Chapter 4). Plug or cap all open fittings.

10 Disconnect the throttle linkage and transaxle shift control cable (if equipped) from the throttle body (see Chapters 4 and 7).

11 Remove the alternator and starter (see Chapter 5).

12 On power steering equipped vehicles, unbolt the power steering pump (see Chapter 10). Tie the pump aside without disconnecting the hoses.

13 On air conditioned models, unbolt the compressor and set it aside (see Chapter 3). **Warning:** *Do not disconnect the refrigerant hoses.*

14 Unbolt the exhaust pipe from the exhaust manifold (see Chapter 2A).

15 Disconnect the water inlet and outlets from the cylinder head (see Chapter 2A and Chapter 3).

16 Remove the right front wheel and the right driveaxle (see Chapter 8).

17 On automatic transaxle-equipped models, detach the torque converter cover from the lower bellhousing (see Chapter 7B).

18 Remove the torque converter-to-driveplate fasteners (see Chapter 7B) and push the converter back slightly into the bellhousing.

19 Attach a lifting sling or chain to bolts or studs on the engine **(see illustration)**. Position a hoist and connect the sling or chain to it. Take up the slack until there is slight tension on the hoist. **Warning:** *Do not place any part of your body under the engine/transaxle when it's supported only by a hoist or other lifting device.*

20 Recheck to be sure nothing except the mounts are still connecting the engine/transaxle to the vehicle. Disconnect anything still remaining.

21 Place a floor jack under the transmission bellhousing area, to support the transaxle when the engine is unbolted and removed. Be sure to place a wood block on the jack head to serve as a cushion.

22 Remove the right engine mount and the front and rear engine mounts with their crossmember (see Chapter 2A), but leave the left (transaxle) mount connected.

23 Slowly raise the engine out of the vehicle **(see illustration)**. It may be necessary to tilt the engine up at the front, using the jack under the transaxle, tilt it up as well. Have an assistant twist the engine sling to angle the engine as needed as it is raised with the hoist.

24 Move the engine away from the vehicle and carefully lower the hoist until the engine is on the floor, supported by wood blocks, or remove the flywheel or driveplate and mount the engine on an engine stand.

2B

6.19 Attach a strong chain to one of the exhaust studs and an intake bolt (arrows)

6.23 The transaxle can be left attached to the vehicle (and supported by a jack) and the engine pulled out separately - guide the engine as it comes up and out to clear any components in the engine compartment

Installation

25 Check the engine/transaxle mounts. If they're worn or damaged, replace them.

26 On manual transaxle-equipped models, inspect the clutch components (see Chapter 8) and, on automatic transaxle models, inspect the converter seal and bushing.

27 On automatic transaxle equipped models, apply a dab of grease to the nose of the torque converter and to the seal lips.

28 Attach the hoist to the engine and carefully guide the engine into place, following the procedure outlined in Chapter 7A or 7B. **Caution:** *Do not use the bolts to force the engine and transaxle into alignment. It may crack or damage major components.* Install the engine-to-transaxle bolts and tighten them to the torque listed in Chapter 7A or 7B Specifications.

29 If you're working on a model equipped with an automatic transaxle, install the torque converter-to-driveplate fasteners and tighten them to the torque listed in the Chapter 7B Specifications.

30 Install the mount bolts and tighten them securely.

31 Reinstall the remaining components and fasteners in the reverse order of removal.

32 Add coolant, oil, power steering and transmission fluids as needed (see Chapter 1).

33 Run the engine and check for proper operation and leaks. Shut off the engine and recheck the fluid levels.

7   Engine rebuilding alternatives

The do-it-yourselfer is faced with a number of options when performing an engine overhaul. The decision to replace the engine block, piston/connecting rod assemblies and crankshaft depends on a number of factors, with the number one consideration being the condition of the block. Other considerations are cost, access to machine shop facilities, parts availability, time required to complete the project and the extent of prior mechanical experience on the part of the do-it-yourselfer.

Some of the rebuilding alternatives include:

Individual parts - If the inspection procedures reveal that the engine block and most engine components are in reusable condition, purchasing individual parts may be the most economical alternative. The block, crankshaft and piston/connecting rod assemblies should all be inspected carefully. Even if the block shows little wear, the cylinder bores should be surface-honed.

Short block - A short block consists of an engine block with a crankshaft and piston/connecting rod assemblies already installed. All new bearings are incorporated and all clearances will be correct. The existing camshafts, valve train components, cylinder head(s) and external parts can be bolted to the short block with little or no machine shop work necessary.

Long block - A long block consists of a short block plus an oil pump, oil pan, cylinder head, camshafts and valve train components, timing chain and sprockets. All components are installed with new bearings, seals and gaskets incorporated throughout. The installation of manifolds and external parts is all that's necessary.

Give careful thought to which alternative is best for you and discuss the situation with local automotive machine shops, auto parts dealers and experienced rebuilders before ordering or purchasing replacement parts.

8   Engine overhaul - disassembly sequence

1 It's much easier to disassemble and work on the engine if it's mounted on a portable engine stand. A stand can often be rented quite cheaply from an equipment rental yard. Before the engine is mounted on a stand, the flywheel/driveplate and rear oil seal retainer should be removed from the engine.

2 If a stand isn't available, it's possible to disassemble the engine with it blocked up on the floor. Be extra careful not to tip or drop the engine when working without a stand.

3 If you're going to obtain a rebuilt engine, all external components must come off first, to be transferred to the replacement engine, just as they will if you're doing a complete engine overhaul yourself. These include:

> Emissions control components
> Distributor, spark plug wires and spark plugs
> Thermostat and housing
> Water pump
> Fuel injection components
> Intake/exhaust manifolds
> Oil filter (always use a new filter)
> Engine mounts
> Clutch and flywheel/driveplate
> Engine rear plate (2 pieces on automatic-transmission models)

Note: *When removing the external components from the engine, pay close attention to details that may be helpful or important during installation. Note the installed position of gaskets, seals, spacers, pins, brackets, springs, washers, bolts and other small items.*

4 If you're obtaining a short-block, which consists of the engine block, crankshaft, pistons and connecting rods all assembled, then the cylinder head, oil pan, front timing cover and oil pump will have to be removed as well, to be reused on the rebuilt short block. See Section 7 for additional information regarding the different possibilities to be considered.

5 If you're planning a complete overhaul, the engine must be disassembled and the components removed in the following order:

> Valve cover
> Oil pan and pick-up tube
> Timing chain cover
> Oil pump
> Timing chain and sprockets
> Camshaft(s) and lifters

> Cylinder head
> Piston/connecting rod assemblies
> Crankshaft rear oil seal retainer
> Crankshaft and main bearings

6 Before beginning the disassembly and overhaul procedures, make sure the following items are available. Also, refer to Section 21 for a list of tools and materials needed for engine reassembly.

> Common hand tools
> Small cardboard boxes or plastic bags for storing parts
> Gasket scraper
> Ridge reamer
> Vibration damper puller
> Micrometers
> Telescoping gauges
> Dial indicator set
> Valve spring compressor
> Cylinder-surfacing hone
> Piston ring-groove cleaning tool
> Electric drill motor
> Tap and die set
> Wire brushes
> Oil gallery brushes
> Cleaning solvent

9   Cylinder head - disassembly

Refer to illustrations 9.2 and 9.4

Note: *New and rebuilt cylinder heads are commonly available for most engines at dealerships and auto parts stores. Due to the fact that some specialized tools are necessary for the disassembly and inspection procedures, and replacement parts may not be readily available, it may be more practical and economical for the home mechanic to purchase a replacement cylinder head rather than taking the time to disassemble, inspect and recondition the original.*

1 Cylinder head disassembly involves removal of the intake and exhaust valves and related components. It's assumed that the lifters and camshafts have already been removed (see Part A as needed).

2 Before the valves are removed, arrange

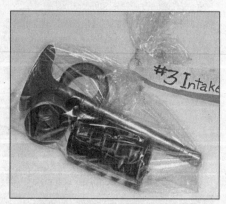

9.2 A small plastic bag, with an appropriate label, can be used to store the valve train components so they can be kept together and reinstalled in the original position

9.4 If the valve won't pull through the guide, deburr the edge of the stem end and the area around the top of the keeper groove with a file or whetstone

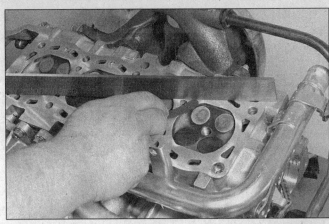

10.12 Check the cylinder head gasket surface for warpage by trying to slip a feeler gauge under the straightedge (see this Chapter's Specifications for the maximum warpage allowed and use a feeler gauge of that thickness)

to label and store them, along with their related components, so they can be kept separate and reinstalled in the same valve guides they are removed from **(see illustration)**.

3   Compress the springs on the first valve with a spring compressor and remove the keepers (see Chapter 2A). Carefully release the valve spring compressor and remove the retainer, the spring and the spring seat (if used).

4   Pull the valve out of the cylinder head and remove the oil seal from the guide. If the valve binds in the guide (won't pull through), push it back into the cylinder head and deburr the area around the keeper groove with a fine file or whetstone **(see illustration)**.

5   Repeat the procedure for the remaining valves. Remember to keep all the parts for each valve together so they can be reinstalled in their original locations.

6   Once the valves and related components have been removed and stored in an organized manner, the cylinder head should be thoroughly cleaned and inspected. If a complete engine overhaul is being done, finish the engine disassembly procedures before beginning the cylinder head cleaning and inspection process.

---

10   Cylinder head - cleaning and inspection

---

1   Thorough cleaning of the cylinder head and related valve train components, followed by a detailed inspection, will enable you to decide how much valve service work must be done during the engine overhaul. **Note:** *If the engine was severely overheated, the cylinder head is probably warped (see Step 12).*

Cleaning

2   Scrape all traces of old gasket material and sealing compound off the cylinder head gasket, intake manifold and exhaust manifold sealing surfaces. Be very careful not to gouge

the cylinder head. Special gasket removal solvents that soften gaskets and make removal much easier are available at auto parts stores.

3   Remove all built-up scale from the coolant passages.

4   Run a stiff wire brush through the various holes to remove deposits that may have formed in them.

5   Run an appropriate-size tap into each of the threaded holes to remove corrosion and thread sealant that may be present. If compressed air is available, use it to clear the holes of debris produced by this operation. **Warning:** *Wear eye protection when using compressed air!*

6   Clean the exhaust and intake manifold stud threads with a wire brush.

7   Clean the cylinder head with solvent and dry it thoroughly. Compressed air will speed the drying process and ensure that all holes and recessed areas are clean. **Note:** *Decarbonizing chemicals are available and may prove very useful when cleaning cylinder heads and valve train components. They are very caustic and should be used with caution. Be sure to follow the instructions on the container.*

8   Clean the lifters with solvent and dry them thoroughly (don't mix them up during the cleaning process). Compressed air will speed the drying process and can be used to clean out the oil passages.

9   Clean all the valve springs, spring seats, keepers and retainers with solvent and dry them thoroughly. Work on the components from one valve at a time to avoid mixing up the parts. Number the valve lifters and their shims and keep them together.

10   Scrape off any heavy deposits that may have formed on the valves, then use a motorized wire brush to remove deposits from the valve heads and stems. Again, make sure the valves don't get mixed up.

Inspection

Note: *Be sure to perform all of the following inspection procedures before concluding that*

machine shop work is required. Make a list of the items that need attention. The inspection procedures for the lifters and camshafts can be found in Chapter 2 Part A.

Cylinder head

Refer to illustrations 10.12 and 10.14

11   Inspect the cylinder head very carefully for cracks, evidence of coolant leakage and other damage. If cracks are found, check with an automotive machine shop concerning repair. If repair isn't possible, a new cylinder head should be obtained.

12   Using a straightedge and feeler gauge, check the cylinder head gasket mating surface for warpage **(see illustration)**. If the warpage exceeds the specified limit, it can be resurfaced at an automotive machine shop.

13   Examine the valve seats in each of the combustion chambers. If they're pitted, cracked or burned, the cylinder head will require valve service that's beyond the scope of the home mechanic.

14   Measure the valve guide inside diameter with a small hole gauge and micrometer **(see**

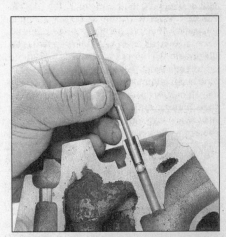

10.14 Use a small hole gauge to determine the inside diameter of the valve guides (the gauge is then measured with a micrometer)

2B

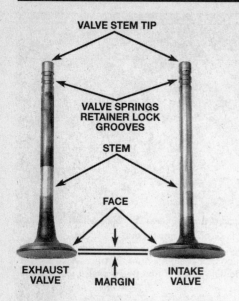

10.15 Check for valve wear at the points shown here

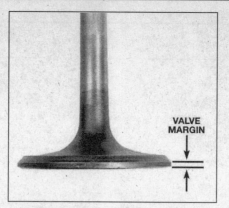

10.16 The margin width on each valve must be as specified (if no margin exists, the valve cannot be reused)

10.17 Measure the free length of each valve spring with a dial or vernier caliper

illustration), then measure the valve stem diameter with a micrometer and subtract it from the valve guide inside diameter to obtain the stem-to-guide clearance. When using a small-hole gauge or telescoping snap gauge, insert the gauge to the middle portion of the guide (where wear should be minimal) and tighten the gauge. Move the gauge up and down in the guide. If the guide isn't worn the clearance should be equal from top to bottom. Loose areas indicate wear. If there's doubt regarding the condition of the valve guides they should be checked by an automotive machine shop (the cost should be minimal).

Valves

Refer to illustrations 10.15 and 10.16

15   Carefully inspect each valve face for uneven wear, deformation, cracks, pits and burned areas. Check the valve stem for scuffing and galling and the neck for cracks. Rotate the valve and check for any obvious indication that it's bent. Look for pits and excessive wear on the end of the stem **(see illustration)**. The presence of any of these conditions indicates the need for valve service by an automotive machine shop.

16   Measure the margin width on each valve **(see illustration)**. Any valve with a margin narrower than specified will have to be replaced with a new one.

Valve components

Refer to illustrations 10.17 and 10.18

17   Check each valve spring for wear (on the ends) and pits. Measure the free height and compare it to the Specifications **(see illustration)**. Any springs that are shorter than specified have sagged and should not be reused. The tension of all springs should be checked with a special fixture before deciding that they're suitable for use in a rebuilt engine (take the springs to an automo-

tive machine shop for this check).

18   Stand each spring on a flat surface and check it for squareness **(see illustration)**. If any of the springs are distorted or sagged, replace all of them with new parts.

19   Check the spring retainers and keepers for obvious wear and cracks. Any questionable parts should be replaced with new ones, as extensive damage will occur if they fail during engine operation.

20   Any damaged or excessively worn parts must be replaced with new ones.

21   If the inspection process indicates that the valve components are in generally poor condition and worn beyond the limits specified, which is usually the case in an engine that's being overhauled, reassemble the valves in the cylinder head and refer to valve servicing recommendations (see Section 11).

11   Valves - servicing

1   Because of the complex nature of the job and the special tools and equipment needed, servicing of the valves, the valve seats and the valve guides, commonly known as a valve job, should be done by a professional.

2   The home mechanic can remove and

10.18 Check each valve spring for squareness - if it is bent it should be replaced

disassemble the cylinder head, do the initial cleaning and inspection, then reassemble and deliver it to a dealer service department or an automotive machine shop for the actual service work. Doing the inspection will enable you to see what condition the cylinder head and valvetrain components are in and will ensure that you know what work and new parts are required when dealing with an automotive machine shop.

3   The dealer service department, or automotive machine shop, will remove the valves and springs, recondition or replace the valves and valve seats, check and replace the valve guides (as necessary), check and replace the valve springs, spring retainers and keepers (as necessary), replace the valve seals with new ones, reassemble the valve components and make sure the installed spring height is correct. The cylinder head gasket surface will also be resurfaced (within specification limits) if it's warped.

4   After the valve job has been performed by a professional, the cylinder head will be in like-new condition. When the cylinder head is returned, be sure to clean it again before installation on the engine to remove any metal particles and abrasive grit that may still be present from the valve service or cylinder head resurfacing operations. Use compressed air, if available, to blow out all the oil holes and passages. **Warning:** *Always wear eye protection when using compressed air!*

12   Cylinder head - reassembly

1   Regardless of whether or not the cylinder head was sent to an automotive repair shop for valve servicing, make sure it's clean before beginning reassembly.

2   If the cylinder head was sent out for valve servicing, the valves and related components will already be in place.

3   If the machine shop did not install them, install new seals on each of the valve guides. Gently tap each valve seal into place until it's seated on the guide (see Chapter 2A). **Caution:** *Don't hammer on the valve seals once they're seated or you may damage them. Don't twist or cock the seals during installa-*

13.1 A ridge reamer is required to remove the ridge from the top of each cylinder - do this before removing the pistons!

13.3 Check the connecting rod side clearance (endplay) with a feeler gauge as shown here

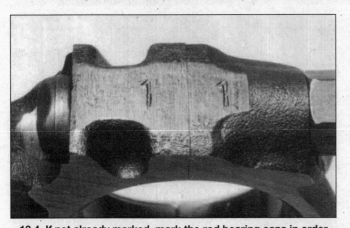

13.4 If not already marked, mark the rod bearing caps in order from the front of the engine to the rear (one mark for the front cap, two for the second one and so on) the marks should always be on the same side of the rod and cap

13.6 To prevent damage to the crankshaft journals and cylinder walls, slip sections of hose over the rod bolts before removing the piston/rod assembly

tion or they won't seat properly on the valve stems.

4   Beginning at one end of the cylinder head, lubricate and install the first valve. Apply moly-base grease or clean engine oil to the valve stem.

5   Drop the spring seat over the valve guide and set the valve spring and retainer in place.

6   Compress the springs with a valve spring compressor and carefully install the keepers in the upper groove, then slowly release the compressor and make sure the keepers seat properly. Apply a small dab of grease to each keeper to hold it in place if necessary.

7   Repeat the procedure for the remaining valves. Be sure to return the components to their original locations - don't mix them up!

13   Pistons/connecting rods - removal

Refer to illustrations 13.1, 13.3, 13.4 and 13.6

Note: Prior to removing the piston/connecting rod assemblies, remove the cylinder head,

the oil pan and the oil pump pick-up tube by referring to the appropriate Sections in Chapter 2 Part A.

1   Use your fingernail to feel if a ridge has formed at the upper limit of ring travel (about 1/4-inch down from the top of each cylinder). If carbon deposits or cylinder wear have produced ridges, they must be completely removed with a special tool (see illustration). Follow the manufacturer's instructions provided with the tool. Failure to remove the ridges before attempting to remove the piston/connecting rod assemblies may result in piston breakage. Caution: When removing a ridge in the cylinder bore, DO NOT damage the top of the cylinder bore or the block deck surface with the ridge removal tool.

2   After the cylinder ridges have been removed, turn the engine upside-down so the crankshaft is facing up.

3   Before the connecting rods are removed, check the endplay (side clearance) with feeler gauges. Slide them between each connecting rod and the crankshaft throw until the play is removed (see illustration). The endplay is equal to the thickness of the feeler gauge(s). If the endplay exceeds the service limit, new connecting rods will be required. If new rods

(or a new crankshaft) are installed, the endplay may fall under the specified minimum (if it does, the rods will have to be machined to restore it - consult an automotive machine shop for advice if necessary). Repeat the procedure for the remaining connecting rods.

4   Check the connecting rods and caps for identification marks. If they aren't plainly marked, use a small center-punch to make the appropriate number of indentations on each rod and cap (1, 2, 3, etc., depending on the cylinder they're associated with) (see illustration).

5   Loosen each of the connecting rod cap nuts 1/2-turn at a time until they can be removed by hand. Remove the number one connecting rod cap and bearing insert. Don't drop the bearing insert out of the cap.

6   Slip a short length of plastic or rubber hose over each connecting rod cap bolt to protect the crankshaft journal and cylinder wall as the piston is removed (see illustration).

7   Remove the bearing insert and push the connecting rod/piston assembly out through the top of the engine. Use a wooden hammer handle to push on the upper bearing surface in the connecting rod. If resistance is felt,

2B

14.1 Checking crankshaft endplay with a dial indicator

14.3 Checking crankshaft endplay with a feeler gauge

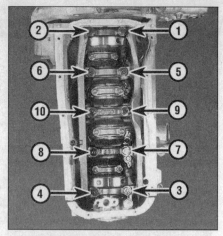

14.4 Main bearing support bolt LOOSENING sequence

15.1a A hammer and a large punch can be used to knock the core plugs sideways in their bores

15.1b Pull the core plugs from the block with pliers

double-check to make sure that all of the ridge was removed from the cylinder.

8   Repeat the procedure for the remaining cylinders.

9   After removal, reassemble the connecting rod caps and bearing inserts in their respective connecting rods and install the cap nuts finger tight. Leaving the old bearing inserts in place until reassembly will help prevent the connecting rod bearing surfaces from being accidentally nicked or gouged.

10   Don't separate the pistons from the connecting rods (see Section 18).

14   Crankshaft - removal

Refer to illustrations 14.1. 14.3 and 14.4
Note: *The crankshaft can be removed only after the engine has been removed from the vehicle. It's assumed that the flywheel or driveplate, crankshaft pulley, timing chain, oil pan, oil pick-up tube, oil pump and piston/connecting rod assemblies have already been removed. The rear main oil seal retainer must be unbolted and separated from the block before proceeding with crankshaft removal.*

1   Before the crankshaft is removed, check the endplay. Mount a dial indicator with the stem in line with the crankshaft and touching the crankshaft end **(see illustration)**.

2   Pry the crankshaft all the way to the rear and zero the dial indicator. Next, pry the crankshaft to the front as far as possible and check the reading on the dial indicator. The distance that it moves is the endplay. If it's greater than specified, check the crankshaft thrust surfaces for wear. If no wear is evident, new thrust bearings should correct the endplay.

3   If a dial indicator isn't available, feeler gauges can be used. Gently pry or push the crankshaft all the way to the front of the engine. Slip feeler gauges between the crankshaft and the front face of the thrust main bearing to determine the clearance **(see illustration)**. The thrust bearing is the number three (center) bearing.

4   This engine uses a large bearing support that incorporates all the main bearing caps in one unit. Loosen the main bearing support bolts 1/4-turn at a time each, in the recommended sequence **(see illustration)**.

5   Gently tap the support with a soft-face hammer, then separate it from the engine

block. If necessary, use the bolts as levers to remove the assembly. Try not to drop the bearing inserts if they come out with the support.

6   Carefully lift the crankshaft out of the engine. It may be a good idea to have an assistant available, since the crankshaft is quite heavy. With the bearing inserts in place in the engine block and main bearing cap assembly, return the support to the engine block and tighten the bolts finger tight.

15   Engine block- cleaning

Refer to illustrations 15.1a, 15.1b, 15.8 and 15. 10
Caution: *The core plugs (also known as freeze or soft plugs) may be difficult or impossible to retrieve if they're driven into the block coolant passages.*

1   Using the wide end of a punch **(see illustration)** tap in on one side of the outer edge of the core plug to turn the plug sideways in the bore. Then, using a pair of pliers, pull the core plug from the engine block **(see illustration)**. Don't worry about the condition of the old core plugs as they are being

15.8 All bolt holes in the block - particularly the main bearing cap and cylinder head bolt holes - should be cleaned and restored with a tap (be sure to remove debris from the holes after this is done)

15.10 A large socket on an extension can be used to drive the new core plugs into the bores

2B

removed because they will be replaced on reassembly with new plugs.

2   Using a gasket scraper, remove all traces of gasket material from the engine block. **Caution:** *Be EXTREMELY CAREFUL not to nick or gouge the gasket sealing surfaces.*

3   Remove the main bearing cap support and separate the bearing inserts from the support and the engine block. Tag the bearings, indicating which cylinder they were removed from and whether they were in the cap or the block, then set them aside.

4   Remove all of the threaded oil gallery plugs from the block. The plugs are usually very tight - they may have to be drilled out and the holes retapped. Use new plugs when the engine is reassembled.

5   If the engine is extremely dirty it should be taken to an automotive machine shop for cleaning.

6   After the block is returned, clean all oil holes and oil galleries one more time. Brushes specifically designed for this purpose are available at most auto parts stores. Flush the passages with warm water until the water runs clear, dry the block thoroughly and wipe all machined surfaces with a light, rust preventive oil. If you have access to compressed air, use it to speed the drying process and to blow out all the oil holes and galleries. **Warning:** *Wear eye protection when using compressed air!*

7   If the block isn't extremely dirty or sludged up, you can do an adequate cleaning job with hot soapy water and a stiff brush. Take plenty of time and do a thorough job. Regardless of the cleaning method used, be sure to clean all oil holes and galleries very thoroughly, dry the block completely and coat all machined surfaces with light oil.

8   The threaded holes in the block must be clean to ensure accurate torque readings during reassembly. Run the proper size tap into each of the holes to remove rust, corrosion, thread sealant or sludge and restore damaged threads **(see illustration)**. If possible, use compressed air to clear the holes of debris produced by this operation. Now is a

good time to clean the threads on the cylinder head bolts and the main bearing cap bolts as well.

9   Reinstall the main bearing cap support and tighten the bolts finger tight.

10   After coating the sealing surfaces of the new core plugs with a non-hardening sealant, such as Permatex no. 2 sealant or equivalent, install them in the engine block **(see illustration)**. Make sure they're driven in straight or leakage could result. Special tools are available for this purpose, but a large socket, with an outside diameter that will just slip into the core plug, a 1/2-inch drive extension and a hammer will work just as well.

11   Apply non-hardening sealant (such as Permatex no. 2 or Teflon pipe sealant) to the new oil gallery plugs and thread them into the holes in the block. Make sure they're tightened securely.

12   If the engine isn't going to be reassembled right away, cover it with a large plastic trash bag to keep it clean.

16   Engine block - inspection

Refer to illustrations 16.2, 16.4a, 16.4b, 16.4c and 16.10

1   Before the block is inspected, it should be cleaned (see Section 15).

2   Visually check the block for cracks, rust and corrosion. Look for stripped threads in the threaded holes. It's also a good idea to have the block checked for hidden cracks by an automotive machine shop that has the special equipment to do this type of work. If defects are found, have the block repaired, if possible, or replaced. The block is marked with a "grade" indication along the cylinder head gasket surface, which indicates the size of the original pistons **(see illustration)**.

3   Check the cylinder bores for scuffing and scoring.

4   Check the cylinders for taper and out-of-round conditions as follows **(see illustrations)**:

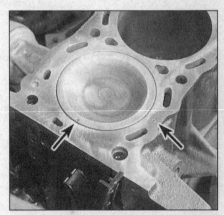

16.2 The top of the block has numbers (right arrow) to indicate the Grade number (size) of each piston - note that the dot on the piston faces forward (left arrow)

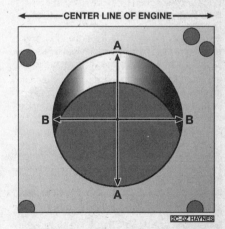

16.4a Measure the diameter of each cylinder at a right angle to the engine centerline (A), and parallel to the engine centerline (B) - out-of-round is the difference between A and B; taper is the difference between A and B at the top of the cylinder and A and B at the bottom of the cylinder

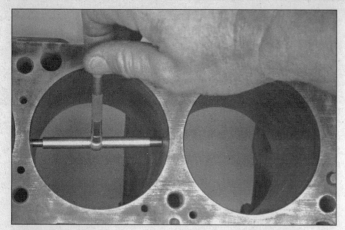

16.4b Measure the diameter of each cylinder - the ability to feel when the telescoping gauge is at the correct point will be developed over time, so work slowly and repeat the check until you're satisfied the bore measurement is accurate

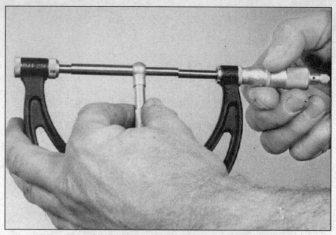

16.4c The gauge is then measured with a micrometer to determine the bore size

5   Measure the diameter of each cylinder at the top (just under the ridge area), center and bottom of the cylinder bore, parallel to the crankshaft axis.

6   Next, measure each cylinder's diameter at the same three locations perpendicular to the crankshaft axis.

7   The taper of each cylinder is the difference between the bore diameter at the top of the cylinder and the diameter at the bottom. The out-of-round specification of the cylinder bore is the difference between the parallel and perpendicular readings. Compare your results to this Chapter's Specifications.

8   Repeat the procedure for the remaining pistons and cylinders.

9   If the cylinder walls are badly scuffed or scored, or if they're out-of-round or tapered beyond the limits given in this Chapter's Specifications, have the engine block rebored and honed at an automotive machine shop. If a rebore is done, oversize pistons and rings will be required.

10   Using a precision straightedge and a feeler gauge, check the block deck (the surface that mates with the cylinder head) for distortion (see illustration). If it's distorted beyond the specified limit, it can be resurfaced by an automotive machine shop.

11   If the cylinders are in reasonably good condition and not worn to the outside of the limits, and if the piston-to-cylinder clearance can be maintained properly, then they don't have to be rebored. Honing is all that's necessary (see Section 17).

17   Cylinder honing

Refer to illustrations 17.3a and 17.3b

1   Prior to engine reassembly, the cylinder bores must be honed so the new piston rings will seat correctly and provide the best possible combustion chamber seal. **Note:** *If you don't have the tools or don't want to tackle the honing operation, most automotive machine shops will do it for a reasonable fee.*

2   Before honing the cylinders, install the main bearing cap support (without bearing inserts) and tighten the bolts to the torque listed in this Chapter's Specifications.

3   Two types of cylinder hones are commonly available - the flex hone or "bottle brush" type and the more traditional surfacing hone with spring-loaded stones. Both will do the job, but for the less experienced mechanic the "bottle brush" hone will probably be easier to use. You'll also need some kerosene or honing oil, rags and an electric drill motor. Proceed as follows:

a) *Mount the hone in the drill motor, compress the stones and slip it into the first cylinder* (see illustration). *Be sure to wear safety goggles or a face shield!*

b) *Lubricate the cylinder with plenty of honing oil, turn on the drill and move the hone up-and-down in the cylinder at a pace that will produce a fine crosshatch pattern on the cylinder walls. Ideally, the crosshatch lines should intersect at approximately a 60-degrees angle (see illustration). Be sure to use plenty of lubricant and don't take off any more material than is absolutely necessary to produce the desired finish.* **Note:** *Piston ring manufacturers may specify a smaller crosshatch angle than the traditional 60-degrees - read and follow any instruc-*

16.10 Check the block deck for distortion with a precision straightedge and a feeler gauge - lay the straightedge across the block, diagonally and from end-to-end when making the check

17.3a A "bottle brush" hone will produce better results if you've never honed cylinders before

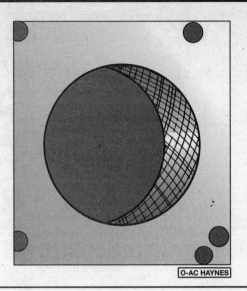

17.3b The cylinder hone should leave a smooth, crosshatch pattern with the lines intersecting at approximately a 60-degree angle

18.4 The piston ring grooves can be cleaned with a special tool, or a section of a broken ring as shown here

tions included with the new rings.

c) *Don't withdraw the hone from the cylinder while it's running. Instead, shut off the drill and continue moving the hone up-and-down in the cylinder until it comes to a complete stop, then compress the stones and withdraw the hone. If you're using a "bottle brush" type hone, stop the drill motor, then turn the chuck in the normal direction of rotation while withdrawing the hone from the cylinder.*

d) *Wipe the oil out of the cylinder and repeat the procedure for the remaining cylinders.*

4   After the honing job is complete, chamfer the top edges of the cylinder bores with a small file so the rings won't catch when the pistons are installed. Be very careful not to nick the cylinder walls with the end of the file.

5   The entire engine block must be washed again very thoroughly with warm, soapy water to remove all traces of the abrasive grit produced during the honing operation. **Note:** *The bores can be considered clean when a lint-free white cloth - dampened with clean engine oil - used to wipe them out doesn't pick up any more honing residue, which will show up as gray areas on the cloth. Be sure to run a brush through all oil holes and galleries and flush them with running water.*

6   After rinsing, dry the block and apply a coat of light rust preventive oil to all machined surfaces. Wrap the block in a plastic trash bag to keep it clean and set it aside until reassembly.

18 Pistons/connecting rods - inspection

Refer to illustrations 18.4, 18.10 and 18.11

1   Before the inspection process can be carried out, the piston/connecting rod assemblies must be cleaned and the original piston rings removed from the pistons. **Note:** *Always use new piston rings when the engine*

is reassembled.

2   Using a piston ring installation tool, carefully remove the rings from the pistons. Be careful not to nick or gouge the pistons in the process.

3   Scrape all traces of carbon from the top of the piston. A hand-held wire brush or a piece of fine emery cloth can be used once the majority of the deposits have been scraped away. Do not, under any circumstances, use a wire brush mounted in a drill motor to remove deposits from the pistons. The piston material is soft and may be eroded away by the wire brush.

4   Use a piston ring groove-cleaning tool to remove carbon deposits from the ring grooves. If a tool isn't available, a piece broken off the old ring will do the job. Be very careful to remove only the carbon deposits - don't remove any metal and do not nick or scratch the sides of the ring grooves **(see illustration)**.

5   Once the deposits have been removed, clean the piston/rod assemblies with solvent and dry them with compressed air (if available). **Warning:** *Wear eye protection when using compressed air!* Make sure the oil return holes in the back sides of the ring grooves and the oil hole in the lower end of each rod are clear.

6   If the pistons and cylinder walls aren't damaged or worn excessively, and if the engine block is not rebored, new pistons won't be necessary. Normal piston wear appears as even vertical wear on the piston thrust surfaces and slight looseness of the top ring in its groove. New piston rings, however, should always be used when an engine is rebuilt.

7   Carefully inspect each piston for cracks around the skirt, at the pin bosses and at the ring lands.

8   Look for scoring and scuffing on the thrust faces of the skirt, holes in the piston crown and burned areas at the edge of the crown. If the skirt is scored or scuffed, the engine may have been suffering from overheating and/or abnormal combustion, which

caused excessively high operating temperatures. The cooling and lubrication systems should be checked thoroughly. A hole in the piston crown is an indication that abnormal combustion (preignition) was occurring. Burned areas at the edge of the piston crown are usually evidence of spark knock (detonation). If any of the above problems exist, the causes must be corrected or the damage will occur again. The causes may include intake air leaks, incorrect fuel/air mixture, incorrect ignition timing and EGR system malfunctions.

9   Corrosion of the piston, in the form of small pits, indicates that coolant is leaking into the combustion chamber and/or the crankcase. Again, the cause must be corrected or the problem may persist in the rebuilt engine.

10   Measure the piston ring side clearance by laying a new piston ring in each ring groove and slipping a feeler gauge in beside it **(see illustration)**. Check the clearance at three or four locations around each groove. Be sure to use the correct ring for each groove - they are different. If the side clearance is greater than specified, new pistons will have to be used.

18.10 Check the ring side clearance with a feeler gauge at several points around the groove

2B

11 Check the piston-to-bore clearance by measuring the bore (see Section 16) and the piston diameter. Make sure the pistons and bores are correctly matched. Measure the piston across the skirt, at a 90-degree angle to the piston pin and at approximately the bottom of the pin bore **(see illustration)**. Subtract the piston diameter from the bore diameter to obtain the clearance. If it's greater than specified, the block will have to be rebored and new pistons and rings installed.

12 Check the piston-to-rod clearance by twisting the piston and rod in opposite directions. Any noticeable play indicates excessive wear, which must be corrected. The piston/connecting rod assemblies should be taken to an automotive machine shop to have the pistons and rods re-sized and new pins installed.

13 If the pistons must be removed from the connecting rods for any reason, they should be taken to an automotive machine shop. While they are there, have the connecting rods checked for bend and twist, since automotive machine shops have special equipment for this purpose. **Note:** *Unless new pistons and/or connecting rods must be installed, do not disassemble the pistons and connecting rods.*

14 Check the connecting rods for cracks and other damage. Temporarily remove the rod caps, lift out the old bearing inserts, wipe the rod and cap bearing surfaces clean and inspect them for nicks, gouges and scratches. After checking the rods, replace the old bearings, slip the caps into place and tighten the nuts finger tight. **Note:** *If the engine is being rebuilt because of a connecting rod knock, be sure to install new rods.*

19 Crankshaft - inspection

Refer to illustrations 19.1, 19.2, and 19.5

1 Remove all burrs from the crankshaft oil holes with a stone or file **(see illustration)**.

2 Clean the crankshaft with solvent and dry it with compressed air (if available). Be

18.11 Measure the piston diameter at a 90-degree angle to the piston pin below the bottom of the piston pin bore

sure to clean the oil holes with a stiff brush and flush them with solvent **(see illustration)**.

3 Check the main and connecting rod bearing journals for uneven wear, scoring, pits and cracks.

4 Check the rest of the crankshaft for cracks and other damage. It should be magnafluxed to reveal hidden cracks, an automotive machine shop will handle the procedure.

5 Using a micrometer, measure the diameter of the main and connecting rod journals and compare the results to this Chapter's Specifications **(see illustration)**. By measuring the diameter at a number of points around each journal's circumference, you'll be able to determine whether or not the journal is out-of-round. Take the measurement at each end of the journal, near the crank throws, to determine if the journal is tapered. Crankshaft runout should be checked also, but large V-blocks and a dial indicator are needed to do it correctly. If you don't have the equipment, have a machine shop check the runout.

6 If the crankshaft journals are damaged, tapered, out-of-round or worn beyond the limits given in the Specifications, have the crankshaft reground by an automotive

19.1 The oil holes should be chamfered so sharp edges don't gouge or scratch the new bearings

machine shop. Be sure to use the correct size bearing inserts if the crankshaft is reconditioned.

7 Check the oil seal journals at each end of the crankshaft for wear. If the seal has worn a groove in the journal, or if it's nicked or scratched, the new seal may leak when the engine is reassembled. In some cases, an automotive machine shop may be able to repair the journal by pressing on a thin sleeve. If repair isn't feasible, a new or different crankshaft should be installed.

8 Examine the main and rod bearing inserts (see Section 20).

9 If the crankshaft requires replacement make certain that you match the original crankshaft with any replacement purchased.

20 Main and connecting rod bearings - inspection and selection

Inspection

Refer to illustrations 20.1 and 20.10

1 Even though the main and connecting

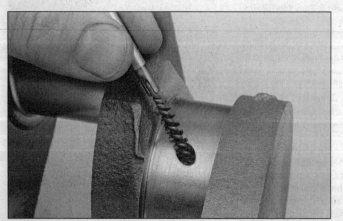

19.2 Use a wire or stiff plastic bristle brush to clean the oil passages in the crankshaft

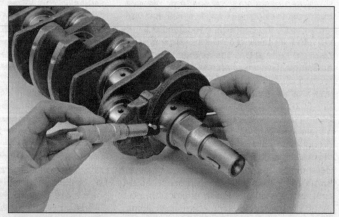

19.5 Measure the diameter of each crankshaft journal at several points to detect taper and out-of-round conditions

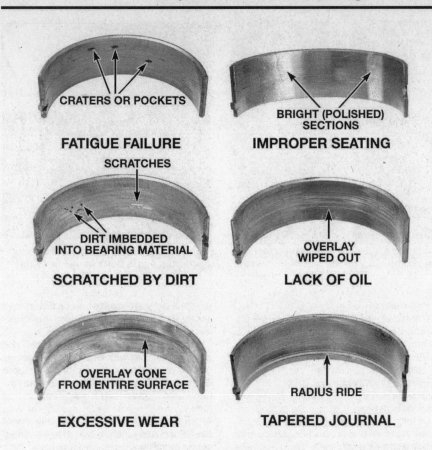

CRATERS OR POCKETS

FATIGUE FAILURE

BRIGHT (POLISHED) SECTIONS

IMPROPER SEATING

SCRATCHES

DIRT IMBEDDED INTO BEARING MATERIAL

SCRATCHED BY DIRT

OVERLAY WIPED OUT

LACK OF OIL

OVERLAY GONE FROM ENTIRE SURFACE

EXCESSIVE WEAR

RADIUS RIDE

TAPERED JOURNAL

20.1 Typical bearing failures

rod bearings should be replaced with new ones during the engine overhaul, the old bearings should be retained for close examination, as they may reveal valuable information about the condition of the engine **(see illustration)**.

2   Bearing failure occurs because of lack of lubrication, the presence of dirt or other foreign particles, corrosion and overloading the engine. Regardless of the cause of bearing failure, it must be corrected before the engine is reassembled to prevent it from happening again.

3   When examining the bearings, remove them from the engine block, the main bearing cap support, the connecting rods and the rod caps and lay them out on a clean surface in the same general position as their location in the engine. This will enable you to match any bearing problems with the corresponding crankshaft journal.

4   Dirt and other foreign particles get into the engine in a variety of ways. It may be left in the engine during assembly, or it may pass through filters or the PCV system. It may get into the oil, and from there into the bearings. Metal chips from machining operations and normal engine wear are often present. Abrasives are sometimes left in engine components after reconditioning, especially when parts are not thoroughly cleaned using the proper cleaning methods. Whatever the source, these foreign objects often end up embedded in the soft bearing material and are easily recognized. Large particles will not embed in the bearing and will score or gouge the bearing and journal. The best prevention for this cause of bearing failure is to clean all parts thoroughly and keep everything spotlessly clean during engine assembly. Frequent and regular engine oil and filter changes are also recommended.

5   Lack of lubrication (or lubrication breakdown) has a number of interrelated causes. Excessive heat (which thins the oil), overloading (which squeezes the oil from the bearing face) and oil leakage or throw off (from excessive bearing clearances, worn oil pump or high engine speeds) all contribute to lubrication breakdown. Blocked oil passages, which usually are the result of misaligned oil holes in a bearing shell, will also oil starve a bearing and destroy it. When lack of lubrication is the cause of bearing failure, the bearing material is wiped or extruded from the steel backing of the bearing. Temperatures may increase to the point where the steel backing turns blue from overheating.

6   Driving habits can have a definite effect on bearing life. Low speed operation in too high a gear (lugging the engine) puts very high loads on bearings, which tends to squeeze out the oil film. These loads cause the bearings to flex, which produces fine cracks in the

bearing face (fatigue failure). Eventually the bearing material will loosen in pieces and tear away from the steel backing. Short trip driving leads to corrosion of bearings because insufficient engine heat is produced to drive off the condensed water and corrosive gases. These products collect in the engine oil, forming acid and sludge. As the oil is carried to the engine bearings, the acid attacks and corrodes the bearing material.

7   Incorrect bearing installation during engine assembly will lead to bearing failure as well. Tight fitting bearings leave insufficient bearing oil clearance and will result in oil starvation. Dirt or foreign particles trapped behind a bearing insert result in high spots on the bearing which lead to failure.

Selection

8   If the original bearings are worn or damaged, or if the oil clearances are incorrect (see Section 23 or 25) and the crankshaft is not damaged or worn excessively, new bearings should be purchased for engine reassembly. However, if the crankshaft has been damaged, it should be reground and new undersize bearings installed. The automotive machine shop that reconditions the crankshaft will provide or help you select the correct-size bearings. Regardless of how the bearing sizes are determined, measure the oil clearance with Plastigage to ensure the bearings are the right size.

9   If you need to use STANDARD size bearings, install bearings that have the same grade number as the original bearings.

10   Crankshafts are marked with two rows of numbers on the front counterweight **(see illustration)**. The top row as shown indicates the grade of the four rod journals (1 to 4 from left to right), and the lower numbers the grade of the five main journals (1 to 5, left-to-right).

11   Remember, the oil clearance is the final judge when selecting new bearing sizes. If you have any questions or are unsure which bearings to use, get help from a dealer parts or service department or other parts supplier.

2B

20.10 The rod journal sizes are marked on the crankshaft in the upper row (A) and the main journals sizes are in the lower row (B)

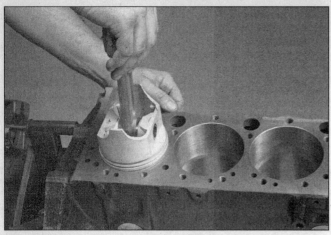

22.3 When checking piston ring end gap, the ring must be square in the cylinder bore (this is done by pushing the ring down with the top of a piston as shown)

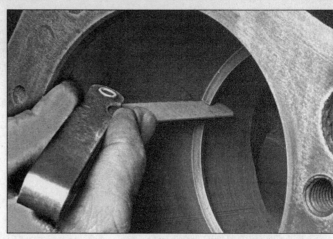

22.4 With the ring square in the cylinder, measure the end gap with a feeler gauge

21   Engine overhaul - reassembly sequence

1   Before beginning engine reassembly, make sure you have all the necessary new parts, gaskets and seals as well as the following items on hand:

Common hand tools
A torque wrench
Piston ring installation tool
Piston ring compressor
Short lengths of rubber or plastic hose to fit over
connecting rod bolts
Plastigage
Feeler gauges
A fine-tooth file
New engine oil
Engine assembly lube or moly-base grease
Gasket sealant
Thread-locking compound

2   In order to save time and avoid problems, engine reassembly must be done in the following general order:

Piston rings
Crankshaft and main bearings
Piston/connecting rod assemblies
Rear main oil seal
Timing chains, sprockets and cover
Cylinder head
Camshafts and lifters
Oil pump
Oil pick-up tube
Oil pan
Valve cover
Intake and exhaust manifolds
Flywheel/driveplate

22   Piston rings - installation

Refer to illustrations 22.3, 22.4, 22.5, 22.8a, 22.8b and 22.11

1   Before installing the new piston rings, the ring end gaps must be checked. It's assumed that the piston ring side clearance has been checked and verified correct (see Section 18).
2   Lay out the piston/connecting rod assemblies and the new ring sets so the ring sets will be matched with the same piston and cylinder during the end gap measurement and engine assembly.
3   Insert the top ring into the first cylinder and square it up with the cylinder walls by pushing it in with the top of the piston (see illustration). The ring should be near the bottom of the cylinder, at the lower limit of ring travel.
4   To measure the end gap, slip feeler gauges between the ends of the ring until a gauge equal to the gap width is found (see illustration). The feeler gauge should slide between the ring ends with a slight amount of drag. Compare the measurement to this Chapter's Specifications. If the gap is larger or smaller than specified, double-check to make sure you have the correct rings before proceeding.
5   If the gap is too small, it must be enlarged or the ring ends may come in contact with each other during engine operation, which can cause serious damage to the engine. The end gap can be increased by filing the ring ends very carefully with a fine file. Mount the file in a vise equipped with soft jaws, slip the ring over the file with the ends

22.5 If the endgap is too small, clamp a file in a vise and file the ring ends (only from the outside toward the inside) to enlarge the gap slightly

22.8a Installing the spacer/expander in the oil control ring groove

22.8b DO NOT use a piston ring installation tool when installing the oil ring side rails

22.11 Installing the compression rings with a ring expander - the mark (arrow) must face up

contacting the file face and slowly move the ring to remove material from the ends. When performing this operation, file only from the outside in **(see illustration)**.

6   Repeat the procedure for each ring that will be installed in the first cylinder and for each ring in the remaining cylinders. Remember to keep rings, pistons and cylinders matched up.

7   Once the ring end gaps have been checked, the rings can be installed on the pistons.

8   The oil control ring (lowest one on the piston) is usually installed first. It's composed of three separate components. Slip the spacer/expander into the groove **(see illustration)**. If an anti-rotation tang is used, make sure it's inserted into the drilled hole in the ring groove. Next, install the lower side rail. Don't use a piston ring installation tool on the oil ring side rails, as they may be damaged. Instead, place one end of the side rail into the groove between the spacer/expander and the ring land, hold it firmly in place and slide a finger around the piston while pushing the rail into the groove **(see illustration)**. Next, install the upper side rail in the same manner.

9   After the three oil ring components have been installed, check to make sure that both the upper and lower side rails can be turned smoothly in the ring groove.

10   The second compression ring is installed next. It's usually stamped with a mark which must face up, toward the top of the piston. **Note:** *Always follow the instructions printed on the ring package or box - different manufacturers may require different approaches. Do not mix up the top and second compression rings, as they have different cross sections.*

11   Use a piston ring installation tool and make sure the identification mark is facing the top of the piston, then slip the ring into the middle groove on the piston. Don't expand the ring any more than necessary to slide it over the piston **(see illustration)**.

12   Install the top ring in the same manner. Make sure the mark is facing up.

13   Repeat the procedure for the remaining pistons and rings.

23 Crankshaft - installation and main bearing oil clearance check

Refer to illustrations 23.10 and 23.14

1   Crankshaft installation is the first major step in engine reassembly. It's assumed at this point that the engine block and crankshaft have been cleaned, inspected and repaired or reconditioned.

2   Position the engine with the bottom facing up.

3   Remove the main bearing support bolts and lift out the support.

4   If they're still in place, remove the old bearing inserts from the block and the main bearing support. Wipe the main bearing surfaces of the block and support with a clean, lint-free cloth. They must be kept spotlessly clean!

Main bearing oil clearance check

5   Clean the back sides of the new main bearing inserts and lay the bearing half with the oil groove and hole in each main bearing saddle in the block (on all engines covered by this manual, the bearings with oil holes go in the block and those without oil holes go in the caps). Lay the other bearing half from each bearing set in the corresponding portion of the main bearing support. Make sure the tab on each bearing insert fits into the recess in the block or support. Also, the oil holes in the block must line up with the oil holes in the bearing insert. **Caution:** *Do not hammer the bearings into place and don't nick or gouge the bearing faces. No lubrication should be used at this time.*

6   The thrust bearings (the only ones in the set with a flange on either side) must be installed in the number three main bearing saddle. Apply a thin film of moly-based grease or engine assembly lube to the back sides of the thrust bearings to hold them in place.

7   Clean the faces of the bearings in the block and the crankshaft main bearing journals with a clean, lint-free cloth. Check or clean the oil holes in the crankshaft, as any

23.10 Lay the Plastigage strips (arrow) on the main bearing journals, parallel to the crankshaft centerline

dirt here can go only one way - straight through the new bearings.

8   Once you're certain the crankshaft is clean, carefully lay it in position in the main bearings. No lubricant should be used at this time.

9   Before the crankshaft can be permanently installed, the main bearing oil clearance must be checked.

10   Trim several pieces of the appropriate size Plastigage (they must be slightly shorter than the width of the main bearings) and place one piece on each crankshaft main bearing journal, parallel with the journal axis **(see illustration)**.

11   Clean the faces of the bearings in the caps and install the main bearing support. Don't disturb the Plastigage. Apply a light coat of oil to the bolt threads and the undersides of the bolt heads, then install them.

12   Following the reverse of the loosening sequence **(see illustration 14.4)**, tighten the main bearing support bolts, in three steps, to the torque listed in this Chapter's Specifications. Don't rotate the crankshaft at any time during this operation!

13   Remove the bolts and carefully lift off

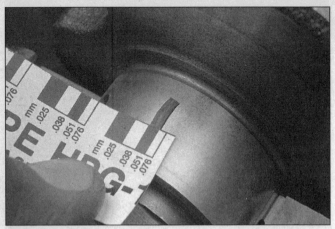

23.14 Compare the width of the crushed Plastigage to the scale on the envelope to determine the main bearing oil clearance (always take the measurement at the widest point of the Plastigage); be sure to use the correct scale - standard and metric ones are included

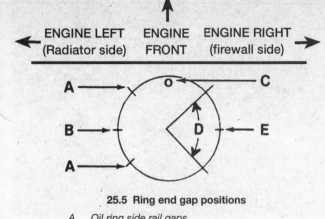

25.5 Ring end gap positions

A   *Oil ring side rail gaps*
B   *Second compression ring gap*
C   *Dimple in the piston (indicating front of engine)*
D   *Oil spacer/expander gap*
E   *Top compression ring gap*

the main bearing caps. Keep them in order. Don't disturb the Plastigage or rotate the crankshaft. If the main bearing support is difficult to remove, tap it gently from side-to-side with a soft-face hammer to loosen it.

14   Compare the width of the crushed Plastigage on each journal to the scale printed on the Plastigage envelope to obtain the main bearing oil clearance **(see illustration)**. Check the Specifications to make sure it's correct.

15   If the clearance is not as specified, the bearing inserts may be the wrong size which means different ones will be required (see Section 20). Before deciding that different inserts are needed, make sure that no dirt or oil was between the bearing inserts and the support or block when the clearance was measured. If the Plastigage is noticeably wider at one end than the other, the journal may be tapered (see Section 19).

16   Carefully scrape all traces of the Plastigage material off the main bearing journals and/or the bearing faces. Don't nick or scratch the bearing faces.

Final crankshaft installation

17   Carefully lift the crankshaft out of the engine. Clean the bearing faces in the block, then apply a thin, uniform layer of clean moly-base grease or engine assembly lube to each of the bearing surfaces. Coat the flanges on the thrust bearings.

18   Lubricate the crankshaft surfaces that contact the oil seals with moly-base grease, engine assembly lube or clean engine oil.

19   Make sure the crankshaft journals are clean, then lay the crankshaft back in place in the block. Clean the faces of the bearings in the bearing support, then apply the same lubricant to them. Install the bearing support.

20   Apply a light coat of oil to the bolt threads and the under-sides of the bolt heads, then install them. Tighten the bolts to 10-to-12 ft-lbs following the recommended sequence

(reverse of the loosening sequence in illustration 14.4), working from the center main out to the ends. Tap the ends of the crankshaft forward and backward with a lead or brass hammer to align the thrust bearings. Retighten all main bearing support bolts to the torque listed in this Chapter's Specifications, following the recommended sequence.

21   Rotate the crankshaft a number of times by hand to check for any obvious binding.

22   Check the crankshaft endplay with a feeler gauge or a dial indicator (see Section 14). The endplay should be correct if the crankshaft thrust faces aren't worn or damaged and new thrust bearings/washers have been installed.

23   Install a new rear main oil seal, then bolt the retainer to the block (see Section 24).

24   Rear main oil seal - installation

Note: *Refer to Chapter 2 Part A for the procedure and illustrations*

1   The crankshaft must be installed first and the main bearing support bolted in place, then the new seal should be installed in the retainer and the retainer bolted to the block. **Note:** *Depending on the design of the engine stand being used, you may not be able to install the rear main seal retainer with the engine on the stand. In this case, install the rear main seal retainer as the last step before engine installation, when the engine is off the stand and on the hoist.*

25   Pistons/connecting rods - installation and rod bearing oil clearance check

Refer to illustrations 25.5, 25.9a, 25.9b, 25.13 and 25.17

1   Before installing the piston/connecting

rod assemblies, the cylinder walls must be perfectly clean, the top edge of each cylinder must be chamfered, and the crankshaft must be in place.

2   Remove the cap from the end of the number one connecting rod (refer to the marks made during removal). Remove the original bearing inserts and wipe the bearing surfaces of the connecting rod and cap with a clean, lint-free cloth. They must be kept spotlessly clean.

Connecting rod bearing oil clearance check

3   Clean the back side of the new upper bearing insert, then lay it in place in the connecting rod. Make sure the tab on the bearing fits into the recess in the rod so the oil holes line up. Don't hammer the bearing insert into place and be very careful not to nick or gouge the bearing face. Don't lubricate the bearing at this time.

4   Clean the back side of the other bearing insert and install it in the rod cap. Again, make sure the tab on the bearing fits into the recess in the cap, and don't apply any lubricant. It's critically important that the mating surfaces of the bearing and connecting rod are perfectly clean and oil free when they're assembled.

5   Position the piston ring gaps at staggered intervals around the piston **(see illustration)**.

6   Slip a section of plastic or rubber hose over each connecting rod cap bolt. **(see illustration 13.6).**

7   Lubricate the piston and rings with clean engine oil and attach a piston ring compressor to the piston. Leave the skirt protruding about 1/4-inch to guide the piston into the cylinder. The rings must be compressed until they're flush with the piston.

8   Rotate the crankshaft until the number one connecting rod journal is at bottom dead center and apply a coat of engine oil to the cylinder walls.

25.9a Position the piston, when installing it, with the mark (arrow) in the piston facing the front of the engine

25.9b With the ring compressor firmly seated against the block, the piston can be driven gently into the cylinder bore with the end of a wooden or plastic hammer handle

9   With the mark on top of the piston facing the front of the engine (timing chain end) **(see illustration)**, gently insert the piston/connecting rod assembly into the number one cylinder bore and rest the bottom edge of the ring compressor on the engine block **(see illustration)**.

10   Tap the top edge of the ring compressor to make sure it's contacting the block around its entire circumference.

11   Gently tap on the top of the piston with the end of a wooden or plastic hammer handle while guiding the end of the connecting rod into place on the crankshaft journal. The piston rings may try to pop out of the ring compressor just before entering the cylinder bore, so keep some pressure on the ring compressor. Work slowly, and if any resistance is felt as the piston enters the cylinder, stop immediately. Find out what's hanging up and fix it before proceeding. Do not, for any reason, force the piston into the cylinder - you might break a ring and/or the piston.

12   Once the piston/connecting rod assembly is installed, the connecting rod bearing oil clearance must be checked before the rod cap is permanently installed.

13   Cut a piece of the appropriate-size Plastigage slightly shorter than the width of the connecting rod bearing and lay it in place on the number one connecting rod journal, parallel with the journal axis **(see illustration)**.

14   Clean the connecting rod cap bearing face, remove the protective hoses from the connecting rod bolts and install the rod cap. Make sure the mating mark on the cap is on the same side as the mark on the connecting rod. Check the cap to make sure the front mark is facing the timing chain end of the engine.

15   Apply a light coat of oil to the undersides of the nuts, then install and tighten them to the torque listed in this Chapter's Specifications. Use a thin-wall socket to avoid erroneous torque readings that can result if the socket is wedged between the rod cap and nut. If the socket tends to wedge itself between the nut and the cap, lift up on it slightly until it no longer contacts the cap. Do not rotate the crankshaft at any time during this operation.

16   Remove the nuts and detach the rod cap, being very careful not to disturb the Plastigage.

17   Compare the width of the crushed Plastigage to the scale printed on the Plastigage envelope to obtain the oil clearance **(see illustration)**. Compare it to this Chapter's Specifications to make sure the clearance is correct.

18   If the clearance is not as specified, the bearing inserts may be the wrong size (which means different ones will be required). Before deciding that different inserts are needed, make sure that no dirt or oil was between the bearing inserts and the connecting rod or cap when the clearance was measured. Also, recheck the journal diameter. If the Plastigage was wider at one end than the other, the journal may be tapered (see Section 19).

Final connecting rod installation

19   Carefully scrape all traces of the Plastigage material off the rod journal and/or bearing face. Be very careful not to scratch the bearing, use your fingernail or the edge of a

25.13 Lay a strip of Plastigage (arrow) on each rod bearing journal, parallel to the crankshaft centerline

25.17 Measure the width of the crushed Plastigage to determine the rod bearing oil clearance (be sure to use the correct scale - standard and metric ones are included)

2B

credit card or similar device.

20   Make sure the bearing faces are perfectly clean, then apply a uniform layer of clean moly-base grease or engine assembly lube to both of them. You'll have to push the piston into the cylinder to expose the face of the bearing insert in the connecting rod, be sure to slip the protective hoses over the rod bolts first.

21   Slide the connecting rod back into place on the journal, remove the protective hoses from the rod cap bolts, install the rod cap and tighten the nuts to the torque listed in this Chapter's Specifications.

22   Repeat the entire procedure for the remaining pistons/connecting rod assemblies.

23   The important points to remember are:

a) Keep the back sides of the bearing inserts and the insides of the connecting rods and caps perfectly clean when assembling them.

b) Make sure you have the correct piston/rod assembly for each cylinder.

c) The mark on the piston top must face the front (timing chain end) of the engine.

d) Lubricate the cylinder walls with clean oil.

e) Lubricate the bearing faces when installing the rod caps after the oil clearance has been checked.

24   After all the piston/connecting rod assemblies have been properly installed, rotate the crankshaft a number of times by hand to check for any obvious binding.

25   As a final step, the connecting rod side clearance (endplay) must be checked (see Section 13).

26   Compare the measured side clearance to this Chapter's Specifications to make sure it's correct. If it was correct before disassembly and the original crankshaft and rods were reinstalled, it should still be right. If new rods or a new crankshaft were installed, the side clearance may be inadequate. If so, the rods will have to be removed and taken to an automotive machine shop for re-sizing.

26 Initial start-up and break-in after overhaul

Warning: *Have a fire extinguisher handy when starting the engine for the first time.*

1   Once the engine has been installed in the vehicle, double-check the engine oil and coolant levels.

2   With the spark plugs out of the engine and the ignition and fuel system disabled (see Section 4), crank the engine until oil pressure registers on the gauge or the light goes out.

3   Install the spark plugs, hook up the plug wires and reconnect the ignition coil wire. Install the fuel pump fuse, if it was removed.

4   Start the engine. It may take a few moments for the fuel system to build up pressure, but the engine should start without a great deal of effort.

5   After the engine starts, it should be allowed to warm up to normal operating temperature. While the engine is warming up, make a thorough check for fuel, oil and coolant leaks.

6   Shut the engine off and recheck the engine oil and coolant levels.

7   Drive the vehicle to an area with no traffic, accelerate from 30 to 50 mph, then allow the vehicle to slow to 30 mph with the throttle closed. Repeat the procedure 10 or 12 times. This will load the piston rings and cause them to seat properly against the cylinder walls. Check again for oil and coolant leaks.

8   Drive the vehicle gently for the first 500 miles (no sustained high speeds) and keep a constant check on the oil level. It is not unusual for an engine to use oil during the break-in period.

9   At approximately 500 to 600 miles, change the oil and filter.

10   For the next few hundred miles, drive the vehicle normally. Do not pamper it or abuse it.

11   After 2000 miles, change the oil and filter again and consider the engine broken in.

Chapter 3
Cooling, heating and air conditioning systems

Contents

Specifications

General

Radiator cap pressure rating	11 to 14 psi
Cooling system test pressure	23 psi
Thermostat rating (opening temperature)	170-degrees F
Refrigerant type	R-134a
Refrigerant capacity	1.54 to 1.76 lbs

Torque specifications

Ft-lbs (unless otherwise indicated)

Thermostat housing bolts	56 to 66 in-lbs
Water pump-to-engine bolts	
6 mm bolts	56 to 66 in-lbs
8 mm bolts	144 to 168 in-lbs

1 General information

Engine cooling system

All vehicles covered by this manual employ a pressurized engine cooling system with thermostatically controlled coolant circulation. An impeller type water pump mounted on the timing chain end of the block pumps coolant through the engine. The coolant flows around each cylinder and toward the transaxle end of the engine. Cast-in coolant passages direct coolant around the intake and exhaust ports, near the spark plug areas and in close proximity to the exhaust valve guides. The heated coolant exits the cylinder head at the timing chain end and goes through the upper radiator hose back to the radiator to be cooled.

A wax-pellet type thermostat is located in a housing near the transaxle end of the engine. During warm up, the closed thermostat prevents coolant from circulating through the radiator. As the engine nears normal operating temperature, the thermostat opens and allows hot coolant to travel through the radiator, where it's cooled before returning to the engine.

The cooling system is sealed by a pressure-type radiator cap, which raises the boiling point of the coolant and increases the cooling efficiency of the radiator. If the system pressure exceeds the cap pressure-relief value, the excess pressure in the system forces the spring-loaded valve inside the cap off its seat and allows the coolant to escape through the overflow tube into a coolant reservoir. When the system cools, the excess coolant is automatically drawn from the reservoir back into the radiator.

The coolant reservoir serves as both the point at which fresh coolant is added to the cooling system to maintain the proper fluid level and as a holding tank for overheated coolant.

This type of cooling system is known as a closed design because coolant that escapes past the pressure cap is saved and re-used.

Heating system

The heating system consists of a blower fan and heater core located in the heater box, the hoses connecting the heater core to the engine cooling system and the heater/air conditioning control head on the dashboard. Hot engine coolant is circulated through the heater core. When the heater mode is activated, a flap opens to expose the heater box to the passenger compartment. A fan switch on the control head activates the blower motor, which forces air through the core, heating the air that enters the passenger compartment.

Air conditioning system

The air conditioning system consists of a condenser mounted in front of the radiator, an evaporator mounted adjacent to the heater core, a compressor mounted on the engine, a filter-drier which contains a high pressure relief valve and the plumbing connecting all of the above components.

A blower fan forces air (either outside

2.5 Inexpensive coolant hydrometers can be purchased at most auto parts stores

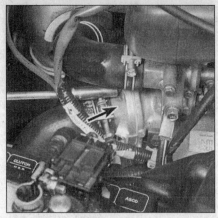

3.10 Remove the bolts from the housing cover (arrow) and separate the cover and thermostat

3.11 Note the position of the thermostat's air bleed valve (arrow) aligned with the casting and how the thermostat is installed (the spring end goes into the engine)

air, or inside air when in the RECIRC mode) through the evaporator core (sort of a radiator-in-reverse), transferring the heat from the air to the refrigerant. The liquid refrigerant boils off into low pressure vapor, taking the heat with it when it leaves the evaporator.

This refrigerant is pumped by the compressor to the condenser, where it cools and returns to a liquid before being sent back through the evaporator.

2   Antifreeze - general information

Refer to illustration 2.5
Warning: *Do not allow antifreeze to come in contact with your skin or painted surfaces of the vehicle. Rinse off spills immediately with plenty of water. Antifreeze is highly toxic if ingested. Never leave antifreeze lying around in an open container or in puddles on the floor; children and pets are attracted by it's sweet smell and may drink it. Check with local authorities about disposing of used antifreeze. Many communities have collection centers which will see that antifreeze is disposed of safely. Never dump used anti-freeze on the ground or into drains.*
Note: *Non-toxic coolant is available at most auto parts stores. Although the coolant is non-toxic, proper disposal is still required.*
1   The cooling system should be filled with a water/ethylene glycol-based antifreeze solution, which will prevent freezing down to at least -20-degrees F, or lower if local climate requires it. It also provides protection against corrosion and increases the coolant boiling point.
2   The cooling system should be drained, flushed and refilled at the specified intervals (see Chapter 1). Old or contaminated antifreeze solutions are likely to cause damage and encourage the formation of corrosion and scale in the system. Use distilled water with the antifreeze.
3   Before adding antifreeze, check all hose connections, because antifreeze tends to leak through very minute openings. Engines don't normally consume coolant, so if the level goes

down, find the cause and correct it.
4   The exact mixture of antifreeze to water which you should use depends on the relative weather conditions. The mixture should contain at least 50-percent antifreeze, but should never contain more than 70-percent antifreeze. Consult the mixture ratio chart on the antifreeze container before adding coolant.

Antifreeze/coolant testing
Warning: *Do not remove the radiator cap to take a sample until the engine has cooled completely. The system is under pressure and extremely hot during and after running the engine, and if the coolant contacts your skin it can cause severe scalding.*
5   Hydrometers used to test the condition of the coolant are available at most auto parts stores **(see illustration)**. They are inexpensive and are very easy to use. To test coolant draw a small amount from the radiator, or coolant reservoir, with the hydrometer until all the balls are submerged (most of them will probably float). **Note:** *It is preferable to take the coolant sample from the radiator. The mixture in the coolant reservoir may be slightly diluted if any water has recently been added.* The strength of the mixture is shown by the number of balls that are floating. Exact temperature protection indicated by the hydrometer is clearly described on the package instructions. Always use antifreeze which meets the vehicle manufacturer's specifications.

3   Thermostat - check and replacement

Note: *See the coolant **Warning** in Section 2.*

Check
1   Before assuming the thermostat is to blame for a cooling system problem, check the coolant level, drivebelt tension (see Chapter 1) and temperature gauge operation.
2   If the engine seems to be taking a long time to warm up (based on heater output or

temperature gauge operation), the thermostat is probably stuck open. Replace the thermostat with a new one.
3   If the engine runs hot, use your hand to check the temperature of the lower radiator hose. If the hose isn't hot, but the engine is, the thermostat is probably stuck closed, preventing the coolant from circulating through the radiator. Replace the thermostat. **Caution:** *Don't drive the vehicle without a thermostat. The engine may never reach the required temperature to allow the computer to go to closed-loop operation. Performance, emissions and fuel economy will probably be poor.*
4   If the radiator-to-thermostat hose is hot, it means that the coolant is flowing and the thermostat is open. Consult the *Troubleshooting* section at the front of this manual for cooling system diagnosis.

Replacement
Refer to illustrations 3.10 and 3.11
5   Disconnect the battery cable from the negative battery terminal.
6   Drain the cooling system (see Chapter 1). If the coolant is relatively new, or is in good condition (see Section 2), save it and re-use it.
7   Follow the lower radiator hose to the engine to locate the thermostat housing.
8   Remove the radiator hose from the housing cover. If the hose is stuck, grasp it near the end with a pair of large adjustable pliers and twist it to break the seal, then pull it off. If the hose is old or deteriorated, cut it off and install a new one. If the hose clamp's nut head is inaccessible, leave the hose attached and unbolt the thermostat cover.
9   If the outer surface of the large fitting that mates with the hose is deteriorated (corroded, pitted, etc.) it may be damaged further by hose removal. If it is, the thermostat housing cover will have to be replaced.
10   Remove the bolts/nuts and detach the housing cover **(see illustration)**. If the cover is stuck, tap it with a soft-face hammer to jar it loose. Be prepared for some coolant to spill

4.3a There are two electrical connectors (arrows) for the two electric fans

4.3b Disconnect the fan electrical connector and use fused jumper wires (on the fan side of the connector) to apply battery voltage and ground

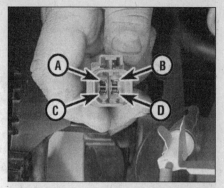

4.3c To test for LOW speed fan operation, apply battery power to A and ground to C - to test for HIGH speed fan operation, connect a jumper wire to provide battery voltage to both terminals A and B, while ground is applied to both terminals C and D

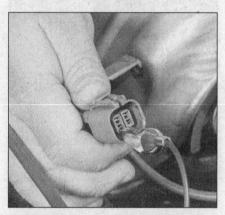

4.4 With the engine hot and running, check for battery voltage at the harness-side of the fan connector

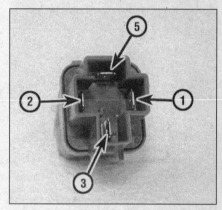

4.6 Terminal designations of cooling fan relay #1 - cooling fan relay #1 is located in the relay strip next to the battery

as the gasket seal is broken.

11   Note the position of the air bleed valve or air bleed hole (see illustration).

12   Install the new thermostat in the housing. Make sure the air bleed faces up and the spring end is directed toward the engine.

13   Position a new gasket or apply RTV sealant to the thermostat cover.

14   Install the cover and bolts. Tighten the bolts evenly to the torque listed in this Chapter's Specifications.

15   Reattach the hose to the fitting and tighten the hose clamp securely. Reconnect the electrical connector for the cooling fan switch.

16   Refill the cooling system (see Chapter 1).

17   Start the engine and allow it to reach normal operating temperature, then check for leaks and proper thermostat operation (as described in Steps 2 through 4).

4   Engine cooling fan(s) and circuit(s) - check and component replacement

Warning: *The models covered by this manual are equipped with airbags. Always disconnect the negative battery cable, then the positive cable and wait 10 minutes before working in the vicinity of the impact sensors, steering column or instrument panel to avoid the possibility of accidental deployment of the airbag, which could cause personal injury (see Chapter 12). The airbag circuits are easily identified by yellow insulation covering the entire wiring harness or just prior to the wire harness connectors. Do not use electrical test equipment on any of these wires or tamper with them in any way.*

1   The two engine cooling fans have two speeds and are controlled by the ECM (computer). The ECM receives input from the coolant temperature sensor, as well as sensors for air conditioning, vehicle speed and engine rpm. The ECM determines when the electric fan(s) should turn on, and at which speed, High or Low.

Motor check

Refer to illustrations 4.3a, 4.3b, 4.3c and 4.4

2   First, check the fuses (see Chapter 12).

3   To test a fan motor, disconnect the electrical connector and use fused jumper wires to connect the fan directly to the battery (see illustrations). If the fan still does not work, replace the motor. The fans operate at two different speeds, each of which can be tested.

4   If the motor tested OK in Step 3, check for trouble codes stored in the computer (see Section 12). If the coolant sensor and vehicle speed sensor check out OK, check the wiring. Check for available battery voltage in the harness-side of the fan connector (see illustration). Voltage should be at this terminal only when the engine is running and hot enough. **Caution:** *Keep your hands away from the fans when testing.*

Engine cooling fan relays

Refer to illustrations 4.6 and 4.7

5   Locate the three cooling fan relays and check the relays for continuity using an ohmmeter.

6   With the blue fan relay #1 at rest (not

energized), there should be **no** continuity between terminals 3 and 5. With the relay energized (by applying voltage to terminals 1 and 2), there should be continuity between terminals 3 and 5 (see illustration). If continuity isn't as specified, replace the relay.

7   Inspect fan relays #2 and #3 (brown) for continuity using an ohmmeter (see illustration). With no current supplied to terminals 1

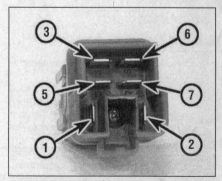

4.7 Terminal designations of cooling fan relays #2 and #3 - they are located in the relay box next to the right (passenger side) fender

3

4.9 The side panels (arrow) of the fan shroud can be pulled straight up and out

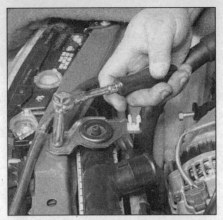

4.10a Remove one of the these brackets at each side of the fan/radiator assembly

4.10b After the brackets are off, remove these two bolts (arrows) holding the fan assembly to the radiator

4.11 To replace the fan motor, unbolt the fan blade nut (arrow) and remove the fan blade

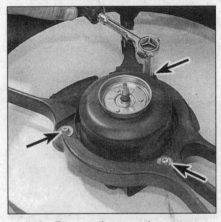

4.12a Remove the mounting screws (arrows) that hold the motor to the fan housing and remove the motor

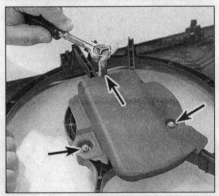

4.12b The right-hand fan motor has a shield over it - remove the screws (arrows) and the shield to access the motor retaining screws

or 2, there should be no continuity between terminals 3 and 5, or between 6 and 7. Using jumper wires, apply battery voltage and ground by connecting the positive jumper wire to terminal number 1 and the negative jumper wire to terminal number 2. With current applied, there now should be continuity between terminals 3 and 5, and between 6 and 7. If continuity isn't as specified, replace the relay(s).

Fan replacement

Refer to illustrations 4.9, 4.10a, 4.10b, 4.11, 4.12a and 4.12b

8   Disconnect the cable from the negative battery terminal.
9   Pull the side plates of the fan shroud (one on each side of the fan assembly) straight up and out **(see illustration)**. Disconnect the electrical connectors if you haven't already **(see illustration 4.3a)**.
10   Unbolt the cooling fan assembly from the radiator and remove it **(see illustrations)**. There are two bolts/brackets at the top and clips at the bottom. Both fans will come out with the shroud. **Note:** *The same brackets at the top retain the radiator to the core support.*
11   Remove the nut and detach the fan

blade assembly from the motor shaft **(see illustration)**.
12   Remove the screws retaining the fan motor to the mounting bracket, then detach the motor **(see illustrations)**.
13   Installation is the reverse of removal.

---

5   Radiator and coolant reservoir - removal and installation

---

Warning: *The models covered by this manual are equipped with airbags. Always disconnect the negative battery cable, then the positive cable and wait 10 minutes before working in the vicinity of the impact sensors, steering column or instrument panel to avoid the possibility of accidental deployment of the airbag, which could cause personal injury (see Chapter 12). The airbag circuits are easily identified by yellow insulation covering the entire wiring harness or just prior to the wire harness connectors. Do not use electrical test equipment on any of these wires or tamper with them in any way.*
Note: *Refer to the coolant* **Warning** *in Section 2.*

Radiator removal

Refer to illustration 5.8

1   Disconnect the cable from the negative battery terminal.
2   Raise the front of the vehicle and support it securely on jackstands.
3   Drain the cooling system (see Chapter 1). If the coolant is relatively new, or is in good condition, save it and reuse it (see Section 2).
4   Loosen the upper and lower hose clamps, then detach the radiator hoses from the fittings. If they're stuck, grasp each hose near the end with a pair of adjustable pliers and twist it to break the seal, then pull it off - be careful not to damage the plastic radiator fittings! If the hoses are old or deteriorated, cut them off and install new ones.
5   Disconnect the coolant reservoir hose from the radiator.
6   Disconnect the cooling fan electrical connectors and remove the fan assembly (see Section 4).
7   Remove the radiator mounting bolts **(see illustration 4.10a)**.
8   If the vehicle is equipped with an automatic transaxle, disconnect the transmission oil cooler lines **(see illustration)** and plug the lines and fittings.

5.8 On automatic transmission models disconnect the cooler lines - have a drain pan ready to catch spilled fluid

5.18 Detach the overflow hose from the radiator, remove this mounting bracket bolt (arrow) and pull the reservoir bottle out of its bracket

6.3 Heavy stains or deposits around the water pump's weep hole (arrow) indicate the pump seal is failing

9   Carefully lift out the radiator. Don't spill coolant on the vehicle or scratch the paint.
10   Inspect the lower mounting pads. Replace them if they are deteriorated, split or broken.
11   With the radiator removed, it can be inspected for leaks and damage. If it needs repair, have a radiator shop or dealer service department perform the work as special techniques are required. **Note:** *Because the core is aluminum and the tanks are plastic, it may be more practical to replace an aging or damaged radiator than repair it.*
12   Bugs and dirt can be removed from the radiator with a soft brush, followed by forcing water from a garden hose through the core from the engine side. Don't bend the cooling fins as this is done.

Radiator installation

13   Installation is the reverse of the removal procedure. Be sure the radiator mounting pads are seated properly at the base of the radiator.
14   After installation, fill the cooling system with the proper mixture of antifreeze and water (see Section 2 and Chapter 1).
15   Start the engine and check for leaks. Allow the engine to reach normal operating temperature, indicated by the upper radiator hose becoming hot. Recheck the coolant level and add more if required.
16   If you're working on an automatic transaxle-equipped vehicle, check and add transmission fluid as needed (see Chapter 1).

Coolant reservoir removal and installation

Refer to illustration 5.18
17   Disconnect the overflow hose from the reservoir, and plug to prevent coolant from spilling if the reservoir is full.
18   Unbolt the coolant reservoir and remove it from the vehicle **(see illustration)**.
19   Pour the coolant into a container. Wash out the reservoir, using soapy water and a long brush to make the coolant level easier to read. Inspect the reservoir for cracks and chafing. Replace it if any damage is found.
20   Installation is the reverse of removal.

6   Water pump - check

Refer to illustration 6.3
1   A failure in the water pump can cause serious engine damage due to overheating.
2   There are three ways to check the operation of the water pump while it's installed on the engine. If the pump is defective, it should be replaced with a new, or rebuilt unit.
3   Water pumps are equipped with weep or vent holes **(see illustration)**. If a failure occurs in the pump seal, coolant will leak from the hole. In most cases you'll need a flashlight to find the hole on the water pump from underneath to check for leaks. **Note:** *Small gray discolorations around the weep hole are normal. Heavy brown deposits and wetness indicate a seal failure in the pump.*
4   If the water pump shaft bearings fail there may be a howling sound at the drivebelt end of the engine while it's running. Shaft wear can be felt if the water pump pulley is rocked up and down. Don't mistake drivebelt slippage, which causes a squealing sound, for water pump bearing failure.
5   Turn the pump shaft by hand and feel for roughness in the bearing. Replace the pump if roughness is felt.

7   Water pump - replacement

Refer to illustrations 7.5, 7.7, 7.8, 7.9a, 7.9b, 7.9c, 7.10 and 7.11
Warning: *Wait until the engine is completely cool before beginning this procedure.*
Note: *Read the coolant Warning in Section 2.*
1   Disconnect the battery cable from the negative battery terminal.
2   Drain the cooling system (see Chapter 1). If the coolant is relatively new, or is in good condition (see Section 2), save it and re-use it.
3   Remove the right hand splash shield (see Chapter 1).
4   Support the vehicle safely on jackstands and remove the right front wheel and engine side cover.
5   Loosen the water pump pulley bolts from underneath the vehicle while the drivebelt is still in place (it keeps the pulley from turning) **(see illustration)**. Remove the drivebelt and the pulley.
6   Unbolt the air conditioning compressor (see Section 14) and secure it aside without disconnecting the refrigerant lines.
7   Refer to Chapter 5 and remove the alternator and alternator mount for access to the water pump **(see illustration)**.

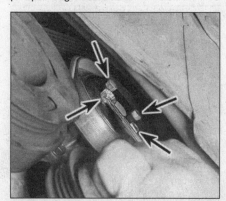

7.5 Loosen the water pump pulley bolts (arrows) with tension on the drivebelt, then remove the belt and the pulley

7.7 The lower alternator mounting bracket (arrow) must be removed for access to one of the water pump bolts

3

7.8a Remove these three water pump bolts (arrows) from underneath . . .

7.8b . . . and these water pump bolts (arrows) from above with the alternator removed

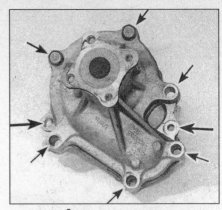

7.9a Water pump mounting bolt details - the six bolts (small arrows) secure the pump to the housing, the other two are housing-to-engine bolts (larger arrows)

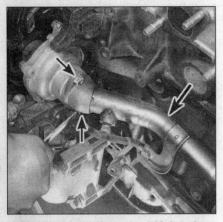

7.9b If removing the pump with housing, remove the two bolts (small arrows) retaining the water pipe (larger arrow) to the back of the water pump housing

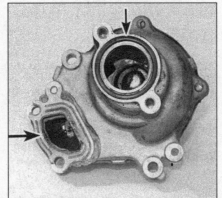

7.10 If the pump was removed with the housing, the water pipe O-ring (top arrow) must be replaced, and the housing-to-block passage (left arrow) must be cleaned and coated with RTV sealant before installation

7.11 Apply a bead of RTV sealant to the water pump mounting surface before bolting it to the housing

8   Remove the water pump bolts and separate the pump from the pump housing (**see illustrations**). If the pump is stuck, gently tap it with a soft faced hammer to break the gasket seal.

9   If necessary, remove the complete water pump housing and pump assembly from the block and the water pipe. Remove the housing-to-block bolts and the two bolts connecting the rear of the water pump to the water pipe (**see illustrations**).

10   If the housing was removed, clean the sealant from the mating surface and replace the O-ring (**see illustration**).

11   Apply a bead of RTV sealant around the perimeter of the water pump and install the pump to the housing (**see illustration**). Tighten the pump-to-housing bolts to the torque listed in this Chapter's Specifications in 1/4-turn increments. Don't over-tighten them or the pump may be damaged or leak.

12   If removed, install the pump/housing assembly to the block and connect the water pipe.

13   Reinstall the alternator, air conditioning compressor (if equipped), and any other components removed for access to the water pump.

14   Refill the cooling system (see Chapter 1). Run the engine and check for leaks.

---

8   Coolant temperature sending unit - check and replacement

Refer to illustration 8.3

Check

1   If the coolant temperature gauge is inoperative, check the fuses first (see Chapter 12).

2   If the temperature indicator shows excessive temperature after running awhile, see the *Troubleshooting* section in the front of the manual.

3   If the temperature gauge indicates Hot shortly after the engine is started cold, and the engine is not up to normal operating temperature, disconnect the connector at the coolant temperature sending unit (**see illustration**). **Note:** *The sending unit has a single wire connector and is located on the intake manifold, on the right (passenger's) side of*

the engine compartment. Do not confuse it with the coolant temperature sensor, which is just above it and has a multiple-wire connector. If the gauge reading drops, replace the sending unit. If the reading remains high, the wire to the gauge may be shorted to ground or the gauge is faulty.

4   If the coolant temperature gauge fails to indicate after the engine has been warmed up (approximately 10 minutes) and the fuses

8.3 Location of the coolant temperature sending unit (arrow)

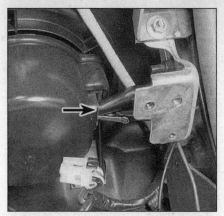

9.3 To test the blower motor, disconnect the electrical connector and attach jumper wires (arrow) to apply power and ground - if it doesn't operate, replace the blower motor

9.8a The glovebox support can be detached by removing these three screws (arrows) on the right side . . .

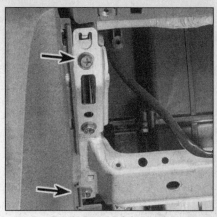

9.8b . . . and these two screws on the left side

checked out okay, shut off the engine. Disconnect the electrical connector at the sending unit and using a jumper wire, connect the connector to a clean ground on the engine. Turn on the ignition without starting the engine. If the gauge now indicates Hot, replace the sending unit.

5   If the gauge still does not work, the circuit may be open or the gauge may be faulty.

6   With the engine still warm and the electrical connector disconnected from the sender, connect an ohmmeter between a good engine ground and the electrical blade on the sender. The resistance should be approximately 70 to 90 ohms at 140-degrees F and should drop to approximately 21 to 24 ohms as the engine reaches 212-degrees F. If the sender doesn't meet these specifications, replace it.

Replacement

Warning: *The engine must be completely cool before removing the sending unit.*

7   With the engine completely cool, remove the cap from the radiator to release any pressure, then reinstall the cap. This reduces coolant loss during sending unit replacement.

8   Disconnect the electrical connector from the sending unit.

9   Prepare the new sending unit for installation by applying a light coat of sealant to the threads.

10   Unscrew the sending unit and quickly install the new one to prevent coolant loss.

11   Tighten the sending unit securely and attach the electrical connector.

12   Refill the cooling system (see Chapter 1) and run the engine. Check for leaks and proper gauge operation.

9   Blower motor and circuit - check, removal and component replacement

Warning: *The models covered by this manual*

are equipped with airbags. Always disconnect the negative battery cable, then the positive cable and wait 10 minutes before working in the vicinity of the impact sensors, steering column or instrument panel to avoid the possibility of accidental deployment of the airbag, which could cause personal injury (see Chapter 12). The airbag circuits are easily identified by yellow insulation covering the entire wiring harness or just prior to the wire harness connectors. Do not use electrical test equipment on any of these wires or tamper with them in any way.*

Blower motor
Check

Refer to illustration 9.3

1   Begin by trying the blower at all speeds, starting at the first (slowest) speed. On standard models, turn the blower speed knob to position 1, and on optional automatic air conditioning models, push the blower button once. There should be only one blade of the fan symbol darkened (two for the next speed, etc.). Try all the speeds. Listen for blower operation at the blower housing, below the glove box.

2   If the blower motor speed does not correspond to the setting selected on the blower switch, or the blower motor does not operate at all, the problem could be a bad fuse, relay, switch, blower motor resistor (or control unit), blower motor or blower motor circuit wiring.

3   Before checking the blower motor or circuit, always check the fuse and relay (if equipped) first (see Chapter 12). If the fan does not operate, disconnect the electrical connector from the blower and using fused jumper wires connect battery voltage and ground to the motor (see illustration). If the blower operates, the problem is elsewhere in the circuit.

4   With the ignition key in the ON position and the blower motor electrical connector connected to the motor, turn the blower switch to the faulty position(s) and, using a test light or voltmeter, check for voltage at the motor electrical connector brown/white wire by backprobing the connector. If the

motor is receiving voltage but not operating, either the motor is defective or the ground circuit is open.

5   To check for a bad ground circuit, backprobe the blower motor electrical connector blue/white wire with a jumper wire connected to ground. If the motor now operates properly, there is a problem with ground circuit. On this system, the blower motor ground circuit consists of the blower resistor (manual), fan control unit (automatic) and blower switch (manual) or control head (automatic). Check for continuity through the various components. If equipped with an automatic system, it may be necessary to have the system checked by a dealer service department or other properly equipped repair facility.

6   If you suspect the blower motor fan is binding, remove the blower motor (see Step 7) to check for free operation of the fan.

Replacement

Refer to illustrations 9.8a, 9.8b, 9.9a, 9.9b and 9.10

7   Disconnect the negative battery cable, then the positive cable and wait 10 minutes before proceeding any further.

8   To access the blower for removal, remove the glove box (see Chapter 11). Then remove the glovebox support bracing **(see illustrations)**.

9   Remove the motor cooling tube, disconnect the electrical connector at the blower

9.9a Remove the rubber cooling (air) tube from the blower motor

9.9b Remove the electrical connector (large arrow) and mounting screws (small arrows)

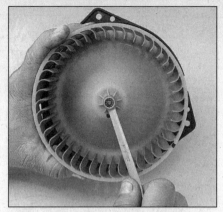

9.10 Remove the nut from the blower shaft and evenly pull the fan from the blower motor

9.12 The blower motor resistor (arrow) is mounted between the blower housing and evaporator housing

9.14 Heater blower motor resistor check - there should be continuity between all terminals

9.16 Blower speed control switch continuity check - there should be continuity between 32 and 104 at all speeds; and between 104, 32 and 31 (switch position 1); 104, 32 and 30 (position 2); 104, 32 and 29 (position 3); and 104, 32 and 28 in position 4

motor and remove the three blower unit retaining screws **(see illustrations)**. Lower the unit from the housing.

10   If you are replacing the motor, detach the fan and transfer it to the new motor **(see illustration)**.

11   Installation is the reverse of removal. Run the blower and check for proper operation.

Blower motor resistor (manual system only)

Refer to illustrations 9.12 and 9.14

12   Locate the heater blower motor resistor **(see illustration)**, usually attached to a heating/air conditioning duct below the blower motor case.

13   Check the wiring and the connectors between the resistor and the motor. Check for loose or corroded connections and damaged wires.

14   Check the blower resistor for continuity with the connector removed **(see illustration)**. There should be continuity between all pins. If not, replace the resistor.

Blower speed control switch (manual system only)

Refer to illustration 9.16

15   Remove the instrument panel bezel (see

Chapter 12). Disconnect the electrical connectors to the heater/air conditioning controls. The blower speed switch is on the right of the heater/air conditioning controls when looking at the back of the panel.

16   Check the continuity between the switch terminals with an ohmmeter as shown **(see illustration)**. There should be continuity between 104 and 32 in all switch positions. At the first blower switch position, there should be continuity from 104 to 32 and 31, and in the second-speed position, from 104 to 32 and 30. In the third-speed position, continuity should exist between 104, 32 and 29, and in the last position, there should be continuity between 104, 32 and 28.

17   If continuity isn't as specified, replace the switch.

10   Heater core - removal and installation

Warning: *The models covered by this manual are equipped with airbags. Always disconnect the negative battery cable, then the positive cable and wait 10 minutes before working in the vicinity of the impact sensors, steering column or instrument panel to avoid the*

possibility of accidental deployment of the airbag, which could cause personal injury (see Chapter 12). The airbag circuits are easily identified by yellow insulation covering the entire wiring harness or just prior to the wire harness connectors. Do not use electrical test equipment on any of these wires or tamper with them in any way.

Removal

Refer to illustrations 10.4, 10.8, 10.11a, 10.11b, 10.11c, 10.11d, 10.12, 10.13, 10.14, 10.15, 10.16a, 10.16b, 10.17, 10.18a, 10.18b and 10.18c

1   Replacement of the heater core is a difficult procedure for the home mechanic, involving removal of the entire dashboard, console, and many wiring connectors. If you attempt it at home, keep track of the assemblies by taking notes and keeping screws and other hardware in small, marked plastic bags for reassembly.

2   Disconnect the negative battery cable, then the positive cable and wait 10 minutes before proceeding any further.

3   Turn the heater control setting to HOT. Drain the cooling system (see Chapter 1). If the coolant is relatively new, or is in good condition (see Section 2), save it and re-use it. **Note:** *See the coolant* **Warning** *in Section 2.*

10.4 Disconnect the heater hoses (arrows) at the firewall by loosening the hose clamps and sliding them off the tubes, then remove the grommet behind them

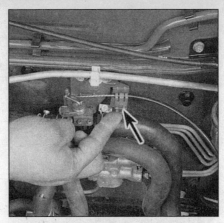

10.8 Release this clip (arrow) and disconnect the water-control valve cable on the engine-side of the firewall

10.11a Remove these two screws retaining the airbag control module and lay the module aside, then begin removing the two metal braces the module was attached to

10.11b Remove the bolts (arrows) where each brace is attached to the floor . . .

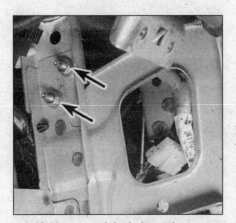

10.11c . . . and the bolts at the top (arrows), attaching the braces to the long tubular cowl brace

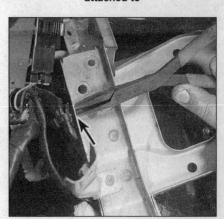

10.11d Use pliers to squeeze the clips (arrow) to release the wiring harness retainers

4   Working in the engine compartment, disconnect the heater hoses where they enter the firewall (see illustration). Caution: *If the heater hoses are stuck firmly, it is better to cut off the hoses than to twist them with pliers and risk breaking the plastic heater core tubes.*

5   Remove the rubber grommet where the heater core tubes go through the firewall.

6   Remove the glove box door, glove box and metal glove box support frame (see illustrations 9.8a and 9.8b).

7   Remove the left and right lower dash panels, center dash fascia, and center dash surround (see Chapter 11).

8   On the engine side of the firewall, at the center, Release the clip that retains the water-control-valve cable and disconnect the cable from the control (see illustration).

9   Push the cable through the firewall so it can be pulled out with the heater core housing.

10   Refer to Chapter 11 and remove the radio, radio bezel, floor console and the complete instrument panel.

11   The layout of the heater/air conditioning components below the dash area is as follows; the blower housing is on the right, below the glovebox area, the evaporator

10.12 Remove the floor heating duct, and the ECM (arrow) under the heater housing

housing is to the left of the blower housing (and connected by screws), and the heater core housing is attached to the left of the evaporator housing, at the center of the vehicle. Remove the steel dash-to-floor braces (at the center of the dash) after the floor console has been removed (see illustrations).

12   Pull out the floor heating duct from the

10.13 Disconnect the air-mix door electrical connector (right arrow) and remove the two housing-to-firewall bolts (left arrows)

bottom front of the heater core housing and refer to Chapter 6 for removal of the ECM (computer) from below the housing (see illustration).

13   At the left side of the heater housing, disconnect the air-mix door electrical connector, and remove the upper and lower

3

10.14 Release the snap (arrows) at the front of this bottom duct and pull the duct down and out - the lower-right housing-to-firewall bolt is behind this duct

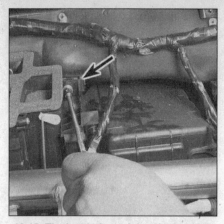

10.15 The last bolt retaining the housing to the firewall is this one (arrow) - it is inaccessible unless the instrument panel has been removed

10.16a Remove the two bolts at each end of the cowl support tube

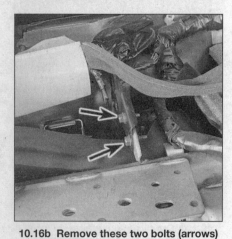

10.16b Remove these two bolts (arrows) retaining the cowl support tube to the bracket

10.17 Pull up on the cowl support tube (upper arrow), while working the heater housing (lower arrow) away from the evaporator housing

10.18a Pry under the end of this clip (arrow) retaining the heater core in the housing, then . . .

housing-to-firewall bolts **(see illustration)**.
14   Pull down the bottom duct from the heater core housing to allow more room under the housing for removal **(see illustration)**. Behind this duct, remove the bottom housing-to-firewall bolt.
15   Remove the remaining housing-to-firewall bolt at the upper right corner of the housing **(see illustration)**.
16   Unbolt the steering column from the cowl support, lower the column and allow it to rest on the front seat. Remove the bolts at each end of the cowl support **(see illustrations)**.
17   While pulling up on the cowl support tube to provide more working room, pull down and left on the heater unit, making sure that all fasteners have been removed. Some twisting is required to separate the heater unit from the evaporator housing to its right **(see illustration)**.
18   With the heater housing on the bench, use a screwdriver to release the strap over the heater core and slide the heater core out of the housing for replacement **(see illustrations)**.

10.18b . . . remove this screw (arrow) to remove the clip

Installation

19   Reinstall the heater core into the heater unit and replace the assembly in the reverse order of removal.
20   Refill the cooling system (see Chapter 1), reconnect the battery and run the

10.18c Remove the heater core from the housing - when replacing the heater core, makes sure the foam sealing strips are in place

engine. Check for leaks and proper system operation. Check the operation of all electrical components of the steering column and dash.

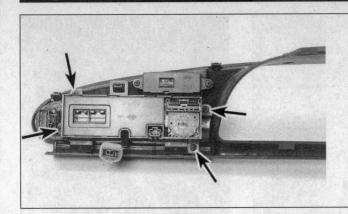

11.2 Remove the control-panel-to-bezel mounting screws (arrows)

11 Air conditioning and heater control assembly - removal and installation

Refer to illustration 11.2

Warning: *The models covered by this manual are equipped with airbags. Always disconnect the negative battery cable, then the positive cable and wait 10 minutes before working in the vicinity of the impact sensors, steering column or instrument panel to avoid the possibility of accidental deployment of the airbag, which could cause personal injury (see Chapter 12). The airbag circuits are easily identified by yellow insulation covering the entire wiring harness or just prior to the wire harness connectors. Do not use electrical test equipment on any of these wires or tamper with them in any way.*

1   Disconnect the negative battery cable, then the positive cable and wait 10 minutes before proceeding any further.

2   Refer to Chapter 12 and remove the instrument cluster bezel. Disconnect the electrical connectors, remove the control panel retaining screws and separate the control panel from the bezel **(see illustration)**.

3   Installation is the reverse of removal. After installation, check the operation of all heating and air conditioning controls.

12 Air conditioning and heating system - check and maintenance

Refer to illustrations 12.1 and 12.7

Warning: *The air conditioning system is under high pressure. Do not loosen any fittings or remove any components until after the system has been discharged. Air conditioning refrigerant should be properly discharged into an EPA-approved container at a dealer service department or an automotive air conditioning repair facility. Always wear eye protection when disconnecting air conditioning system fittings.*

1   The following maintenance checks should be performed on a regular basis to ensure that the air conditioning system continues to operate at peak efficiency.

   a) *Check the compressor drivebelt. If it's worn or deteriorated, replace it (see Chapter 1).*
   b) *Check the drivebelt tension and, if necessary, adjust it (see Chapter 1).*
   c) *Check the system hoses. Look for cracks, bubbles, hard spots and deterioration. Inspect the hoses and all fittings for oil bubbles and seepage. If there's any evidence of wear, damage or leaks, replace the hose(s).*
   d) *Inspect the condenser fins for leaves, bugs and other debris. Use a "fin comb" or compressed air to clean the condenser.*
   e) *Make sure the system has the correct refrigerant charge.*
   f) *Check the evaporator housing drain tube for blockage* **(see illustration)**.

2   It's a good idea to operate the system for about 10 minutes at least once a month, particularly during the winter. Long term non-use can cause hardening, and subsequent failure, of the seals.

3   Because of the complexity of the air conditioning system and the special equipment necessary to service it, in-depth troubleshooting and repairs are not included in this manual. However, simple checks and component replacement procedures are provided in this Chapter. For more complete information on the air conditioning system, refer to the Haynes *Automotive Heating and Air Conditioning Manual.*

4   The most common cause of poor cooling is simply a low system refrigerant charge. If a noticeable drop in cool air output occurs, one of the following quick checks will help you determine if the refrigerant level is low.

5   Warm the engine up to normal operating temperature.

6   Place the air conditioning temperature selector at the coldest setting and the blower at the highest setting. Open the doors (to make sure the air conditioning system doesn't cycle off as soon as it cools the passenger compartment).

7   Feel the evaporator inlet and outlet pipes at the firewall **(see illustration)**. The smaller pipe leading from the condenser outlet to the evaporator should be cold, and the larger evaporator outlet line should be slightly colder (3 to 10 degrees F). If the evaporator outlet is considerably warmer than the inlet,

12.1 The air conditioning evaporator core drain tube (arrow) is located above the right side steering gear boot

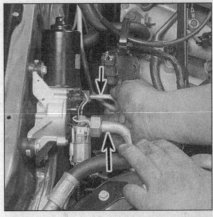

12.7 Feel the two pipes (arrows) leading to the evaporator at the firewall to make a determination if the system needs a charge

the system needs a charge. Insert a thermometer in the center air distribution outlet in the dash while operating the air conditioning system - the temperature of the output air should be approximately 35 to 40 degrees F below the ambient air temperature (down to approximately 40 degrees F). If the ambient (outside) air temperature is very high, say 110 degrees F, the duct air temperature may be as high as 60 degrees F, but generally the air conditioning is 30 to 50 degrees F cooler than the ambient air. If the air isn't as cold as it used to be, the system probably needs a charge. Further inspection or testing of the system is beyond the scope of this manual and should be left to a professional.

8   If the compressor clutch doesn't engage when the air conditioning is turned on, check the air conditioning relay for continuity (see Section 4 and test the relay in the same manner as the fan relays).

Adding refrigerant

Refer to illustrations 12.9 and 12.12

Caution: *All models covered by this manual are equipped with "environmentally-friendly" R-134a refrigerant. The refrigerant, O-rings,*

3

seals and compressor oil used with R-134a is NOT compatible with the elements of older R-12 systems. Use only R-134a compatible refrigerant, oil and components when working on these systems.

9   Buy an automotive charging kit at an auto parts store. A charging kit includes a 12-ounce can of R-134a refrigerant, a tap valve and a short section of hose that can be attached between the tap valve and the system low side service valve **(see illustration)**. Because one can of refrigerant may not be sufficient to bring the system charge up to the proper level, it's a good idea to buy a couple of additional cans. Try to find at least one can that contains red refrigerant dye. If the system is leaking, the red dye will leak out with the refrigerant and help you pinpoint the location of the leak.

10   Connect the charging kit by following the manufacturer's instructions.

11   Back off the valve handle on the charging kit and screw the kit onto the refrigerant can, making sure first that the O-ring or rubber seal inside the threaded portion of the kit is in place. **Warning:** *Wear protective eye wear when dealing with pressurized refrigerant cans.*

12   Remove the dust cap from the low-side charging port and attach the quick-connect fitting on the kit hose **(see illustration)**. **Warning:** *DO NOT hook the charging kit hose to the system high side! The fittings on the charging kit are designed to fit **only** on the low side of the system.*

13   Warm the engine to normal operating temperature and turn on the air conditioning. Keep the charging kit hose away from the fan and other moving parts.

14   Turn the valve handle on the kit until the stem pierces the can, then back the handle out to release the refrigerant. You should be able to hear the rush of gas. Add refrigerant to the low side of the system until both the outlet and the evaporator inlet pipe feel about the same temperature. Allow stabilization time between each addition. **Warning:** *Never add more than two cans of refrigerant to the system.* The can may tend to frost up, slowing the procedure. Wrap a shop towel wet with hot water around the bottom of the can to keep it from frosting.

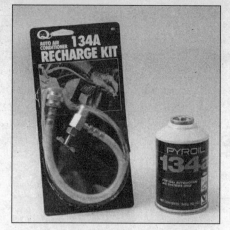

12.9 A basic charging kit for R-134a systems is available at most auto parts stores - it must say R-134a (not R-12) and so must the can of refrigerant

15   If you have an accurate thermometer, you can place it in the center air conditioning duct inside the vehicle to monitor the air temperature. A charged system that is working properly, should output air down to approximately 40 degrees F.

16   When the can is empty, turn the valve handle to the closed position and release the connection from the low-side port. Replace the dust cap.

17   Remove the charging kit from the can and store the kit for future use with the piercing valve in the UP position, to prevent inadvertently piercing the can on the next use.

Automatic heating/air conditioning system diagnosis

Note: *The 1998 and later models are not equipped with the self-diagnosis feature and the trouble codes are no longer used.*

18   Some models are equipped with an optional automatic climate-control system. This system has its own computer that receives input from sensors and controls the functions of the heating/air conditioning system. The computer also offers a self-diagnostic capability the check all aspects of the sys-

12.12 Add R-134a refrigerant to the low-side port only - the procedure is easier if you wrap the can with a warm, wet towel to prevent icing

tem in a few minutes.

19   Start the engine and push in the OFF button on the heater/air conditioning control panel for at least five seconds. This must be done within 10 seconds of having started the engine. This begins the self-diagnostic procedure, which goes through tests in the following order: sensors, mode and intake door motors, actuators, and temperatures at the sensors. At the end of each test portion, the unit will display a code, either a trouble code or a code to indicate that that portion of the system is working OK. To advance to the next test, push the HOT (arrow up) button. To go back to a previous test, push the COLD (arrow down) button. **Note:** *The testing should be done outdoors in daylight, for the "sunload" sensor to test properly.*

20   All of the LED's should light during the initial phase of the test. Next, press the HOT button and the computer will test the sensors, with 2 (the number of the test) displayed. If all the sensors are operating properly, code 20 will be displayed. If there is a problem, a trouble code will be displayed (from 21 through 26). Check the following code chart to find which sensor is at fault:

Code number	Problem	Code number	Problem
21	Open or short in ambient sensor	35	Problem with Floor/Defrost mode
22	Open or short in in-car sensor	36	Problem with Defrost mode
25	Open or short in sunload sensor	37	Problem with recirc (intake door motor)
26	PBR (potentio balance resistor, on mix-door motor)	38	Problem at 20% fresh position (intake door motor)
31	Problem with Vent mode	39	Problem at Fresh position (intake door motor)
32	Problem with Bi-level mode		
34	Problem with Floor mode		

21   Press the HOT button again and 3 (test 3) should be displayed. When this test is finished, either a trouble code (31 through 39) will show or code 30 (mode and intake door functions OK).

22   For test 4, put the fresh vent lever to OFF. Code 41 will show first, putting the mode door in the vent position. Watch the vent door mechanism or listen for its operation. During the test 4 sequence, no trouble

codes will be displayed, you have to watch or listen for the actuator's operation. Pushing the defrost button once will change to the next test until all 6 tests have been made. Refer to the following chart for the test order:

Code		Actuator position				
	Mode door	Intake door	Air mix door	Fresh vent door	Blower voltage	Compressor
41	Vent	Recirc	Full cold	Open	4-5 volts	On
42	Bi-level	Recirc	Full cold	Open	9-11 volts	On
43	Bi-level	20% fresh	Full hot	Closed	7-9 volts	On
44	Floor/defr 1	Fresh	Full hot	Closed	7-9 volts	Off
45	Floor/defr 2	Fresh	Full hot	Closed	7-9 volts	Off
46	Defrost	Fresh	Full hot	Closed	10-12 volts	On

23 The final test 5 sequence (after pushing the HOT button again), displays the temperatures at the sensors. First displayed is the ambient sensor, then push the defrost button once to advance the display. Next is the in-car sensor, then the display returns to code 5 again. If either of the temperatures displayed are substantially different from actual ambient or in-car temperatures, check the sensor circuits.

Eliminating air-conditioner odors

24 Unpleasant odors that often develop in air conditioning systems are caused by the growth of a fungus, usually on the surface of the evaporator core. The warm, humid environment there is a perfect breeding ground for mildew to develop.
25 The evaporator core on most vehicles is difficult to access, and the dealerships have a lengthy, expensive process for eliminating the fungus by opening up the evaporator case and using a powerful disinfectant and rinse on the core until the fungus is gone. You can service your own system at home, but it takes something much stronger than basic household germ-killers or deodorizers.
26 Aerosol disinfectants for automotive air conditioning systems are available in most auto parts stores, but remember when shopping for them that the most effective treatments are also the most expensive. The basic procedure for using these sprays is to start by running the system in the RECIRC mode for ten minutes with the blower on its highest speed. Use the highest heat mode to dry out the system and keep the compressor from engaging by disconnecting the wiring connector at the compressor (see Section 14).
27 The disinfectant can is usually equipped with a long spray hose. Find a point on the evaporator case that would allow direct spray over the entire core and drill a small hole in the plastic housing. **Caution:** *Use a stop if necessary to prevent the drill from going in far enough to damage the core.* Point the nozzle inside the hole and spray, according to the manufacturer's recommendations. After use, close up the small hole with a dab of flexible body sealer.
28 Once the evaporator has been cleaned, the best way to prevent the mildew from returning is to always run your air conditioning system on the exterior-air source, don't use the MAX/AC setting. The drier outside air retards the growth of fungus. Also make sure your evaporator housing drain tube is clear (see Step 1).
29 Using the blower at high speed to dry out the evaporator core is also good idea. A commercial kit you can install, is available at many auto parts stores. It installs easily and automatically runs your blower on HIGH for five minutes, one hour after the car has been shut off.

Heating systems

30 If the air coming out of the heater vents isn't hot, the problem could stem from any of the following causes:

a) *The thermostat is stuck open, preventing the engine coolant from warming up enough to carry heat to the heater core. Replace the thermostat (see Section 3).*
b) *A heater hose is blocked, preventing the flow of coolant through the heater core. Feel both heater hoses at the firewall. They should be hot. If one of them is cold, there is an obstruction in one of the hoses or in the heater core, or the heater control valve is shut. Detach the hoses and back flush the heater core with a water hose. If the heater core is clear but circulation is impeded, remove the two hoses and flush them out with a water hose.*
c) *If flushing fails to remove the blockage from the heater core, the core must be replaced.* (see Section 10).

31 If the blower motor speed does not correspond to the setting selected on the blower switch, the problem could be a bad fuse, circuit, switch, blower motor resistor or motor (see Section 9).
32 If there isn't any air coming out of the vents:

a) *Turn the ignition ON and activate the fan control. Place your ear at the heating/air conditioning register (vent) and listen. Most motors are audible. Can you hear the motor running?*
b) *If you can't (and have already verified that the blower switch and the blower motor resistor are good), the blower motor itself is probably bad (see Section 9 for testing).*

33 If the carpet under the heater core is damp, or if antifreeze vapor or steam is coming through the vents, the heater core is leak-

13.2 The receiver/drier (arrow) is mounted to the left of the radiator, in front of the battery

ing. Remove it (see Section 10) and install a new unit (most radiator shops will not repair a leaking heater core).

13 Air conditioning receiver/drier - removal and installation

Refer to illustration 13.2
Warning: *The air conditioning system is under high pressure. Do not loosen any fittings or remove any components until after the system has been discharged. Air conditioning refrigerant should be properly discharged into an EPA-approved container at a dealer service department or an automotive air conditioning repair facility. Always wear eye protection when disconnecting air conditioning system fittings.*
1 Have the refrigerant discharged at a dealer service department or an automotive air conditioning repair facility.
2 The receiver/drier, which acts as a reservoir and filter for the refrigerant, is located at the driver's side of the radiator, just in front of the battery **(see illustration)**. **Note:** *It may be easier to work on the receiver/drier if the battery is removed first.*
3 Detach the two refrigerant lines from the receiver/drier, and disconnect the electrical connector on the triple-pressure switch.
4 Immediately cap the open fittings to prevent the entry of dirt and moisture.

3

14.3 Disconnect the two refrigerant lines (arrows) at the compressor (view from underneath)

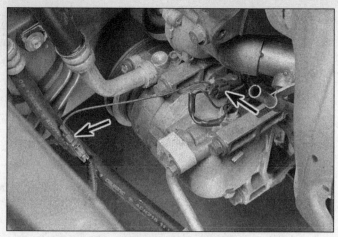

14.4 Disconnect the electrical connectors (arrows) at the compressor

5   Unbolt the receiver/drier mounting bolt, at the clamp, and lift it from the vehicle.
6   Install new O-rings on the lines and lubricate them with clean refrigerant oil.
7   Installation is the reverse of removal. **Note:** *Do not remove the sealing caps until you are ready to reconnect the lines. Do not mistake the inlet (marked IN) and the outlet connections.*
8   If a new receiver/drier is to be install, drain the oil from the old unit into a measured container. When installing the new drier, drain any oil from the new unit, then add new oil in the same amount as removed from the old one. The amount should be about 0.2 ounces. **Caution:** *Use only refrigerant oil compatible with the R-134a system in your vehicle.*
9   Have the system evacuated, charged and leak tested by the shop that discharged it.

14   Air conditioning compressor - removal and installation

Refer to illustrations 14.3, 14.4 and 14.6
Warning: *The air conditioning system is under high pressure. Do not loosen any fittings or remove any components until after the system has been discharged. Air conditioning refrigerant should be properly discharged into an EPA-approved container at a dealer service department or an automotive air conditioning repair facility. Always wear eye protection when disconnecting air conditioning system fittings.*
1   Have the refrigerant discharged at a dealer service department or an automotive air conditioning repair facility.
2   Disconnect the cable from the negative battery terminal.
3   Detach the refrigerant lines from the compressor **(see illustration)** and immediately cap the open fittings to prevent the entry of dirt and moisture.
4   Disconnect the electrical connector from the compressor **(see illustration).**

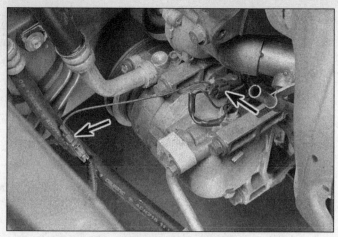

14.6 Remove the mounting bolts (arrows) and the compressor from the mounting bracket - there are two more bolts accessible from above the compressor

5   Remove the drivebelt (see Chapter 1).
6   Remove the compressor mounting bolts **(see illustration)** and remove the compressor from the engine compartment. **Note:** *Keep the compressor level during handling and storage. If the compressor seized or you find metal particles in the refrigerant lines, the system must be flushed out by an air conditioning technician and the receiver/drier must be replaced* (see Section 13).
7   Prior to installation, turn the center of the clutch six times to evenly disperse any oil that has collected in the head.
8   Install the compressor in the reverse order of removal.
9   If you are installing a new compressor, the cycling clutch assembly may have to be transferred to the new compressor. **Note:** *The removal of the clutch assembly will require the use of several special tools. You may want to have the clutch assembly transferred to the new compressor by an air conditioning shop or dealer service department. Refer to the compressor manufacturer's instructions for adding refrigerant oil to the system.* **Caution:** *Use only refrigerant oil*

compatible with the R-134a system in your vehicle.
10   Have the system evacuated, charged and leak tested by the shop that discharged it.

15   Air conditioning condenser - removal and installation

Refer to illustration 15.3
Warning 1: *The models covered by this manual are equipped with airbags. Always disconnect the negative battery cable, then the positive cable and wait 10 minutes before working in the vicinity of the impact sensors, steering column or instrument panel to avoid the possibility of accidental deployment of the airbag, which could cause personal injury (see Chapter 12). The airbag circuits are easily identified by yellow insulation covering the entire wiring harness or just prior to the wire harness connectors. Do not use electrical test equipment on any of these wires or tamper with them in any way.*
Warning 2: *The air conditioning system is under high pressure. Do not loosen any fittings or remove any components until after the system has been discharged. Air conditioning refrigerant should be properly discharged into an EPA-approved container at a dealer service department or an automotive air conditioning repair facility. Always wear eye protection when disconnecting air conditioning system fittings.*
1   Have the refrigerant discharged at a dealer service department or an automotive air conditioning repair facility.
2   Remove the fan assembly and radiator (see Sections 4 and 5).
3   Disconnect the refrigerant lines from the condenser **(see illustration).**
4   Immediately cap the open fittings to prevent the entry of dirt and moisture.
5   Unbolt the condenser mounts and lift it out of the vehicle. Store it upright to prevent oil loss.
6   Installation is the reverse of removal.

15.3 Disconnect the refrigerant lines at the condenser (upper arrow), there is a fitting on each side of the condenser - the lower arrow indicates the mounting bolt, there is one on each side

16.3 Unbolt the two refrigerant lines (arrows) at the engine side of the firewall - use a back-up wrench when disconnecting the fittings

7   If a new condenser is installed, add 2.5 ounces (75 cc) of refrigerant oil to the system. **Caution:** *Use only refrigerant oil compatible with the R-134a system in your vehicle.*

8   Have the system evacuated, charged and leak tested by the shop that discharged it.

16   Air conditioning evaporator - removal and installation

Refer to illustrations 16.3 and 16.6

Warning 1: *The models covered by this manual are equipped with airbags. Always disconnect the negative battery cable, then the positive cable and wait 10 minutes before working in the vicinity of the impact sensors, steering column or instrument panel to avoid the possibility of accidental deployment of the airbag, which could cause personal injury (see Chapter 12). The airbag circuits are easily identified by yellow insulation covering the entire wiring harness or just prior to the wire harness connectors. Do not use electrical test equipment on any of these wires or tamper with them in any way.*

Warning 2: *The air conditioning system is under high pressure. Do not loosen any fittings or remove any components until after the system has been discharged. Air conditioning refrigerant should be properly discharged into an EPA-approved container at a*

dealer service department or an automotive air conditioning repair facility. Always wear eye protection when disconnecting air conditioning system fittings.

1   Have the refrigerant discharged at a dealer service department or an automotive air conditioning repair facility.

2   Disconnect the negative battery cable, then the positive cable and wait 10 minutes before proceeding any further.

3   Working in the engine compartment, disconnect the suction and liquid lines **(see illustration)**. Use a back-up wrench to avoid twisting and damaging the lines.

4   Immediately cap the open fittings to prevent the entry of dirt and moisture and remove the inlet and outlet grommets.

5   Refer to Section 10 for removal of the heater core housing. The heater core housing and evaporator housing are mounted together, and once the heater core housing is removed, there is ample access to the evaporator housing.

6   Disconnect all electrical connectors and tubing from the evaporator assembly, remove the mounting nuts and bolts and remove the unit from the vehicle **(see illustration)**.

7   To remove the evaporator, separate the two halves of the housing and pull out the evaporator core.

8   Check the evaporator fins for blockage; if they are dirty clean them with compressed air - never use water for this purpose!

9   Check fittings for cracks and signs of

16.6 After removing the heater unit, remove the evaporator unit (arrow)

wear; replace parts as necessary.

10   Installation is the reverse of the removal procedure. Be sure to replace all O-rings removed during disassembly with new ones.

11   If a new evaporator was installed, add 2.5 ounces (75 cc) of refrigerant oil to the system. **Caution:** *Use only refrigerant oil compatible with the R-134a system in your vehicle.*

12   Have the system evacuated, charged and leak tested by the repair facility that discharged it.

3

Notes

Chapter 4
Fuel and exhaust systems

Contents

Specifications

Fuel pressure

Fuel system pressure (at idle)	
Vacuum hose attached	32 to 38 psi
Vacuum hose detached	38 to 46 psi
Fuel system hold pressure (after 5 minutes)	30 to 40 psi
Fuel pump pressure (maximum)	65 psi

Injector resistance .... 10 to 14 ohms

Torque specifications

Throttle body mounting bolts (use cross pattern)	
First tightening sequence	72 to 96 in-lbs
Second tightening sequence	156 to 192 in-lbs
Injector cover mounting screws	48 to 60 in-lbs
Fuel pressure regulator mounting screws	48 to 60 in-lbs
Air intake plenum mounting bolts	144 to 168 in-lbs
EGR valve-to-base mounting assembly	144 to 168 in-lbs
EGR base mounting assembly to intake manifold	144 to 168 in-lbs
Fuel rail mounting bolts	144 to 168 in-lbs
Exhaust pipe-to-exhaust manifold bolts	15 to 25 ft-lbs

1   General information

The fuel system on these models consists of a fuel tank, an electric fuel pump (located in the fuel tank), a fuel pump relay, fuel injector(s), an air induction system and a throttle body unit. The fuel injection system uses separate injectors positioned over the intake valves, mounted in the intake manifold. 1996 and later models are equipped with a system that routes air through a cut valve into each fuel injector chamber to aid in atomizing the air/fuel mixture for a more complete combustion and less emissions. This system is called the Air Assisted Injector system (refer to Section 14).

Multi Port Fuel Injection (MPFI) system

Multi Port Fuel Injection uses timed impulses to inject the fuel directly into the intake port of each cylinder. The injectors are controlled by the Engine Control Module (ECM). The ECM monitors various engine parameters and delivers the exact amount of fuel required into the intake ports. The throttle body serves only to control the amount of air passing into the system. Because each cylinder is equipped with its own injector, much better control of the fuel/air mixture ratio is possible.

Fuel pump and lines

Fuel is circulated from the fuel tank to the fuel injection system, and back to the fuel tank, through a pair of metal lines running along the underside of the vehicle. An electric fuel pump is located inside the fuel tank. A vapor return system routes all vapors back to the fuel tank through a separate return line.

The fuel pump will operate as long as the engine is cranking or running and the ECM is receiving ignition reference pulses

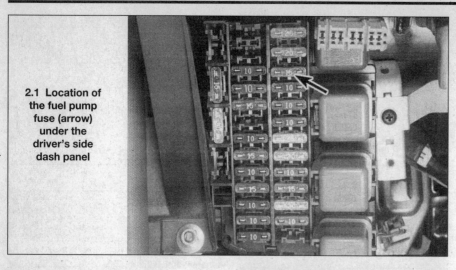

2.1 Location of the fuel pump fuse (arrow) under the driver's side dash panel

from the electronic ignition system. If there are no reference pulses, the fuel pump will shut off after two or three seconds.

Exhaust system

The exhaust system includes an exhaust manifold fitted with an exhaust oxygen sensor, a catalytic converter, an exhaust pipe, and a muffler.

The catalytic converter is an emission control device added to the exhaust system to reduce pollutants. A single-bed converter is used in combination with a three-way (reduction) catalyst. Refer to Chapter 6 for more information regarding the catalytic converter.

2   Fuel pressure relief procedure

Refer to illustration 2.1
Warning: *Gasoline is extremely flammable, so take extra precautions when you work on any part of the fuel system. Don't smoke or allow open flames or bare light bulbs near the work area, and don't work in a garage where a natural gas-type appliance (such as a water heater or a clothes dryer) with a pilot light is present. Since gasoline is carcinogenic, wear latex gloves when there's a possibility of being exposed to fuel, and, if you spill any fuel on your skin, rinse it off immediately with soap and water. Mop up any spills immediately and do not store fuel-soaked rags where they could ignite. The fuel system is under constant pressure, so, if any fuel lines are to be disconnected, the fuel pressure in the system must be relieved first. When you perform any kind of work on the fuel system, wear safety glasses and have a Class B type fire extinguisher on hand.*
Note: *After the fuel pressure has been relieved. it's a good idea to lay a shop towel over any fuel connection to be disassembled, to absorb the residual fuel that may leak out when servicing the fuel system.*
1   Remove the fuel pump fuse from the fuse panel **(see illustration)**.
2   Start the engine and allow it to run until it stops. This should take only a few seconds.

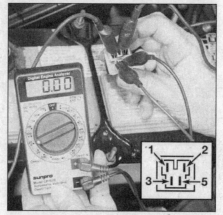

3.3b Apply battery voltage using jumper wires attached from the battery terminals to terminals 1 and 2 on the fuel pump relay and check that continuity exists between terminals 3 and 5

Disconnect the cable from the negative terminal of the battery before working on the fuel system.
3   The fuel system pressure is now relieved. When you're finished working on the fuel system, simply install the fuel pump fuse back into the fuse panel and connect the negative cable to the battery.
4   It is a good idea to back-up any fuel line that will be disconnected using a shop rag to catch fuel that might spill out.

3   Fuel pump/fuel pressure - check

Warning: *Gasoline is extremely flammable, so take extra precautions when you work on any part of the fuel system. See the* **Warning** *in Section 2.*
Note: *To perform the fuel pressure test, you will need to obtain a fuel pressure gauge and adapter set (fuel line fittings).*

Preliminary inspection

Refer to illustrations 3.3a, 3.3b and 3.4
1   Should the fuel system fail to deliver the

3.3a Check for battery voltage on the fuel pump relay connector located in the engine compartment near the air cleaner assembly

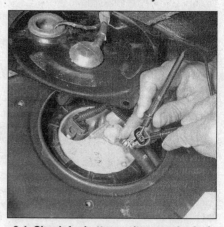

3.4 Check for battery voltage to the fuel pump at the fuel tank with the ignition key ON

proper amount of fuel, or any fuel at all, inspect it as follows. Remove the fuel filler cap. Have an assistant turn the ignition key to the ON position (engine not running) while you listen at the fuel filler opening. You should hear a whirring sound that lasts for a couple of seconds.
2   If you don't hear anything, check the fuel pump fuse (see Section 3). If the fuse is blown, replace it and see if it blows again. If it does, trace the fuel pump circuit for a short. If it isn't blown, check the fuel pump relay.
3   Remove the relay and check for battery voltage to the fuel pump relay connector **(see illustration)**. If there is battery voltage present, check the relay for proper operation. With battery voltage applied to terminals 1 and 2, continuity should exist between 3 and 5 **(see illustration)**. Without battery voltage applied, there should be no continuity between terminals 3 and 5.
4   If battery voltage is present, remove the rear seat (see Chapter 11) and check for battery voltage at the fuel pump electrical connector **(see illustration)**. If there is no voltage, check the fuel pump circuit. If there is voltage present, replace the pump (see Section 4).

3.6a Remove the clamp and carefully slide the fuel line off the fuel rail

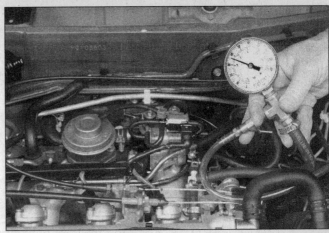

3.6b Using a T-fitting, install the fuel pressure gauge between the fuel filter and the fuel rail

3.9 If the fuel pressure is low, squeeze the fuel return line - if the fuel pressure increases, the pressure regulator is defective. The tool shown here is a pair of flat-nosed pliers with no sharp edges (be sure to wrap a rag around the hose to prevent damage)

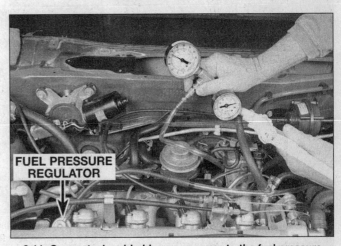

FUEL PRESSURE REGULATOR

3.11 Connect a hand-held vacuum pump to the fuel pressure regulator - apply vacuum and confirm that as vacuum is applied, the fuel pressure decreases

Operating pressure check

Refer to illustrations 3.6a, 3.6b, 3.9, 3.11 and 3.13

5   Relieve the fuel system pressure (see Section 2). Detach the cable from the negative battery terminal.

6   Remove the fuel line from the fuel rail and attach a fuel pressure gauge **(see illustrations)** between the fuel filter and the fuel rail. Tighten the hose clamps securely.

7   Attach the cable to the negative battery terminal. Start the engine.

8   Check the fuel pressure at idle, comparing your reading with the value listed in this Chapter's Specifications. Disconnect the vacuum hose and watch the gauge - the pressure should jump up considerably as soon as the hose is disconnected. If it doesn't, check for a vacuum signal to the fuel pressure regulator (see Step 13).

9   If the fuel pressure is low, pinch the fuel return line shut **(see illustration)** and watch the gauge. If the pressure doesn't rise, the fuel pump is defective or there is a restriction in the fuel feed line (possibly a clogged fuel

filter). If the pressure rises sharply, replace the pressure regulator.

10   If the fuel pressure is too high, turn the engine off. Disconnect the fuel return line and blow through it to check for a blockage. If there is no blockage, replace the fuel pressure regulator.

11   Hook up a hand-held vacuum pump to the port on the fuel pressure regulator **(see illustration)**.

12   Read the fuel pressure gauge with vacuum applied to the pressure regulator and also with no vacuum applied. The fuel pressure should decrease as vacuum increases (and increase as vacuum decreases).

13   Connect a vacuum gauge to the pressure regulator vacuum hose. Start the engine and check for vacuum **(see illustration)**. If there isn't vacuum present, check for a clogged hose or vacuum port or a vacuum leak. If the amount of vacuum is adequate, replace the fuel pressure regulator.

14   Turn the ignition switch to OFF, wait five minutes and recheck the pressure on the gauge. Compare the reading with the hold pressure listed in this Chapter's Specifica-

3.13 Detach the vacuum line from the fuel pressure regulator and see if vacuum is present when the engine is running

tions. If the hold pressure is less than specified:

a) *The fuel lines may be leaking.*

b) *The fuel pressure regulator may be*

4

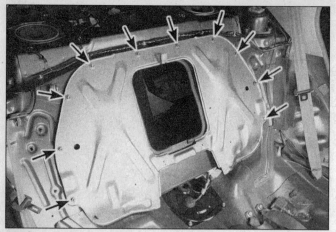

4.5 Remove the trunk shield bolts (arrows)

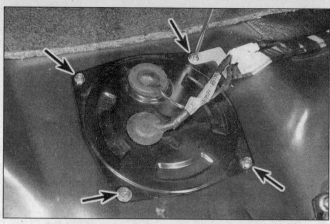

4.6 Remove the fuel pump/fuel level sending unit access cover screws (arrows)

4.7 Disconnect the electrical connectors from the fuel pump assembly

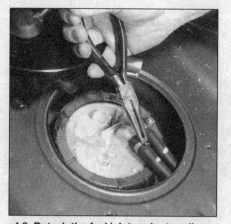

4.8 Detach the fuel inlet and return lines from the fuel pump assembly

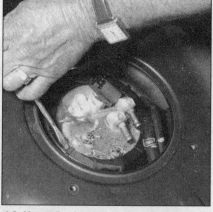

4.9 Use a flat-bladed screwdriver and tap the lock ring counterclockwise

allowing the fuel pressure to bleed through to the return line
c) A fuel injector (or injectors) may be leaking.
d) The fuel pump may be defective.

4 Fuel pump - removal and installation

Refer to illustrations 4.5, 4.6, 4.7, 4.8, 4.9, 4.10, 4.11, 4.12, 4.14a, 4.14b, 4.15 and 4.16
Warning: Gasoline is extremely flammable, so take extra precautions when you work on any part of the fuel system. See the **Warning** in Section 2.
1   Unless the vehicle has been driven far enough to completely empty the tank, it's a good idea to siphon the residual fuel out before removing the fuel pump from the vehicle. **Warning:** DO NOT start the siphoning action by mouth! Use a siphoning kit (available at most auto parts stores).
2   Relieve the fuel pressure (see to Section 2).
3   Detach the cable from the negative terminal of the battery.
4   Remove the rear seat from the vehicle (see Chapter 11).

5   Remove the trunk shield **(see illustration)**.
6   Remove the access cover for the fuel pump/fuel level sending unit assembly **(see illustration)**.
7   Disconnect the electrical connectors from the fuel pump/fuel level sending unit **(see illustration)**.
8   Disconnect the fuel inlet and return line from the fuel level sending unit/fuel pump

assembly **(see illustration)**.
9   Using a screwdriver and hammer, tap the lock ring on the assembly counterclockwise **(see illustration)**. Remove the lock ring from the assembly.
10   Lift the fuel gauge assembly and disconnect the fuel pump and fuel level sending unit electrical connectors **(see illustration)**.
11   Disconnect the fuel lines from the assembly **(see illustration)**.

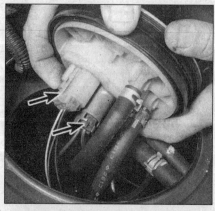

4.10 Unplug the electrical connectors from the bottom of the fuel pump assembly and . . .

4.11 . . . detach the fuel lines

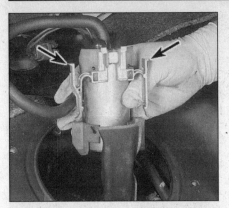

4.12 Reach inside the fuel pump access hole on top of the fuel tank and squeeze the tabs (arrows), releasing the fuel pump from the housing and lift the pump from the fuel tank

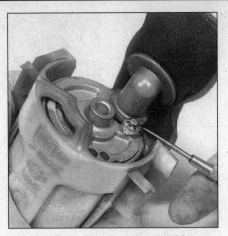

4.14a Using a fine-tipped screwdriver or pick, remove the retaining clip and . . .

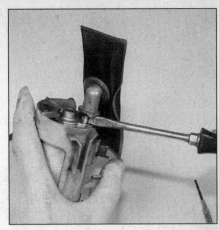

4.14b . . . carefully pry the filter sock from the fuel pump

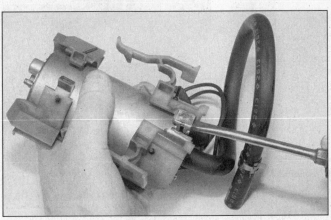

4.15 Lift the locking tab and separate the holder

4.16 Slide the fuel pump from the holder

4

12   Reach into the fuel tank access hole on top of the fuel tank and squeeze the locking tabs together to release the pump from the bracket assembly **(see illustration)**.

13   Lift the fuel pump from the fuel tank.

14   Remove the retaining clip **(see illustration)** and separate the filter sock from the fuel pump **(see illustration)**.

15   To remove the fuel pump from the holder, pry the locking tab while pulling the case to separate the holder **(see illustration)**.

16   Separate the fuel pump from the holder **(see illustration)**.

17   Installation is the reverse of removal. Be sure to replace the O-ring with a new one before installing the cover.

---

5   **Fuel level sending unit - check and replacement**

Warning: *Gasoline is extremely flammable, so take extra precautions when you work on any part of the fuel system. See the* **Warning** *in Section 2.*

Note: *These models are equipped with a separate fuel level sending unit and fuel pump that are accessible from the single opening at the top of the tank.*

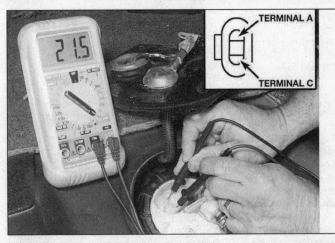

5.4 Connect the probes of the ohmmeter onto terminals A and C and check the fuel level sending unit resistance

Check

Refer to illustration 5.4

1   Before performing any tests on the fuel level sending unit, fill the tank with fuel.

2   Raise the vehicle and support it securely with jackstands.

3   Lift the rear seat and remove the access cover plate (see Section 4) to locate the fuel level sending unit electrical connector positioned on the fuel tank.

4   Connect the probes of an ohmmeter to the electrical terminals **(see illustration)** and check the resistance. Use the 200-ohm or low scale on the ohmmeter.

5   With the fuel tank completely full, the resistance should be approximately 4.0 to 5.0 ohms. With the tank half full, the resistance should be approximately 27.0 to 35.0 ohms. With the fuel tank empty, the resistance of the sending unit should be approximately 80 to 83.0 ohms.

6   If the readings are incorrect, replace the sending unit. **Note:** *A more accurate check of*

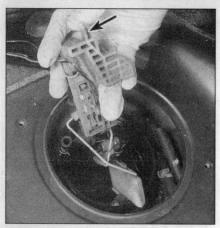

5.13 Reach inside the fuel tank opening, press the tab down (arrow) and slide the fuel level sending unit forward and up

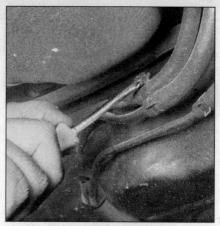

7.6 Remove the fuel feed, return and vapor lines

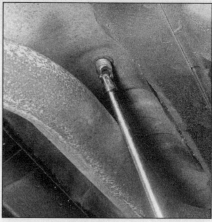

7.8a Remove the bolt that retains the heat shield directly forward of the fuel tank

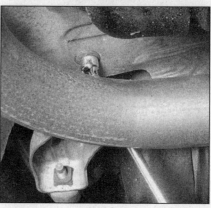

7.8b Remove the heat shield corner bolt

the sending unit can be made by removing it from the fuel tank and checking its resistance while manually operating the float arm.
7   If the readings are incorrect, replace the sending unit.

Replacement

Refer to illustration 5.13
8   Siphon the remaining fuel into an approved gasoline container. **Warning:** *DO NOT start the siphoning action by mouth! Use a siphoning kit (available at most auto parts stores).*
9   Remove the rear seat from the vehicle (see Chapter 11). Remove the fuel pump access cover (see Section 4).
10   Disconnect the fuel pump and the fuel level sending unit electrical connectors from the assembly (see Section 4).
11   Remove the fuel lines from the assembly (see Section 6).
12   Remove the lock ring from the fuel gauge assembly (see Section 4).
13   Lift the sending unit from the tank **(see illustration)**. Carefully angle the sending unit out of the opening without damaging the fuel level float located at the bottom of the assembly.
14   Installation is the reverse of removal. Be sure to replace the O-ring with a new one before installing the cover.

---

6   Fuel lines and fittings - repair and replacement

Warning: *Gasoline is extremely flammable, so take extra precautions when you work on any part of the fuel system. See the* **Warning** *in Section 2.*

Inspection

1   Once in a while, you will have to raise the vehicle to service or replace some component (an exhaust pipe hanger, for example). Whenever you work under the vehicle, always inspect the fuel lines and fittings for possible damage or deterioration.

2   Check all hoses and pipes for cracks, kinks, deformation or obstructions.
3   Make sure all hose and pipe clips attach their associated hoses or pipes securely to the underside of the vehicle.
4   Verify all hose clamps attaching rubber hoses to metal fuel lines or pipes are snug enough to assure a tight fit between the hoses and the metal pipes.

Replacement

5   If you must replace any damaged sections, use hoses approved for use in fuel systems or pipes made from steel only (it's best to use original-equipment pipe that's already flared and bent to the proper shape). Do not install substitutes constructed from inferior or inappropriate material, as this could result in a fuel leak and a fire.
6   Always, before detaching or disassembling any part of the fuel line system, note the routing of all hoses and pipes and the orientation of all clamps and clips to assure that replacement sections are installed in exactly the same manner.
7   Before detaching any part of the fuel system, be sure to relieve the fuel tank pressure by removing the fuel filler cap.
8   Always use new hose clamps after loosening or removing them.
9   While you're under the vehicle, it's a good idea to check the following related components:

   a) *Check the condition of the fuel filter - make sure that it's not clogged or damaged (see Chapter 1).*
   b) *Inspect the evaporative emission control (EVAP) system. Verify that all hoses are attached and in good condition (see Chapter 6).*

---

7   Fuel tank - removal and installation

Refer to illustrations 7.6, 7.8a, 7.8b, 7.8c, 7.8d and 7.11
Warning: *Gasoline is extremely flammable,*

so take extra precautions when you work on any part of the fuel system. See the **Warning** in Section 2.*
Note: *The following procedure is much easier to perform if the fuel tank is empty. Some tanks have a drain plug for this purpose. If the tank does not have a drain plug, drain the fuel into an approved fuel container using a commercially available siphoning kit (NEVER start a siphoning action by mouth) or wait until the fuel tank is nearly empty, if possible.*
1   Remove the fuel tank filler cap to relieve fuel tank pressure.
2   Relieve the fuel system pressure (see Section 2).
3   Detach the cable from the negative terminal of the battery.
4   Raise the vehicle and place it securely on jackstands.
5   If there is still fuel in the tank, siphon it out from the fuel feed line. Remember - NEVER start the siphoning action by mouth! Use a siphoning kit, which can be purchased at most auto parts stores.
6   Disconnect the fuel lines and the vapor return line **(see illustration)**. **Note:** *The fuel feed and return lines and the vapor return line are three different diameters, so reattachment is simplified. If you have any doubts, however, clearly label the three lines and the fittings. Be sure to plug the hoses to prevent leakage and*

7.8c Remove the rear heat shield circlip from the stud

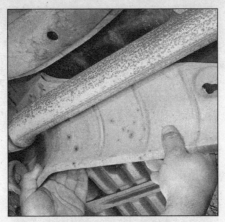

7.8d Slide the heat shield off the studs and maneuver it out

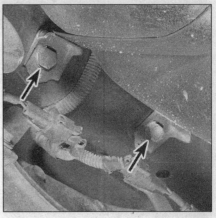

7.11 Remove the bolts from the fuel tank retaining straps (arrows)

contamination of the fuel system.

7   Disconnect the fuel filler tube from the filler pipe.

8   If equipped, remove the bolts from the heat shield located on the front side of the fuel tank **(see illustrations)**.

9   Support the fuel tank with a floor jack. Position a wood block between the jack head and the fuel tank to protect the tank.

10   If equipped, disconnect the electrical connectors for the fuel pump and/or fuel level sending unit located under the rear seat (see Section 4).

11   Remove the fuel tank strap bolts **(see illustration)**.

12   Remove the tank from the vehicle.

13   Installation is the reverse of removal.

8   Fuel tank - cleaning and repair

1   All repairs to the fuel tank or filler neck should be carried out by a professional who has experience in this critical and potentially dangerous work. Even after cleaning and flushing of the fuel system, explosive fumes can remain and ignite during repair of the tank.

2   If the fuel tank is removed from the vehicle, it should not be placed in an area where sparks or open flames could ignite the fumes coming out of the tank. Be especially careful inside garages where a natural gas-type appliance is located, because the pilot light could cause an explosion.

9   Air cleaner assembly - removal and installation

Refer to illustration 9.3

1   Detach the cable from the negative terminal of the battery.

2   Remove the air filter from the air cleaner housing (see Chapter 1).

3   Remove the mounting bolts from the air cleaner housing **(see illustration)**.

4   Loosen the clamp on the air intake duct and separate the duct from the housing.

5   Installation is the reverse of removal.

9.3 Remove the bolts (arrows) from the air cleaner housing

10   Accelerator cable - removal, installation and adjustment

Removal

Refer to illustrations 10.1, 10.3 and 10.4

1   Detach the accelerator cable from the throttle lever **(see illustration)**.

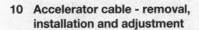

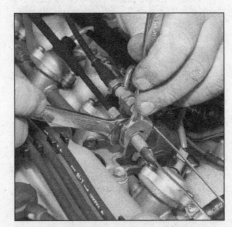

10.3 Unscrew the nut nearest the exposed cable, then pass the cable through the bracket

10.1 Rotate the throttle lever and detach the cable end

2   If equipped, remove the cruise control cable from the cable bracket.

3   Separate the accelerator cable from the cable bracket **(see illustration)**.

4   Detach the nylon collar from the upper end of the accelerator pedal arm **(see illustration)**.

5   Remove the cable through the firewall from the engine compartment.

10.4 Pull the accelerator cable forward and then up through the slot (arrow) in the accelerator pedal arm

4

12.6 Clean the interior of the throttle body using carburetor cleaner spray and shop towels

12.7 Use a stethoscope to determine if the injectors are working properly - they should make a steady clicking sound that rises and falls with engine speed changes

Installation

6   Installation is the reverse of removal. Be sure the cable is routed correctly.

7   If necessary, at the engine compartment side of the firewall, apply sealant around the accelerator cable to prevent water from entering the passenger compartment.

Adjustment

8   To adjust the cable:

a) Lift up on the cable to remove any slack.

b) Turn the adjusting nut until it is 1/8-inch away from the cable bracket.

c) Tighten the locknut and check cable deflection at the throttle linkage. Deflection should be 3/8 to 1/2-inch. If deflection is not within specifications, loosen the locknut and turn the adjusting nut until the deflection is as specified.

d) After you have adjusted the throttle cable, have an assistant help you verify that the throttle valve opens all the way when you depress the accelerator pedal to the floor and that it returns to the idle position when you release the accelerator. Verify the cable operates smoothly. It must not bind or stick.

e) If the vehicle is equipped with an automatic transaxle, adjust the transaxle throttle valve cable (see Chapter 7B).

11   Electronic fuel injection system - general information

The Multi Port Fuel Injection (MPFI) system is a multi-point, pulse timed, speed density controlled fuel injection system. On the MPFI system, fuel is metered into each intake port in accordance with engine demand through injectors mounted on the intake manifold. The two-piece air intake manifold includes an air intake plenum.

This system incorporates an on-board Engine Control Module (ECM) computer that accepts inputs from various engine sensors to compute the required fuel flow rate necessary to maintain a prescribed air/fuel ratio throughout the entire engine operational range. The computer then outputs a command to the fuel injectors to meter the correct quantity of fuel. The system automatically senses and compensates for changes in altitude, load and speed.

The fuel delivery system includes an electric in-tank fuel pump which forces pressurized fuel through a series of metal and plastic lines and an inline fuel filter to the fuel charging manifold assembly.

A constant fuel pressure drop is maintained across the injector nozzles by a pressure regulator. The regulator is positioned downstream from the fuel injectors. Excess fuel passes through the regulator and returns to the fuel tank through a fuel return line.

12   Electronic fuel injection system - check

Warning: *Gasoline is extremely flammable, so take extra precautions when you work on any part of the fuel system. See the* **Warning** *in Section 2.*
Note: *The following procedure is based on the assumption that the fuel pump is working and the fuel pressure is adequate (see Section 3).*

Preliminary checks

1   Check all electrical connectors that are related to the system. Loose electrical connectors and poor grounds can cause many problems that resemble more serious malfunctions.

2   Check to see that the battery is fully charged, as the control unit and sensors depend on an accurate supply voltage in order to properly meter the fuel.

3   Check the air filter element - a dirty or partially blocked filter will severely impede performance and economy (see Chapter 1).

4   If a blown fuse is found, replace it and see if it blows again. If it does, search for a grounded wire in the harness to the fuel pump (see Chapter 12).

System checks

Refer to illustrations 12.6, 12.7, 12.8 and 12.9

5   Check the condition of the vacuum hoses connected to the intake manifold.

12.8 Install a special injector test light ("noid light") into the electrical connector and confirm that it blinks when the engine is cranked

6   Remove the air intake duct from the throttle body and check for dirt, carbon or other residue build-up in the throttle body, particularly around the throttle plate. If it's dirty, clean it with aerosol carburetor cleaner, a rag and a toothbrush, if necessary **(see illustration)**.

7   With the engine running, place an automotive stethoscope against each injector, one at a time, and listen for a clicking sound, indicating operation **(see illustration)**. If you don't have a stethoscope, you can place the tip of a long screwdriver against the injector and listen through the handle.

8   If an injector isn't functioning (not clicking), purchase a special injector test light (sometimes called a "noid" light) and install it into the injector electrical connector **(see illustration)**. Start the engine and check to see if the noid light flashes. If it does, the injector is receiving proper voltage. If it doesn't flash, further diagnosis should be performed by a dealer service department or other repair shop.

9   With the engine OFF and the fuel injector electrical connectors disconnected, measure the resistance of each injector **(see illustration)**. Compare your measurements with the injector resistance listed in this Chapter's Specifications. If any injector is open or has an abnormally high resistance, replace it with a new one.

12.9 Measure the resistance of each injector

13.9 Remove the mounting nuts (arrows) that hold the plenum to the manifold - the numbers represent the bolt TIGHTENING sequence

13   Multi Port Fuel Injection (MPFI) system - component check and replacement

Warning: *Gasoline is extremely flammable, so take extra precautions when you work on any part of the fuel system. See the* **Warning** *in Section 2.*

Air intake plenum with throttle body

Note: *This procedure requires removal of the cylinder head.*

Removal
Refer to illustration 13.9
1   Detach the cable from the negative terminal of the battery.
2   Disconnect the electrical connectors at the IAC/AAC valve, throttle position sensor (TPS) and EGR/EVAP canister control solenoid (see Chapter 6).
3   Detach the accelerator cable (see Section 10) and transmission linkage (see Chapter 7) from the throttle body assembly.
4   Remove the accelerator cable from the intake manifold (see Section 10).
5   Clearly label, then detach, the vacuum lines from air intake plenum, the EGR valve, the BPT valve, the throttle body and the fuel pressure regulator.
6   Detach the PCV (see Chapter 6).
7   If equipped, detach the canister purge line or lines from the throttle body.
8   Remove the cylinder head/intake manifold assembly from the engine (see Chapter 2, Part A).
9   Remove the air intake plenum and the throttle body as an assembly from the upper intake manifold **(see illustration)**.

Installation
10   Be sure to clean and inspect the mounting faces of the lower intake manifold (see Chapter 2A) and the air intake plenum before positioning the new gasket(s) onto the lower intake mounting face. Install the air intake

13.14 Two of the throttle body-to-engine brace bolts are located directly above the starter and two are located on the throttle body lower section (starter removed for clarity)

plenum and throttle body assembly onto the intake manifold. Ensure the gasket remains in place. Install the upper intake manifold retaining bolts and tighten them in the correct sequence **(see illustration 13.9)** to the torque listed in this Chapter's Specifications. The remainder of installation is the reverse of removal.

Throttle body

Removal
Refer to illustrations 13.14 and 13.15
11   Detach the cable from the negative terminal of the battery.
12   Detach the Throttle Position Sensor (TPS) connector (see Chapter 6).
13   Disconnect the linkage from the throttle body and carefully mark and remove the vacuum hoses.
14   Remove the throttle body-to-engine brace **(see illustration)**.
15   Unscrew the four throttle body mounting nuts **(see illustration)** and remove the throttle body.

13.15 Remove the four large Allen head bolts (arrows) to remove the throttle body from the intake manifold

16   Remove and discard the gasket between the throttle body and air intake plenum.

Installation
17   Clean the gasket mating surfaces. If scraping is necessary, be careful not to damage the gasket surfaces or allow material to drop into the manifold. Installation is the reverse of removal. Be sure to use a new gasket and tighten the throttle body mounting nuts to the torque listed in this Chapter's Specifications.

Throttle Position Sensor (TPS)
18   Refer to Chapter 6, Section 4 for the check and replacement procedures for the TPS.

Fuel rail assembly
Refer to illustration 13.23, 13.24, 13.25 and 13.27

Removal
19   Relieve the fuel pressure (see Section 2).
20   Detach the cable from the negative terminal of the battery.

4

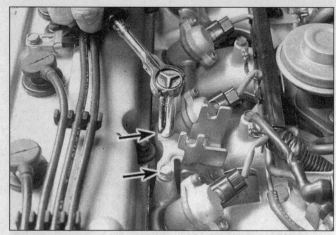

13.23 Remove the accelerator cable bracket bolts (arrows)

13.24 Remove the fuel rail mounting bolts (arrows) and . . .

13.25 . . . lift the fuel rail assembly from the engine compartment

13.27 Pry the injector rubber grommets from the intake manifold and install new ones

21   Remove the air intake plenum assembly (see Steps 1 through 9).
22   Disconnect the fuel feed and return lines from the fuel rail assembly.
23   Remove the accelerator and cruise control cable bracket from the cylinder head **(see illustration)**.
24   Remove the fuel rail retaining bolts **(see illustration)**.
25   Disconnect each fuel injector electrical connector. Carefully lift the fuel rail and fuel injectors from the engine **(see illustration)**.
26   If necessary, remove the injectors from the fuel rail (see Steps 43 and 44).

Installation

Note: *It's a good idea to replace the rubber grommets in the cylinder head whenever the fuel rail is removed.*
27   Install the fuel injectors on the fuel rail, if removed. Remove the old rubber grommets from the intake manifold and install new ones **(see illustration)**.
28   Install the fuel rail and injector assembly onto the engine and install the retaining bolts. Tighten the bolts to the torque listed in this Chapter's Specifications.
29   The remainder of installation is the reverse of removal.

Fuel pressure regulator
Check

Note: *This procedure assumes the fuel filter is in good condition.*
30   Check the fuel pressure and perform the necessary steps to diagnose problems with the pressure regulator (see Section 3).
31   Start the engine and check for leakage around the fuel rail, fuel lines and the fuel pressure regulator.
32   If the fuel pressure regulator is faulty or leaking, replace it with a new part.

Replacement

33   Relieve the fuel pressure from the system (see Section 2).
34   Disconnect the cable from the negative terminal of the battery.
35   Remove the fuel rail from the engine (see Steps 19 through 24).
36   Clean any dirt from around the fuel pressure regulator.
37   Loosen the hose clamp and remove the bolts from the fuel pressure regulator. Detach the regulator from the fuel rail.
38   Install new O-rings on the pressure regulator and lubricate them with a light coat of oil.
39   Installation is the reverse of removal.

Tighten the pressure regulator mounting bolts and the hose clamp securely.

Fuel injectors

Refer to illustrations 13.41, 13.42, 13.43a and 13.43b

Removal

40   Relieve the system fuel pressure (see Section 2).
41   Remove the screws that retain the injector in the fuel rail **(see illustration)**.
42   Carefully pry the injector using a flat-bladed screwdriver and pull up and gently rock the injector from side-to-side **(see illustration)**.
43   Remove both O-rings from the injector and install new ones **(see illustrations)**.
44   Inspect the plastic injector "hat" (covering the injector pintle) and washer for signs of deterioration. Replace as required. If the hat is missing, look for it in the intake manifold.

Installation

45   Lubricate the new O-rings with light grade oil and install two on each injector. **Caution:** *Do not use silicone grease. It will clog the injectors.*

13.41 Remove the injector cover mounting screws and cover to expose the fuel injector

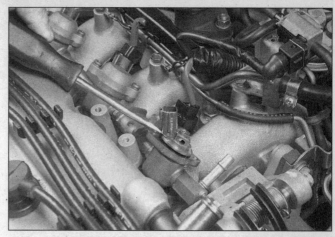

13.42 Carefully pry the fuel injector out of the fuel rail

46   Using a light twisting motion, install the injector(s).

47   The remainder of installation is the reverse of removal.

Idle Air Control/Auxiliary Air Control (IAC/AAC) valve

Refer to illustrations 13.50, 13.51a, 13.51b, 13.52a and 13.52b

General Information

48   The IAC/AAC valve is attached to the air intake plenum below the intake manifold. The ECM (computer) actuates the IAC/AAC valve by an on/off pulse signal. The longer the on-duty signal is left on, the larger the amount of air that will be allowed to flow through the IAC/AAC valve. This valve works in conjunction with the Fast Idle Control Device (FICD) solenoid. Before diagnosing the idle control components, make sure there are no obvious intake leaks (broken hoses, leaking gaskets etc.) where they are mounted.

Check

49   With the engine running, disconnect the

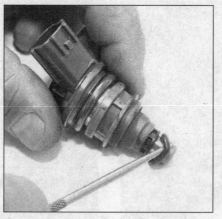

13.43a Using a pick or small-tipped screwdriver, remove the O-ring from the injector nozzle

IAC/AAC valve electrical connector and confirm that the idle speed drops immediately. If it does not, continue checking the idle control system.

50   First, check the IAC/AAC valve resis-

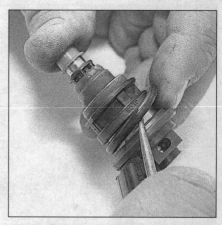

13.43b Remove the large O-ring from the top of the injector

tance **(see illustration)**. Probe the terminals of the IAC/AAC valve with an ohmmeter and record the resistance. It should be approximately 10 ohms. If the IAC/AAC valve resistance is out of range, replace it.

51   Next, remove the valve **(see illustration**

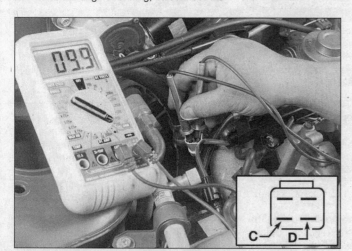

13.50 Disconnect the IAC/AAC valve connector at the right (passenger side) rear of the intake manifold and check the resistance of the valve on terminals C and D - it should be 10 ohms

13.51a Remove the four mounting bolts from the IAC/AAC valve assembly and . . .

13.51b . . . spray carburetor cleaner into the passageway to clean any deposits that may have accumulated

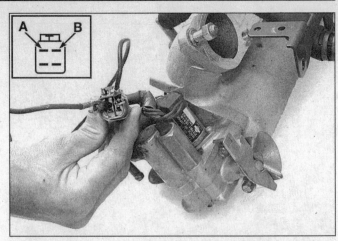

13.52a Using jumper wires, apply battery voltage to terminals A and B and listen for a loud "click" when the solenoid activates

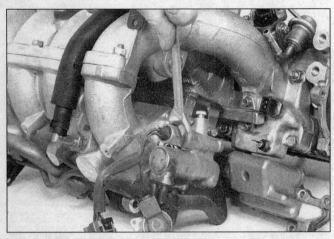

13.52b Remove the FICD using an open-end wrench

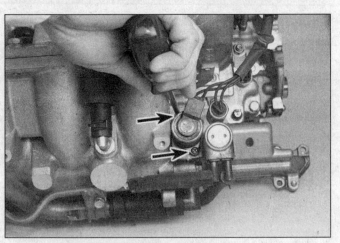

13.54 Remove the IAC valve mounting screws (arrows) and separate the unit from the air intake plenum

13.54) and check the pintle for excessive carbon deposits. If necessary, clean it with carburetor cleaner spray. Also clean the valve housing to remove any deposits **(see illustrations)**.
52   Also, check the FICD solenoid. Using jumper wires from the battery terminals, apply battery voltage to the FICD solenoid. There should be a distinct "click" **(see illustration)**. Listen carefully for the solenoid activation. If there is no sound, replace the FICD **(see illustration)**.

Removal

Refer to illustrations 13.54 and 13.55
53   Disconnect the electrical connector from the AAC valve and the FICD solenoid.
54   Remove the valve attaching screws and separate the assembly from the air intake plenum **(see illustration)**.
55   Check the condition of the O-ring. Replace it with a new one if it's cracked or otherwise deteriorated **(see illustration)**.
56   Clean the sealing surface and the bore of the throttle body assembly to ensure a good seal. **Caution:** *The AAC valve is an electrical component and must not be soaked in any liquid cleaner, as damage may result.*

13.55 Replace the O-ring for the IAC valve

Installation

57   Install a new gasket on the IAC/AAC valve assembly.
58   Install the IAC/AAC valve and tighten the screws securely.
59   Plug in the electrical connector to the IAC/AAC valve assembly.

Air regulator

Refer to illustration 13.62

General information

60   The air regulator valve provides air bypass when the engine is cold for fast idle. It consists of wax, a piston and a spring built

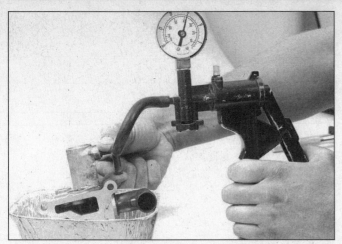

13.62 Immerse the IAC/AAC assembly in a pan of heated water to make sure that air does not flow through the bypass chamber

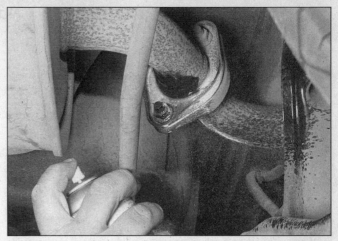

15.4 Be sure to use penetrating spray on the exhaust flange nuts before attempting to remove them

into the regulator assembly. When the temperature of the wax is low, the shutter opens and allows air to bypass from the air intake duct into the intake manifold. As the engine warms, the wax expands and does not allow any air to circulate into the intake manifold. This condition continues until the engine is stopped and the temperature of the engine block cools down.

Check

61   Remove the IAC/AAC assembly **(see illustration 13.51a)**, remove the IAC valve and then the FICD from the assembly.
62   Blow air through the inlet and confirm that air exits the outlet pipe. Next, immerse the assembly in a pan of heated water **(see illustration)** (above 176-degrees F) and make sure air does not flow through.

Replacement

63   Remove the IAC/AAC assembly **(see illustration 13.51a)**.
64   Remove the IAC valve **(see illustration 13.54)** and the FICD **(see illustration 13.52b)** from the IAC/AAC assembly.
65   The air regulator is an integral part of the assembly and cannot be replaced separately. Replace the IAC/AAC assembly as a complete unit.
66   Installation is the reverse of removal.

14   Air assisted injector system

General information

1   The air assisted injector system is installed on 1996 and 1997 models. The air assisted injector system incorporates special injectors that use an air cut valve located near the fuel injector to control the flow of air into the injector solenoid chamber for the purpose of atomizing the air/fuel mixture for more complete combustion and less emissions. Air is allowed into the injectors when the engine and cut valve are cold. During warm-up, the air is eventually cut off.

2   The air cut valve consists of wax, a piston and a spring valve. When the engine coolant is cold, the wax is compressed, opening the air bypass port. As the engine coolant warms, the wax expands and closes the air by-pass port.

Check

3   Remove the air cut valve from the engine (see Steps 6 and 7).
4   Immerse the tip of the air cut valve into warm water (176-degrees F or more) and apply air pressure to the inlet valve. Confirm that air does not pass through.
5   Allow the air cut valve to cool completely. Pressurize the inlet valve and confirm that air flows freely.

Replacement

6   Remove the cylinder head/intake manifold assembly to gain access to the air assisted injector system (see Chapter 2). The air cut valve can be reached under the intake manifold above the starter assembly but it is very difficult to maneuver any tools.
7   Remove the hoses from the air cut valve.
8   Installation is the reverse of removal.

15   Exhaust system servicing - general information

Refer to illustration 15.4
Warning: *Inspection and repair of exhaust system components should be done only after enough time has elapsed after driving the vehicle to allow the system components to cool completely. Also, when working under the vehicle, make sure it is securely supported on jackstands.*

1   The exhaust system consists of the exhaust manifold(s), the catalytic converter(s), the muffler, the tailpipe and all connecting pipes, brackets, hangers and clamps. The exhaust system is attached to the body with mounting brackets and rubber hangers.

If any of the parts are improperly installed, excessive noise and vibration will be transmitted to the body.
2   Conduct regular inspections of the exhaust system to keep it safe and quiet. Look for any damaged or bent parts, open seams, holes, loose connections, excessive corrosion or other defects which could allow exhaust fumes to enter the vehicle. Deteriorated exhaust system components should not be repaired; they should be replaced with new parts.
3   If the exhaust system components are extremely corroded or rusted together, welding equipment will probably be required to remove them. The convenient way to accomplish this is to have a muffler repair shop remove the corroded sections with a cutting torch. If, however, you want to save money by doing it yourself (and you don't have a welding outfit with a cutting torch), simply cut off the old components with a hacksaw. If you have compressed air, special pneumatic cutting chisels can also be used. If you do decide to tackle the job at home, be sure to wear safety goggles to protect your eyes from metal chips and work gloves to protect your hands.
4   Here are some simple guidelines to follow when repairing the exhaust system:

a) *Work from the back to the front when removing exhaust system components.*
b) *Apply penetrating oil to the exhaust system component fasteners to make them easier to remove* **(see illustration)**.
c) *Use new gaskets, hangers and clamps when installing exhaust system components.*
d) *Apply anti-seize compound to the threads of all exhaust system fasteners during reassembly.*
e) *Be sure to allow sufficient clearance between newly installed parts and all points on the underbody to avoid overheating the floor pan and possibly damaging the interior carpet and insulation. Pay particularly close attention to the catalytic converter and heat shield.*

4

Notes

Chapter 5
Engine electrical systems

Contents

Specifications

General

Alternator brush length	
New	1/2 inch
Minimum	1/4 inch
Battery voltage	
Engine off	12 volts
Engine running	14 to 15 volts
Firing order	1-3-4-2

Ignition system

Ignition coil resistance	
Primary resistance	1.0 to 2.0 ohms
Secondary resistance	10.0 to 12.8 K-ohms
Primary winding-to-case resistance	Infinite
Ignition coil-to-distributor cap wire resistance	5,000 ohms per foot
Ignition timing	18 to 22 degrees BTDC

1 General information

The engine electrical systems include all ignition, charging and starting components. Because of their engine-related functions, these components are considered separately from chassis electrical devices like the lights, instruments, etc.

Be very careful when working on the engine electrical components. They are easily damaged if checked, connected or handled improperly. The alternator is driven by an engine drivebelt which could cause serious injury if your hands, hair or clothes become entangled in it with the engine running. Both the starter and alternator are connected directly to the battery and could arc or even cause a fire if mishandled, overloaded or shorted out.

Never leave the ignition switch on for long periods of time with the engine off. Don't disconnect the battery cables while the engine is running. Correct polarity must be maintained when connecting battery cables from another source, such as another vehicle, during jump starting. Always disconnect the negative cable first and hook it up last or the battery may be shorted by the tool being used to loosen the cable clamps.

Additional safety related information on the engine electrical systems can be found in *Safety first* near the front of this manual. It should be referred to before beginning any operation included in this Chapter.

2 Battery - emergency jump starting

Refer to the *Booster battery (jump) starting* procedure at the front of this manual.

3 Battery - removal and installation

Refer to illustration 3.1

1   Disconnect the negative battery cable, then the positive battery cable, from the battery **(see illustration)**. **Caution:** *Always disconnect the negative cable first and hook it up last or the battery may be shorted by the*

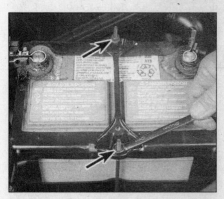

3.1 Detach the nuts from the battery hold-down clamps (arrows) and then remove the cable from the negative battery terminal followed by the positive battery cable

tool being used to loosen the cable clamps.

2   Locate the battery hold-down clamp straddling the top of the battery. Remove the nuts and the hold-down clamp.

3   Lift out the battery. Special battery lifting straps that attach to the battery posts are available at auto parts stores, lifting and moving the battery is much easier if you use one.

4   Installation is the reverse of removal.

4   Battery cables - check and replacement

1   Periodically inspect the entire length of each battery cable for damage, cracked or burned insulation and corrosion. Poor battery cable connections can cause starting problems and decreased engine performance.

2   Check the cable-to-terminal connections at the ends of the cables for cracks, loose wire strands and corrosion. The presence of white, fluffy deposits under the insulation at the cable terminal connection is a sign that the cable is corroded and should be replaced. Check the terminals for distortion, missing mounting bolts and corrosion.

3   When replacing the cables, always disconnect the negative cable first and hook it up last or the battery may be shorted by the tool used to loosen the cable clamps. Even if only the positive cable is being replaced, be sure to disconnect the negative cable from the battery first.

4   Disconnect and remove the cable. Make sure the replacement cable is the same length and diameter.

5   Clean the threads of the relay or ground connection with a wire brush to remove rust and corrosion. Apply a light coat of petroleum jelly to the threads to prevent future corrosion.

6   Attach the cable to the relay or ground connection and tighten the mounting nut/bolt securely.

7   Before connecting the new cable to the battery, make sure that it reaches the battery post without having to be stretched. Clean the battery posts thoroughly and apply a light coat of petroleum jelly to prevent corrosion.

8   Connect the positive cable first, followed by the negative cable.

5   Ignition system - general information

1   The ignition system is designed to ignite the fuel/air charge entering each cylinder at just the right moment. It does this by producing a high voltage spark between the electrodes of each spark plug.

2   The vehicles covered by this manual are equipped with an electronic ignition system. This system consists of the camshaft position sensor (located in the distributor), the power transistor, the ignition coil, an ignition circuit resistor and the primary and secondary wiring.

3   The Electronic Control Module (ECM)

6.1 To use a calibrated ignition tester, simply disconnect a spark plug wire and connect it to the tester, clip the tester to a convenient ground (like a valve cover bolt) and operate the starter - if there is enough power to fire the plug, sparks will be visible between the electrode tip and the tester body

controls the spark timing advance characteristics. Vacuum advance and vacuum hoses have been eliminated from the design.

4   The camshaft position sensor is the basis of this computer controlled system. It monitors engine speed and piston position and relays this data to the computer which in turn controls the fuel injection duration (fuel injector on/off time) and ignition timing. The camshaft position sensor consists of a rotor plate and a wave forming circuit. The rotor plate has 360 slits for each degree, or one percent signal (engine speed signal) and four slits for the 180-degree camshaft position signal. Light Emitting Diodes (LED) and photo diodes are built into the wave forming circuit. When the rotor plate passes the space between the LED and the photo diode, the slits on the rotor plate continually cut the beam of light sent to the photo diode from the LED. They are then converted into on-off pulses by the wave forming circuit and then sent to the ECM.

5   The camshaft position sensor-type distributor must be replaced as a complete unit. There are no replacement parts available for the distributor except the cap, rotor and seal cover.

6   Ignition system - check

Refer to illustrations 6.1, 6.7, 6.9, 6.10, 6.11a, 6.11b, 6.12 and 6.13

Warning: *Because of the high voltage generated by the ignition system, extreme care should be taken whenever an operation is performed involving ignition components.*

1   If the engine turns over but won't start, disconnect the spark plug wire from any spark plug and attach it to a calibrated ignition system tester (available at most auto parts stores). Connect the clip on the tester to a bolt or metal bracket on the engine **(see**

6.7 Check for battery voltage to the ignition coil on the brown wire

illustration). If you're unable to obtain a calibrated ignition tester, remove the wire from one of the spark plugs and using an insulated tool, pull back the boot and hold the end of the wire about 1/4-inch from a good ground.

2   Crank the engine and watch the end of the tester or spark plug wire to see if bright blue, well-defined sparks occur.

3   If sparks occur, sufficient voltage is reaching the plug to fire it. However, the plugs themselves may be fouled, so remove and check them as described in Chapter 1. Repeat the check at the remaining plug wires to verify that the distributor cap, rotor and all the spark plug wires are good.

4   If no sparks or intermittent sparks occur, remove the distributor cap and check the cap and rotor as described in Chapter 1. If moisture is present, dry out the cap and rotor, then reinstall the cap and repeat the spark test.

5   If there's still no spark, detach the coil secondary wire from the distributor cap and hook it up to the tester (reattach the plug wire to the spark plug), then repeat the spark check. Again, if you don't have a tester, hold the end of the wire about 1/4-inch from a good ground. If there's no spark, the coil-to-cap wire may be bad (check the resistance with an ohmmeter and compare it to the ignition coil-to-distributor cap wire resistance found in this Chapter's Specifications.

6   If sparks now occur, the distributor cap, rotor or plug wire(s) may be defective.

7   If no sparks occur and the coil-to-cap wire is good, check the primary wire connections at the coil to make sure they're clean and tight. Check for voltage to the coil **(see illustration)**. Check the coil primary and secondary resistance (see Section 7).

8   Check to make sure the ignition switch receives battery voltage. Also check for battery voltage to the Electronic Control Module (ECM). This can be observed by monitoring the red light on the ECM (see Chapter 6) after the ignition key is turned ON. If there is no power to the ECM, check the ECM relay which is located directly in front of the ECM (see Chapter 12).

9   Check to make sure the IGN SW fuse is not blown **(see illustration)** and there is volt-

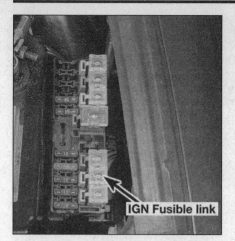

6.9 Location of the 30 amp ignition switch fusible link (arrow)

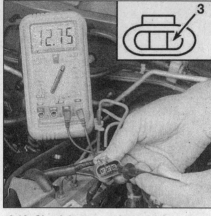

6.10 Check battery voltage to the power transistor on terminal number 3 (green wire)

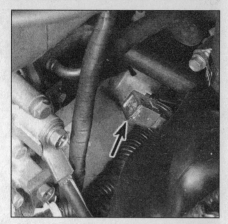

6.11a The ignition resistor is located in the wiring harness below the air intake duct (arrow)

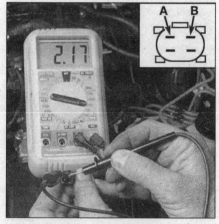

6.11b Remove the resistor from its square case and check the resistance between the terminals A and B. There should be approximately 2.2 K-ohms

age to the fuse.
10   Check for battery voltage to the power transistor (see illustration). If battery voltage is present, check the power transistor (see Section 10).
11   Check the ignition circuit resistor (see

6.12 Check the ground circuit for the power transistor on terminal number 2 of the power transistor wiring harness connector and engine ground. Continuity should exist

illustrations). Resistance should be 2.2 K-ohms.
12   Check the ground circuit for the power transistor on terminal number 2 of the wiring harness connector and engine ground (see

illustration). Continuity should exist.
13   Check for continuity between terminal number 2 on the ignition coil connector and terminal number 3 on the power transistor connector (see illustration). Continuity should exist. Refer to the wiring diagram at the end of Chapter 12 for additional information on the electrical schematic for the ignition system.

7   Ignition coil - check and replacement

Check
Refer to illustrations 7.1 and 7.2

Primary and secondary coil resistance
1   With the ignition off, disconnect the electrical connector from the coil. Connect an ohmmeter across the two primary terminals of the coil (see illustration). The resistance should be as listed in this Chapter's Specifications for the primary resistance. If not, replace the coil.
2   Remove the distributor cap-to-coil wire

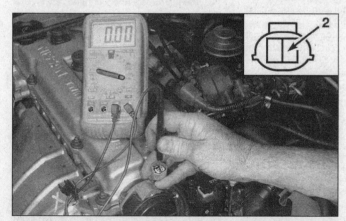

6.13 Check for continuity between terminal 2 on the ignition coil connector and terminal number 3 on the power transistor. Continuity should exist

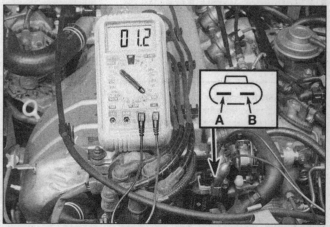

7.1 Checking the coil primary resistance

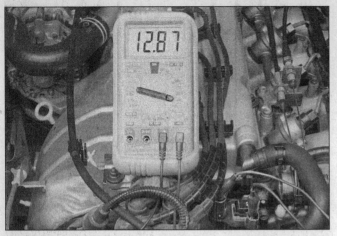

7.2 Checking the coil secondary resistance

7.9 To replace the ignition coil, disconnect the coil primary connection, detach the coil secondary lead and remove both coil bracket bolts (arrows)

and connect an ohmmeter between the negative primary terminal (terminal B) and the secondary cap-to-coil wire terminal **(see illustration)**. The resistance should be as listed in this Chapter's Specifications for the secondary resistance. If not, replace the coil.

Ignition coil primary winding-to-case resistance

3   Measure the resistance from the positive primary terminal (terminal A) to the case of the ignition coil.
4   If the indicated resistance is less than the resistance listed in this Chapter's Specifications, replace the ignition coil.
5   Reconnect the ignition coil wires.

Replacement

Refer to illustration 7.9

6   Detach the cable from the negative terminal of the battery.
7   Detach the connector from the primary terminals on the coil.
8   Disconnect the coil secondary lead.
9   Remove both bracket bolts and detach the coil **(see illustration)**.
10   Installation is the reverse of removal.

---

8   Distributor - removal and installation

Removal

Refer to illustrations 8.5a and 8.5b

1   Detach the cable from the negative terminal of the battery.
2   Disconnect the electrical connector from the distributor. Follow the wires as they exit the distributor to find the connector.
3   Note the raised "1" on the distributor cap. This marks the location for the number one cylinder spark plug wire terminal. **Note:** *Some distributor caps may have been replaced with aftermarket units that are not marked.*
4   Remove the distributor cap (see Chapter 1). Using a socket and breaker bar on the crankshaft pulley bolt, rotate the engine until

8.5a Apply an alignment mark on the perimeter of the distributor body inline with the rotor tip (arrows)

the rotor is pointing toward the number one spark plug terminal (see the TDC locating procedure in Chapter 2).
5   Make a mark on the edge of the distributor base directly below the rotor tip and inline with it **(see illustration)**. Also, mark the distributor base and the engine block to ensure that the distributor can be reinstalled correctly **(see illustration)**.
6   Remove the distributor hold-down bolt and clamp, then pull the distributor straight out to remove it. **Caution:** *DO NOT turn the engine while the distributor is removed, or the alignment marks will be useless.*

Installation

7   Insert the distributor into the engine in exactly the same relationship to the block that it was in when removed.
8   To mesh the helical gears on the camshaft and the distributor, it may be necessary to turn the rotor slightly. Make sure the distributor is seated completely and the alignment marks made previously are aligned. If not, remove the distributor and reposition it. **Note:** *If the crankshaft has been moved while the distributor is out, locate Top*

8.5b Mark the base of the distributor body and the engine block to clearly define the position of the distributor

Dead Center (TDC) for the number one piston (see Chapter 2) and position the distributor and rotor accordingly.
9   Place the hold-down clamp in position and loosely install the bolt.
10   Install the distributor cap and tighten the screws securely.
11   Plug in the electrical connector.
12   Reattach the spark plug wires to the plugs (if removed).
13   Connect the cable to the negative terminal of the battery.
14   Check and, if necessary, adjust the ignition timing (see Section 9) and tighten the distributor hold-down bolt securely.

---

9   Ignition timing - check and adjustment

Refer to illustrations 9.2, 9.3 and 9.6
Note: *The following procedure should apply to most vehicles covered by this manual. However, if the information specified on your vehicle's VECI label differs from this procedure, use the information on the VECI label.*
1   Connect a tachometer according to the

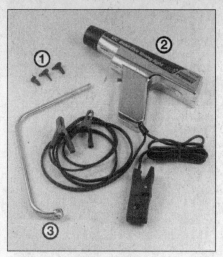

9.2 Tools needed to check and adjust the ignition timing

1 **Vacuum plugs** - *Vacuum hoses will, in most cases, have to be disconnected and plugged. Molded plugs in various shapes and sizes are available for this*

2 **Inductive pick-up timing light** - *Flashes a bright, concentrated beam of light when the number one spark plug fires. Connect the leads according to the instructions supplied with the light*

3 **Distributor wrench** - *On some models, the hold-down bolt for the distributor is difficult to reach and turn with conventional wrenches or sockets. A special wrench like this must be used*

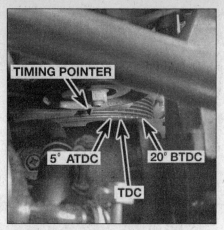

9.3 Location of the timing marks on the crankshaft pulley

9.6 Location of the idle speed adjustment screw

manufacturer's specifications.

2 With the ignition switch off, connect a timing light according to the manufacturer's specifications **(see illustration)**. Install the inductive pick-up onto the number one cylinder spark plug wire.

3 Locate the timing marks on the timing cover and the crankshaft pulley **(see illustration)**.

4 Start the engine and allow it to warm up to normal operating temperature.

5 Turn the engine off and disconnect the electrical connector from the throttle position sensor (see Chapter 6).

6 Start the engine and check the ignition timing. Verify that the engine idle is correct (600 to 700 rpm with an automatic transaxle or 550 to 650 rpm with a manual transaxle). If the idle speed is not correct, adjust the idle speed screw **(see illustration)**. Aim the timing light at the timing scale on the front engine cover. Refer to the timing Specifications listed at the beginning of this chapter. If necessary, loosen the distributor hold-down bolt (see Section 8) and slowly rotate the distributor until the timing marks align. Tighten the hold-down bolt and recheck the timing.

7 Turn the engine off, reconnect the TPS and observe the idle speed. Idle speed should increase approximately 50 to 100 rpm. If the idle speed is incorrect or cannot be adjusted, refer to the TPS adjustment in Chapter 6. In the event the idle speed cannot be adjusted properly, check the IAC-AAC assembly and verify that there are no defective components or intake leaks that will cause the idle to fluctuate abnormally (see

Chapter 4).

8 Turn the engine off and remove the tachometer and the timing light.

10 Power transistor - check and replacement

Caution: *The power transistor is a delicate and relatively expensive electrical component. Failure to follow the step-by-step procedures could result in damage to the module and/or other electronic devices, including the ECM. Additionally, all ECM controlled devices are protected by a Federally mandated emissions warranty. Check with the dealer concerning this warranty before attempting to diagnose and replace this unit yourself.*

Check

Refer to illustrations 10.3a and 10.3b

1 Detach the cable from the negative terminal of the battery.

2 Disconnect the electrical connector from the power transistor.

3 Using an ohmmeter, check for continuity across the designated terminals **(see illustrations)**.

5

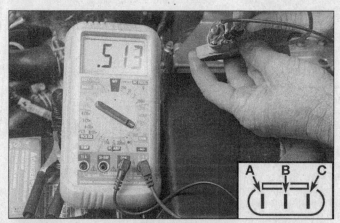

10.3a Connect the negative probe of the ohmmeter to terminal A and the positive probe of the ohmmeter to terminal B. The meter should indicate continuity

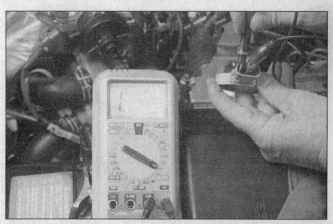

10.3b Connect the positive probe of the ohmmeter to terminal A and the negative probe of the ohmmeter to terminal B. The meter should indicate NO continuity (infinity)

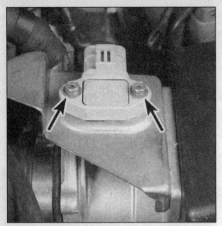

10.7 Remove the power transistor mounting bolts (arrows) and lift the unit from the bracket

12.3 Once the engine has been started, confirm that the charging voltage increases to approximately 14 to 15 volts

13.2 Remove the electrical connections (arrows) from the backside of the alternator

4   If the test results are incorrect, replace the power transistor with a new part.

Replacement

Refer to illustration 10.7

5   Detach the cable from the negative terminal of the battery.
6   Disconnect the electrical connector from the power transistor.
7   Remove the bolts that retain the power transistor to the bracket assembly **(see illustration)**.
8   Installation is the reverse of removal.

11 Charging system - general information and precautions

The charging system includes the alternator, an internal voltage regulator, a charge indicator or warning light, the battery and the wiring between all the components. The charging system supplies electrical power for the ignition system, the lights, the radio, etc. The alternator is driven by a drivebelt at the front of the engine. Refer to the wiring diagrams at the end of Chapter 12 for additional information.

The purpose of the voltage regulator is to limit the alternator's voltage to a preset value. This prevents power surges, circuit overloads, etc., during peak voltage output.

The charging system doesn't ordinarily require periodic maintenance. However, the drivebelt, battery and wires and connections should be inspected at the intervals outlined in Chapter 1.

Be very careful when making electrical circuit connections to a vehicle equipped with an alternator and note the following:

a) *When reconnecting wires to the alternator from the battery, be sure to note the polarity.*
b) *Before using arc welding equipment to repair any part of the vehicle, disconnect the wires from the alternator and the battery terminals.*

c) *Never start the engine with a battery charger connected.*
d) *Always disconnect both battery cables before using a battery charger (negative cable first, positive cable last).*

12 Charging system - check

Refer to illustration 12.3

1   If a malfunction occurs in the charging circuit, do not immediately assume that the alternator is causing the problem. First check the following items:

a) *The battery cables where they connect to the battery. Make sure the connections are clean and tight.*
b) *The battery electrolyte specific gravity. If it is low, charge the battery.*
c) *Check the external alternator wiring and connections.*
d) *Check the fuses and fusible links (see Chapter 12).*
e) *Check the drivebelt condition and tension (see Chapter 1).*
f) *Check the alternator mounting bolts for tightness.*
g) *Run the engine and check the alternator for abnormal noise.*

2   Using a voltmeter, check the battery voltage with the engine off. It should be approximately 12-volts.
3   Start the engine and check the battery voltage again. It should now be approximately 14 to 15-volts **(see illustration)**.
4   If the indicated voltage reading is less or more than the specified charging voltage, the problem may be within the alternator.
5   Due to the special equipment necessary to test or service the alternator, it is recommended that if a fault is suspected the vehicle be taken to a shop with the proper equipment. But if the home mechanic feels confident in the use of an ohmmeter, and in some cases a soldering iron, the component check and replacement procedures for the most

common alternator type are included in Section 14.
6   Some models are equipped with an ammeter on the instrument panel that indicates charge or discharge - current passing in or out of the battery. With all electrical equipment switched ON, and the engine idling, the gauge needle may show a discharge condition. At fast idle or normal driving speeds the needle should stay on the charge side of the gauge, with the charged state of the battery determining just how far over (the lower the battery state of charge, the farther the needle should swing toward the charge side).
7   Some models are equipped with a voltmeter on the instrument panel that indicates battery voltage with the key ON (engine not running), and alternator output when the engine is running.
8   The charge light on the instrument panel illuminates with the key ON and engine not running, and should go out when the engine runs.
9   If the gauge does not show a charge when it should or the alternator light (if equipped) remains on, there is a fault in the system. Before inspecting the brushes or replacing the alternator, the battery condition, alternator belt tension and electrical cable connections should be checked.

13 Alternator - removal and installation

Refer to illustration 13.2

1   Detach the cable from the negative terminal of the battery.
2   Disconnect the electrical connectors from the alternator **(see illustration)**.
3   Loosen the alternator bolts and detach the drivebelt.
4   Remove the adjustment and pivot bolts and separate the alternator from the engine.
5   Installation is the reverse of removal.
6   After the alternator is installed, adjust the drivebelt tension (see Chapter 1).

14.2 Paint a mark across the front cover, stator and rear cover to aid in reassembly

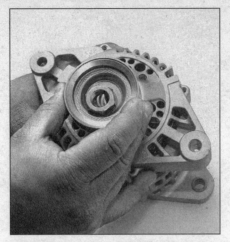

14.3 Remove the pulley nut and remove the pulley

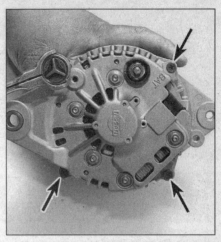

14.4a Remove the four bolts from the rear cover (arrows)

14 Alternator components - check and replacement

Disassembly

Refer to illustrations 14.2, 14.3, 14.4a, 14.4b, 14.4c, 14.5a, 14.5b, 14.6a, 14.6b, 14.6c and 14.7

1   Remove the alternator from the vehicle

(Section 13).
2   Scribe or paint marks on the front and rear end frame housings of the alternator to facilitate reassembly **(see illustration)**.
3   Remove the nut retaining the pulley to the rotor shaft and remove the pulley **(see illustration)**.
4   Remove the four through-bolts holding

the front and rear covers together, then separate the rear cover assembly from the front cover. Remove the rotor **(see illustrations)**.
5   Remove the nuts retaining the stator, then separate the stator and diode assembly from the end frame **(see illustrations)**.
6   Remove the screws attaching the regulator and the brush holder to the diode

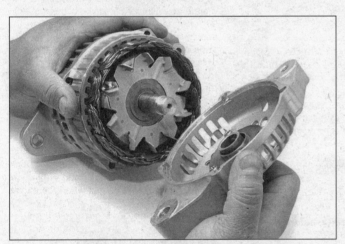

14.4b Separate the front cover from the alternator body

14.4c Remove the rotor from the alternator assembly

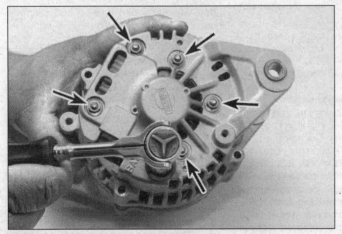

14.5a Remove the mounting nuts (arrows)

14.5b Separate the rear cover from the stator and diode assembly

5

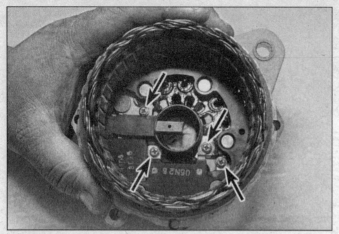

14.6a Remove the diode assembly mounting screws (arrows)

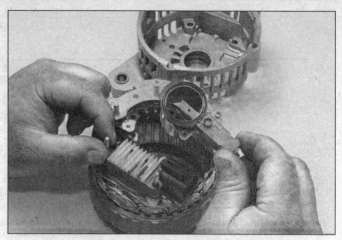

14.6b Remove the brush assembly holder from the diode assembly

14.6c Remove the voltage regulator

14.7 Remove the four set screws and lift the diode assembly from the stator

14.8a Continuity should exist between the rotor slip rings

14.8b No continuity should exist between the slip ring(s) and rotor shaft

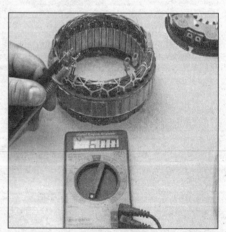

14.9 Check for continuity on each stator lead. There should be no breaks in the windings, therefore continuity should exist between each terminal

assembly and remove the brush holder and regulator **(see illustrations)**.

7   Remove the diode assembly from the stator **(see illustration)**. On some types of alternators it will be necessary to use a solder gun and heat sink to melt the solder joints that connect the stator leads to the diode assembly.

Component checks

Refer to illustrations 14.8a, 14.8b and 14.9

8   Check the rotor for an open between the two slip rings **(see illustration)**. There should be continuity between the slip rings. Check for grounds between each slip ring and the rotor shaft **(see illustration)**. There should be

no continuity (infinite resistance) between the rotor shaft and either slip ring. If the rotor fails either test, or if the slip rings are excessively worn, the rotor is defective.

9   Check for opens between the center terminal and each end terminal of the stator

14.13 Install a paper clip into the backside of the alternator to hold the brushes in place during installation - after installation, simply pull the paper clip out

windings **(see illustration)**. If either reading is high (infinite resistance), the stator is defective. Check for a grounded stator winding between each stator terminal and the frame. If there's continuity between any stator winding and the frame, the stator is defective.

10   Because the diode assembly varies between models and load requirements, have the diode assembly checked at a dealer service department or qualified automobile electric repair facility.

11   Measure the length of the brushes and replace them if they are at or near the minimum brush length found in this Chapter's Specifications. **Note:** *On some models the brush leads are soldered onto the voltage regulator assembly.*

Reassembly

Refer to illustration 14.13

12   Install the components in the reverse order of removal, noting the following:

13   Before installing the brush holder, push the brushes into the holder and slip a straightened paper clip or other suitable pin through the hole in the brush holder to hold the brushes in a retracted position. After the front and rear end frames have been bolted together, remove the paper clip **(see illustration)**.

15   Starting system - general information and precautions

1   The function of the starting system is to crank the engine to start it. The system is composed of the starter motor, starter solenoid, battery, ignition switch, clutch start switch (manual transaxle models), neutral start/back-up light switch (automatic transaxle models), diode box and connecting wires.

2   Turning the ignition key to the Start position actuates the starter relay through the starter control circuit. The starter solenoid then connects the battery to the starter. The

battery supplies the electrical energy to the starter motor, which does the actual work of cranking the engine.

3   All vehicles are equipped with a starter/solenoid assembly that is mounted to the transmission bellhousing.

4   All vehicles are equipped with a clutch start switch or a neutral start switch in the starter control circuit, which prevents operation of the starter unless the shift lever is in Neutral or Park (automatic) or the clutch is depressed (manual).

5   Never operate the starter motor for more than 15 seconds at a time without pausing to allow it to cool for at least two minutes. Excessive cranking can cause overheating, which can seriously damage the starter.

16   Starter motor and circuit - in-vehicle check

Note: *Before diagnosing starter problems, make sure the battery is fully charged.*

General check

1   If the starter motor doesn't turn at all when the switch is operated, make sure the shift lever is in Neutral or Park or the clutch is fully depressed.

2   Make sure the battery is charged and that all cables at the battery and starter solenoid terminals are secure.

3   If the starter motor spins but the engine doesn't turn over, the drive assembly in the starter motor is slipping and the starter motor must be replaced (see Section 17).

4   If, when the switch is actuated, the starter motor doesn't operate at all but the starter solenoid operates (clicks), the problem lies with either the battery, the starter solenoid contacts or the starter motor connections.

5   If the starter solenoid doesn't click when the ignition switch is actuated, either the starter solenoid circuit is open or the solenoid itself is defective. Check the starter solenoid circuit (see the wiring diagrams at the end of Chapter 12) or replace the solenoid (see Section 18).

6   To check the starter solenoid circuit, remove the push-on connector from the solenoid wire. Make sure that the connection is clean and secure and the relay bracket is grounded. If the connections are good, check the operation of the solenoid with a jumper wire. To do this, place the transmission in Park or Neutral and apply the parking brake. Remove the push-on connector from the solenoid. Connect a jumper wire between the battery positive terminal and the exposed terminal on the solenoid. If the starter motor now operates, the starter solenoid is okay. The problem is in the ignition switch, Neutral start switch or in the starting circuit wiring (look for open or loose connections).

7   If the starter motor still doesn't operate, replace the starter solenoid (see Section 18).

8   If the starter motor cranks the engine at

an abnormally slow speed, first make sure the battery is fully charged and all terminal connections are clean and tight. Also check the connections at the starter solenoid and battery ground. Eyelet terminals should not be easily rotated by hand. Also check for a short to ground. If the engine is partially seized, or has the wrong viscosity oil in it, it will crank slowly.

Starter cranking circuit test

Note: *To determine the location of excessive resistance in the starter circuit, perform the following simple series of tests.*

9   Disconnect the ignition coil wire from the distributor cap and ground it on the engine.

10   Connect a remote control starter switch from the battery terminal of the starter solenoid to the S terminal of the solenoid.

11   Connect a voltmeter positive lead to the starter motor terminal of the starter solenoid, then connect the negative lead to ground.

12   Actuate the ignition switch and take the voltmeter readings as soon as a steady figure is indicated. Do not allow the starter motor to turn for more than 15 seconds at a time. A reading of 9-volts or more, with the starter motor turning at normal cranking speed, is normal. If the reading is 9-volts or more but the cranking speed is slow, the motor is faulty. If the reading is less than 9-volts and the cranking speed is slow, the solenoid contacts are probably burned.

17   Starter motor - removal and installation

Refer to illustrations 17.3 and 17.4

1   Detach the cable from the negative terminal of the battery.

2   Raise the vehicle and support it securely on jackstands.

3   Disconnect the battery cable and the solenoid connector from the starter motor **(see illustration)**.

17.3 Disconnect the solenoid electrical connector from the harness and the cable from the starter terminal connection (arrows)

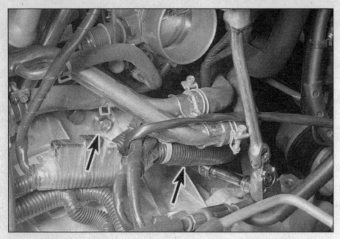

17.4 Remove the starter mounting bolts (arrows) from the bellhousing

18.1 Remove the terminal nut from the solenoid

4   Remove the starter motor mounting bolts and detach the starter from the engine **(see illustration)**. Depending upon the year and engine type, the starter/solenoid assembly may be mounted above the transaxle or below the intake manifold.

5   If necessary, turn the wheels to one side to provide removal access.

6   Installation is the reverse of removal.

18 Starter solenoid - replacement

Refer to illustrations 18.1, 18.2a and 18.2b

1   Remove the nut and disconnect the wire from the solenoid electrical terminal **(see illustration)**.

2   Remove the two bolts from the solenoid and separate the solenoid from the starter assembly **(see illustrations)**.

3   Installation is the reverse of removal.

18.2a Remove the mounting bolts (arrows) that retain the solenoid to the gear case

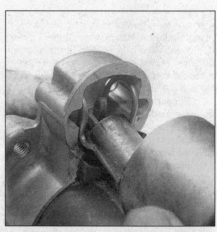

18.2b Remove the solenoid from the gear case - note the position of the plunger, lever and the return spring

Chapter 6
Emissions and engine control systems

Contents

1   General information

Refer to illustrations 1.1, 1.7a and 1.7b

To prevent pollution of the atmosphere from incompletely burned and evaporating gases, and to maintain good driveability and fuel economy, a number of emission control systems are incorporated **(see illustration)**. They include the:

EGR with backpressure transducer (BPT)
Evaporative Emission Control (EVAP) system
Multi Port Fuel Injection (MPFI) system
Positive Crankcase Ventilation (PCV) system
Exhaust Gas Recirculation (EGR) system
Air Assisted Fuel Injection (Chapter 4)
Catalytic converter

All of these systems are linked, directly or indirectly, to the emission control system.

The Sections in this Chapter include general descriptions, checking procedures within the scope of the home mechanic and component replacement procedures (when possible) for each of the systems listed above.

Before assuming that an emissions control system is malfunctioning, check the fuel and ignition systems carefully. The diagnosis of some emission control devices requires specialized tools, equipment and training. If checking and servicing become too difficult or if a procedure is beyond your ability, consult a dealer service department or other qualified repair facility. Remember, the most frequent cause of emissions problems is simply a loose or broken vacuum hose or wire, so always check the hose and wiring connections first.

This doesn't mean, however, that emission control systems are particularly difficult to maintain and repair. You can quickly and easily perform many checks and do most of the regular maintenance at home with common tune-up and hand tools. **Note:** *Because of a Federally mandated extended warranty which covers the emission control system components, check with your dealer about*

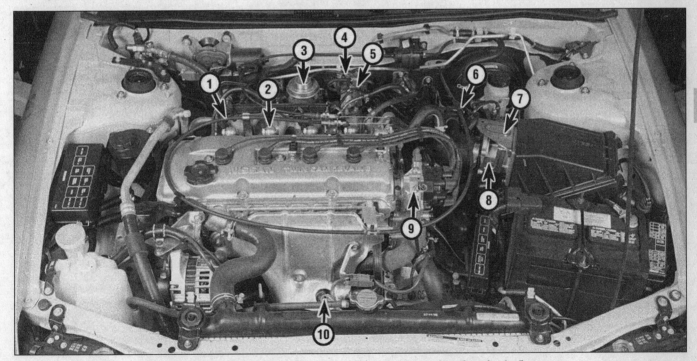

1.1 Emission and engine control system component locations (typical)

1	Idle Air Control/Auxiliary Air Control (IAC/AAC) valve (located below the intake manifold)
2	Fuel injector
3	EGR valve
4	Backpressure Transducer (BPT) valve
5	EGR and canister purge solenoid

6	Charcoal canister (below brake master cylinder)
7	Power transistor
8	Mass Airflow (MAF) sensor
9	Camshaft sensor (located in distributor)
10	Oxygen sensor

6

1.7a The Vehicle Emission Control Information (VECI) label is located under the hood and contains tune-up and emissions control information on your vehicle

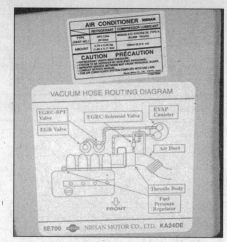

1.7b Vacuum hose routing diagram (1996 model shown)

warranty coverage before working on any emissions-related systems. Once the warranty has expired, you may wish to perform some of the component checks and/or replacement procedures in this Chapter to save money.

Pay close attention to any special precautions outlined in this Chapter. It should be noted that the illustrations of the various systems may not exactly match the system installed on the vehicle you're working on because of changes made by the manufacturer during production or from year to year.

A Vehicle Emissions Control Information (VECI) label is located in the engine compartment **(see illustrations)**. This label contains important emissions specifications and adjustment information, as well as a vacuum hose schematic with emissions components identified. When servicing the engine or emissions systems, the VECI label in your particular vehicle should always be checked for up-to-date information.

2   On Board Diagnosis (OBD) system and trouble codes

General information

1   The Multi Port Fuel Injection (MPFI) system controls the fuel injection system, the spark advance system, the self diagnosis system, the cooling fans, etc. by means of the Engine Control Module (ECM).
2   The ECM receives signals from various sensors which monitor changing engine operations such as intake air volume, intake air temperature, coolant temperature, engine RPM, acceleration/deceleration, exhaust temperature, etc. These signals are utilized by the ECM to determine the correct injection duration and ignition timing.
3   The Sections in this Chapter include general descriptions and checking procedures, within the scope of the home mechanic and component replacement procedures (when possible). Before assuming the fuel and ignition systems are malfunctioning, check the emission control system thoroughly. The emission system and the fuel system are closely interrelated, but can be checked separately. The diagnosis of some of the fuel and emission control devices

requires specialized tools, equipment and training. If checking and servicing become too difficult or if a procedure is beyond your ability, consult a dealer service department or other qualified repair facility. Remember, the most frequent cause of fuel and emissions problems is simply a loose or broken vacuum hose or wire, so always check the hose and wiring connections first. **Note:** *Because of federally mandated extended warranty which covers the emission control system components (and any other components which have a primary purpose other than emission control but have significant effects on emissions), check with your dealer about warranty coverage before working on any emission related systems. Once the warranty has expired, you may wish to perform some of the component checks and/or replacement procedures in this Chapter to save you money.*

Precautions

4   Always disconnect the power by either turning off the ignition switch or disconnecting the battery terminals before disconnecting engine control systems wiring connectors.
5   When installing a battery, be particularly careful to avoid reversing the positive and negative cables.
6   Do not subject engine control systems or emission related components to severe impact during removal or installation.
7   Do not be careless during troubleshooting. Even slight terminal contact can invalidate a testing procedure and even damage one of the numerous transistor circuits.
8   Never attempt to work on the ECM or open the ECM cover. The ECM is protected by a government mandated extended warranty that will be nullified if you tamper with it.
9   If you are inspecting electronic control system components during rainy weather, make sure water does not enter any part. When washing the engine compartment, do not spray these parts or their connectors with water.

Self diagnosis
Refer to illustration 2.12
10   The self diagnosis mode is useful to diagnose malfunctions in major sensors and actuators of the Electronic Fuel Injection system. There are two different modes available

for diagnosing driveability problems.
11   Mode I is the initial test mode of the vehicle with the ignition key ON and the engine not running. The test lamp on the ECM (computer) will light and remain lit without pulsing. Also, the CHECK ENGINE light on the dash will remain ON. This indicates to the home mechanic that the ECM is receiving power and the bulb is not blown or defective. **Note:** *The CHECK ENGINE light will remain on in the event of a malfunction or a stored trouble code. However, there are a few codes that will not illuminate the CHECK ENGINE light but the trouble code is stored in the ECM memory. It is a good idea to check for trouble codes if you experience driveability problems.*
12   Mode II is used to access the self diagnosis system. To access the correct mode for self diagnosis, turn the ignition switch to the ON (engine not running) position. Turn the diagnostic mode selector on the ECM fully clockwise **(see illustration)**. Wait two seconds and turn the mode selector fully counterclockwise. Wait until the inspection lamp flashes (at least two seconds).
13   Carefully observe the lights on the ECM. If everything in the self diagnosis system is functioning properly, the computer will flash a code 55. This code will be represented by five long flashes on the red LED followed by five quick flashes. If the computer has actual trouble codes stored, carefully observe the flashes and record the exact number onto paper. For example, code 43 (throttle position sensor or TPS circuit) is indicated by four long flashes on the red LED followed by three quick flashes. Refer to the trouble code chart below for the exact codes and component failures. **Note:** *1995 through 1997 models are equipped with the OBD-II self diagnosis system. However, translation codes are available for the home mechanic. The code extraction process is the same and many of the codes are the same. Consult the code chart for the exact code number and component failure. For example, code 1005, EGRC solenoid, will be 10 long flashes followed by five short flashes. Observe the flashing lights carefully*

and write down the digits as they are flashed to avoid confusion. Multiple codes will be separated by a distinct long pause between digits.

14   If the ignition key is turned OFF during the code extraction process and possibly turned back on, the diagnosis will automatically return to Mode I. Restart the procedure to extract the codes. **Note:** *Switching the diagnostic modes is not possible if the engine is running.*

15   The accompanying list of diagnostic trouble codes is a compilation of all of the trouble codes that may be encountered. Not all codes pertain to all models and on later models not all codes will illuminate the Check Engine light when set. The codes listed under the "Check Engine Light Flash Code" column are codes that may be displayed by the Check Engine light.

Clearing codes

16   After the tests have been performed and the repairs completed, erase the memory. With the ignition key ON (engine not running), turn the diagnostic mode selector on the ECM fully clockwise and then fully counterclockwise. This will erase any memory ECM has stored concerning a particular component. **Note:** *Be sure to turn the mode selector fully counterclockwise for normal vehicle operation if the codes were not erased from the memory.*

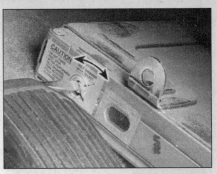

2.12 To start the self diagnosis system inspection, with the ignition key ON (engine not running), turn the mode selector clockwise and hold it there for two seconds then completely counterclockwise and observe the flashing light on the computer

Trouble code chart - 1993 and 1994 models

Trouble codes	Circuit or system	Probable cause
Code 11* (1 long flash, 1 short flash)	Camshaft position sensor/circuit	Check the camshaft position sensor or circuit (see Section 4).
Code 12 (1 long flash, 2 short flashes)	Mass airflow sensor/circuit	The mass airflow sensor source or ground circuit(s) may be shorted or open. Check the mass airflow sensor or circuit (see Section 4).
Code 13 (1 long flash, 3 short flashes)	Coolant temperature sensor	The coolant sensor or circuit may be shorted or open. Check the coolant temperature sensor and circuit (see Section 4).
Code 14 (1 long flash, 4 short flashes)	Vehicle speed sensor.	The vehicle speed sensor signal circuit is open (see Section 4).
Code 21* (2 long flashes, 1 short flash)	Ignition signal	The ignition signal in the primary circuit is not entered during engine cranking or running (see a dealer service department).
Code 31 (3 long flashes, 1 short flash)	ECM control unit	The ECM input signal is beyond "normal" range (see a dealer service department).
Code 32 (3 long flashes, 2 short flashes)	EGR function	The EGR control valve does not operate (see Section 5).
Code 33 (3 long flashes, 3 short flashes)	Oxygen sensor	The oxygen sensor circuit is open (see Section 4).
Code 34* (3 long flashes, 4 short flashes)	Knock sensor	The knock sensor circuit is open or shorted (see Section 4).
Code 35 (3 long flashes, 5 short flashes)	Exhaust gas temperature sensor	The exhaust gas temperature sensor circuit is open or shorted (see Section 4).
Code 43 (4 long flashes, 3 short flashes)	Throttle position sensor (TPS)	The TPS circuit is open or shorted (see Section 4).
Code 45 (4 long flashes, 5 short flashes)	Injector leak	The injector(s) have fuel leaks (see Chapter 4).
Code 54* (5 long flashes, 4 short flashes)	Signal circuit from A/T to ECM faulty	The ECM signal from the automatic transaxle is malfunctioning (see a dealer service department).
Code 55* (5 long flashes, 5 short flashes)	EFI system	Normal operation.

*These codes do not illuminate the CHECK ENGINE light on the dash, but it is possible to access the OBD system and read the codes on the flashing light on the computer

Trouble code chart - 1995 through 1997 models

Trouble codes	Circuit or system	Probable cause
Code 0505 (5 long flashes, 5 short flashes)	No self diagnostic codes present	
Code 0101 (1 long flash, 1 short flash)	Camshaft position sensor	Check the camshaft sensor or the circuit (see Section 4).
Code 0102 (1 long flash, 2 short flashes)	Mass airflow sensor/circuit	The mass airflow sensor source or ground circuit(s) may be shorted or open. Check the mass airflow sensor or circuit (see Section 4).
Code 0103 (1 long flash, 3 short flashes)	Coolant temperature sensor	The coolant sensor or circuit may be shorted or open. Check the coolant temperature sensor and circuit (see Section 4).
Code 0104 (1 long flash, 4 short flashes)	Vehicle speed sensor.	The vehicle speed sensor signal circuit is open (see Section 4).
Code 0114 (1 long flash, 14 short flashes)	Fuel system rich	Check the injectors, fuel pressure, fuel pressure regulator, oxygen sensor and/or the MAF sensor (see Section 4).

6

Trouble code chart - 1995 and later models (continued)

Trouble codes	Circuit or system	Probable cause
Code 0115 (1 long flash, 15 short flashes)	Fuel system lean	Check the injectors, fuel pressure, fuel pressure regulator, oxygen sensor and/or the MAF sensor (see Section 4).
Code 0201 (2 long flashes, 1 short flash)	Ignition signal	The ignition signal in the primary circuit is not entered during engine cranking or running (see a dealer service department).
Code 0205 (2 long flashes, 5 short flashes)	IACV/AAC valve	Check the IACV/AAC valve and circuit (see Chapter 4).
Code 0301 (3 long flashes, 1 short flash)	ECM control unit	The ECM input signal is beyond "normal" range (see a dealer service department).
Code 0302 (3 long flashes, 2 short flashes)	EGR Function	The EGR control valve does not operate (see Section 5).
Code 0306 (3 long flashes, 6 short flashes)	EGR BPT valve	Check the EGR BPT valve (see Section 5).
Code 0307 (3 long flashes, 3 short flashes)	Closed loop operation	ECM does not enter closed loop operation. Check front oxygen sensor and/or heater.
Code 0304 (3 long flashes, 4 short flashes)	Knock sensor	The knock sensor circuit is open or shorted (see Section 4).
Code 0305 (3 long flashes, 5 short flashes)	Exhaust gas temperature sensor	The exhaust gas temperature sensor circuit is open or shorted (see Section 4).
Code 0401 (4 long flashes, 1 short flash)	Intake air temperature sensor	IAT circuit is open or shorted (see Section 4).
Code 0403 (4 long flashes, 3 short flashes)	Throttle position sensor (TPS)	The TPS circuit is open or shorted (see Section 4).
Code 0503 (5 long flashes, 3 short flashes)	Front O2 sensor	Defective O2 sensor, shorted O2 sensor circuit, defective injectors or incorrect fuel pressure.
Code 0605 (6 long flashes, 5 short flashes)	Cylinder 4 misfire	Check number 4 spark plug, ignition wire, compression or injector.
Code 0606 (6 long flashes, 6 short flashes)	Cylinder 3 misfire	Check number 3 spark plug, ignition wire, compression or injector.
Code 0607 (6 long flashes, 7 short flashes)	Cylinder 2 misfire	Check number 2 spark plug, ignition wire, compression or injector.
Code 0608 (6 long flashes, 8 short flashes)	Cylinder 1 misfire	Check number 1 spark plug, ignition wire, compression or injector.
Code 0701 (7 long flashes, 1 short flash)	Multiple cylinder misfire	Check the spark plugs, ignition wires, injectors and compression readings for each cylinder.
Code 0702 (7 long flashes, 2 short flashes)	Catalytic converter	Possible three way catalytic converter defective. Check for injector problems, damaged exhaust tube or intake manifold leak.
Code 0707 (7 long flashes, 7 short flashes)	Rear O2 sensor	Check the rear oxygen sensor and circuit (see Section 4).
Code 0802 (8 long flashes, 2 short flashes)	Crankshaft sensor	Check the crankshaft sensor and circuit (see Section 4).
Code 0804 (8 long flashes, 4 short flashes)	A/T diagnosis comm line	Possible dead battery. Harness connectors between the ECM and automatic transaxle damaged.
Code 0901 (9 long flashes, 1 short flash)	Front O2 sensor heater	Check the front O2 sensor heater and circuit (see Section 4).
Code 0902 (9 long flashes, 2 short flashes)	Rear O2 sensor heater	Check the rear O2 sensor heater and circuit (see Section 4).
Code 0905 (9 long flashes, 5 short flashes)	Crankshaft position sensor	Possible damaged flywheel. Check the crank sensor and/or circuit (see Section 4).
Code 0908 (9 long flashes, 8 short flashes)	Coolant temperature sensor	Check the ECT sensor and circuit (see Section 4).
Code 1003 (10 long flashes, 3 short flashes)	Clutch Start switch	Check the Clutch Start switch circuit (see Chapter 7).
Code 1005 (10 long flashes, 5 short flashes)	EGR solenoid valve	The EGR valve and EVAP canister purge control solenoid valve circuit is open or shorted (see Sections 5 and 6).
Code 1101 (11 long flashes, 1 short flash)	Neutral/Start back-up light switch	Check the Neutral/Start back-up light switch and circuit (see Chapter 7).
Code 1102 (11 long flashes, 2 short flashes)	A/T vehicle speed sensor	Check the vehicle speed sensor and circuit for the automatic transaxle (see Chapter 7).
Code 1103 (11 long flashes, 3 short flashes)	A/T first signal	Have the vehicle diagnosed by a dealer service department.
Code 1104 (11 long flashes, 4 short flashes)	A/T second signal	Have the vehicle diagnosed by a dealer service department.
Code 1105 (11 long flashes, 5 short flashes)	A/T third signal	Have the vehicle diagnosed by a dealer service department.

Trouble codes	Circuit or system	Probable cause
Code 1106 (11 long flashes, 6 short flashes)	A/T fourth signal	Have the vehicle diagnosed by a dealer service department.
Code 1108 (11 long flashes, 8 short flashes)	Shift solenoid A	Have the vehicle diagnosed by a dealer service department.
Code 1201 (12 long flashes, 1 short flash)	Shift solenoid B	Have the vehicle diagnosed by a dealer service department.
Code 1203 (12 long flashes, 3 short flashes)	Overrun clutch solenoid	Have the vehicle diagnosed by a dealer service department.
Code 1204 (12 long flashes, 4 short flashes)	TCC solenoid	Have the vehicle diagnosed by a dealer service department.
Code 1205 (12 long flashes, 5 short flashes)	Line pressure solenoid	Have the vehicle diagnosed by a dealer service department.
Code 1206 (12 long flashes, 6 short flashes)	TPS for A/T	Have the vehicle diagnosed by a dealer service department.
Code 1207 (12 long flashes, 7 short flashes)	Speed signal A/T	Have the vehicle diagnosed by a dealer service department.
Code 1208 (12 long flashes, 8 short flashes)	Fluid temperature sensor	Have the vehicle diagnosed by a dealer service department.
Code 1308 (13 long flashes, 8 short flashes)	Cooling fan	Have the vehicle diagnosed by a dealer service department.

OBD-II trouble codes - 1998 and later models

Scan tool trouble code	Code identification
P0100	Mass air flow sensor
P0105	Absolute pressure sensor
P0110	Intake air temperature sensor
P0115	Engine coolant temperature sensor (ESTC)
P0120	Throttle position sensor
P0125	Engine coolant temperature sensor (ECT)
P0130	Front heated oxygen sensor circuit
P0131	Front heated oxygen sensor lean shift monitoring
P0132	Front heated oxygen sensor rich shift monitoring
P0133	Front heated oxygen sensor response monitoring
P0134	Front heated oxygen sensor high voltage
P0135	Front heated oxygen sensor heater
P0137	Rear heated oxygen sensor minimum voltage monitoring
P0138	Rear heated oxygen sensor maximum voltage monitoring
P0139	Rear heated oxygen sensor response monitoring
P0140	Rear heated oxygen sensor high voltage
P0141	Rear heated oxygen sensor heater
P0171	Fuel injection system function lean side
P0172	Fuel injection system function rich side
P0180	Fuel tank temperature sensor
P0304	Multiple cylinder misfire
P0325	Knock sensor
P0335	Crankshaft position sensor
P0340	Camshaft position sensor
P0400	EGR function
P0402	EGRC-BPT valve function
P0420	Three way catalyst function
P0440	Evaporative emission control system small leak, negative pressure
P0443	Evaporative emission control system canister purge volume control solenoid valve
P0446	Evaporative emission control system vent control valve circuit
P0450	Evaporative emission control system pressure sensor
P0460	Fuel level sensor slosh

6

OBD-II trouble codes - 1998 and later models (continued)

Scan tool trouble code	Code identification
P0461	Fuel level sensor
P0464	Fuel level sensor circuit
P0500	Vehicle speed sensor
P0505	Idle air control valve-auxiliary air control valve
P0510	Closed throttle position switch
P0600	A/T control
P0605	Engine control module
P1105	Manifold absolute pressure/barometric pressure switch solenoid valve
P1148	Closed loop control
P1320	Ignition signal
P1336	Crankshaft position sensor
P1400	EGRC solenoid valve
P1401	EGR temperature sensor
P1402	EGR function open
P1440	Evaporative emission control system small leak, positive pressure
P1441	Evaporative emission control system very small leak, positive pressure
P1444	Evaporative emission canister purge volume control solenoid valve
P1446	Evaporative emission canister vent control valve closed
P1447	Evaporative emission control system purge control monitoring
P1448	Vent control valve
P1464	Fuel level sensor circuit
P1490	Vacuum cut valve bypass valve
P1491	Vacuum cut valve bypass valve
P1605	A/T diagnosis communication line
P1760	Park/Neutral position switch

3   Engine Control Module (ECM) - removal and installation

Refer to illustration 3.4

1   The Engine Control Module (ECM) is located behind the center console, to the right of the accelerator pedal.

2   Disconnect the negative battery cable from the battery. **Warning:** *The models covered by this manual have airbags. Always disconnect the negative battery cable, then the positive cable and wait 10 minutes before working in the vicinity of the impact sensors, steering column or instrument panel to avoid the possibility of accidental deployment of the airbag, which could cause personal injury (see Chapter 12). The airbag circuits are easily identified by yellow insulation covering the entire wiring harness or just prior to the wire harness connectors. Do not use electrical test equipment on any of these wires or tamper with them in any way.*

3   Remove the side trim panels to expose the ECM on both sides of the center console (see Section 2).

4   Disconnect the harness connector from the ECM **(see illustration)**.

5   Remove the retaining nuts from the ECM bracket assembly.

6   Working inside the passenger's compartment, carefully slide the ECM out the passenger's side **(see illustration)**. **Note:** *Avoid any static electricity damage to the computer by using gloves and a special anti-static pad to store the ECM on once it is removed.*

4   Information sensors

Coolant temperature sensor
General description

1   The coolant temperature sensor is a thermistor (a resistor which varies the value of its resistance in accordance with temperature changes). The change in the resistance values will directly affect the voltage signal from the coolant sensor to the ECM. As the sensor temperature DECREASES, the resistance values will INCREASE. As the sensor temperature INCREASES, the resistance values will DECREASE. A failure in the coolant sensor circuit should set a Code 13. This code indicates a failure in the coolant temperature circuit, so in most cases the appropriate solution to the problem will be either repair of a wire or replacement of the sensor.

3.4 Remove the bolt and unplug the harness connector from the ECM

4.2 Check the resistance of the coolant temperature sensor (arrow) with the engine completely cold and then with the engine at operating temperature; resistance should decrease as temperature increases

4.3 Working on the harness side, check the reference voltage from the ECM to the coolant temperature sensor with the ignition key ON and the engine not running; it should be approximately 5.0 volts

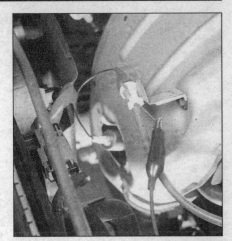

4.8a Backprobe the oxygen sensor and observe the millivolt signal the sensor produces as it goes from cold to fully warm (single wire O2 sensor shown)

Check

Refer to illustrations 4.2 and 4.3

2   Check the resistance value of the coolant temperature sensor while it is completely cold (50 to 65-degrees F = 2,300 to 3,000 ohms). Next, start the engine and warm it up until it reaches operating temperature **(see illustration)**. The resistance should be lower (176 to 180-degrees F = 300 to 330 ohms). **Note:** *Access to the coolant temperature sensor makes it difficult to position test probes on the terminals. If necessary, remove the sensor and perform the tests in a pan of heated water to simulate the conditions.*

3   If the resistance values on the sensor are correct, check the reference voltage to the sensor from the ECM **(see illustration)**. It should be approximately 5.0 volts. The reference voltage wire on 1993 and 1994 models is blue/orange. On 1995 and later models it is black/orange.

Replacement

Warning: *Wait until the engine is completely cool before removing the sensor.*

4   Before installing the new sensor, wrap the threads with Teflon sealing tape to prevent leakage and thread corrosion.

5   Disconnect the electrical connector and remove the sensor. **Caution:** *Handle the coolant sensor with care. Damage to this sensor will affect the operation of the entire fuel injection system.* Install the sensor and tighten it securely. Check the level of the coolant and add some, if necessary (see Chapter 1).

Oxygen sensor

General information

6   The oxygen sensor, located in the exhaust manifold, monitors the oxygen content of the exhaust gas stream. The oxygen content in the exhaust reacts with the oxygen

sensor to produce a voltage output which varies from 0.1-volt (high oxygen, lean mixture) to 0.9-volts (low oxygen, rich mixture). The ECM constantly monitors this variable voltage output to determine the ratio of oxygen to fuel in the mixture. The ECM alters the air/fuel mixture ratio by controlling the pulse width (open time) of the fuel injectors. A mixture ratio of 14.7 parts air to 1 part fuel is the ideal mixture ratio for minimizing exhaust emissions, thus allowing the catalytic converter to operate at maximum efficiency. It is this ratio of 14.7 to 1 which the ECM and the oxygen sensor attempt to maintain at all times. **Note 1:** *1993 and 1994 models are equipped with a single wire oxygen sensor and 1995 and later models are equipped with a heated oxygen sensor (three wire) type. These later models are equipped with a precatalytic and post catalytic converter O2 sensor. Manufacturer terminology specifies them as the front heated O2 sensor and the rear heated O2 sensor.* **Note 2:** *1995 and later models can be equipped with the single wire or the heated oxygen sensor because of the transitional stage of OBD-I and OBD-II.*

7   The oxygen sensor produces no voltage when it is below its normal operating temperature of about 600-degrees F. During this initial period before warm-up, the ECM operates in open loop mode.

Check

Refer to illustrations 4.8a, 4.8b, 4.11a, 4.11b, 4.12a and 4.12b

8   Locate the oxygen sensor electrical connector and backprobe the oxygen sensor connector with a straight-pin **(see illustration)**. Connect the positive probe of a voltmeter onto the pin and the negative probe to ground. On heated oxygen sensor models, backprobe the yellow wire **(see illustration)** and observe the sensor voltage signal. **Note:** *The signal voltage on 1995 and later rear*

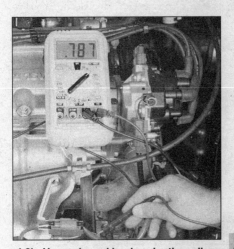

4.8b Use a pin and backprobe the yellow wire to check the SIGNAL voltage on a heated oxygen sensor

6

heated O2 sensor will not display the wide range of change in the voltage values. The O2 sensor detects very small amounts of O2 because the emissions have been catalyzed at this point. If the rear heated O2 sensor displays the same voltage values as the front heated O2 sensor, then either the catalytic converter is defective or the O2 sensor is malfunctioning.

9   Monitor the voltage signal as the engine goes from cold to warm.

10 The oxygen sensor will produce a steady voltage signal at first (open loop) of approximately 0.1 to 0.2 volts with the engine cold. After a period of approximately two minutes, the engine will reach operating temperature and the oxygen sensor will start to fluctuate between 0.1 to 0.9 volts (closed loop). If the oxygen sensor fails to reach the closed loop mode or there is a very long period of time until it does switch into closed loop mode, replace the oxygen sensor with a new part.

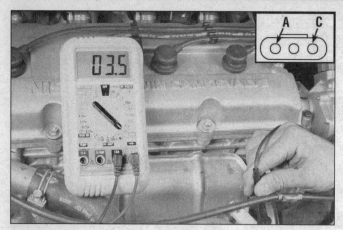

4.11a Connect the probes of the ohmmeter to terminals A and C and check the resistance of the oxygen sensor heater

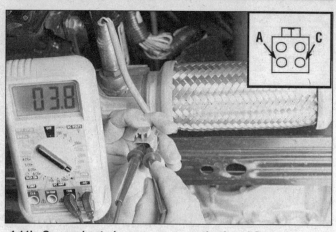

4.11b On rear heated oxygen sensors, the A and C terminals are located on the bottom of the sensor connector

11   Also inspect the oxygen sensor heater. Disconnect the oxygen sensor electrical connector and connect an ohmmeter between terminals A and C **(see illustrations)**. It should measure approximately 2.3 to 4.3 ohms.

12   Check for proper supply voltage to the heater. Measure the voltage on the harness side of the oxygen sensor electrical connector, terminal C **(see illustrations)**. There should be battery voltage with the ignition key ON (engine not running). If there is no voltage, check the circuit between the ignition switch, fuse box and the sensor (see Chapter 12).

13   The proper operation of the oxygen sensor depends on four conditions:

a) *Electrical - The low voltages generated by the sensor depend upon good, clean connections which should be checked whenever a malfunction of the sensor is suspected or indicated.*

b) *Outside air supply - The sensor is designed to allow air circulation to the internal portion of the sensor. Whenever the sensor is removed and installed or replaced, make sure the air passages are not restricted.*

c) *Proper operating temperature - The ECM will not react to the sensor signal until the sensor reaches approximately 600-degrees F. This factor must be taken into consideration when evaluating the performance of the sensor.*

d) *Unleaded fuel - The use of unleaded fuel is essential for proper operation of the sensor. Make sure the fuel you are using is of this type.*

14   In addition to observing the above conditions, special care must be taken whenever the sensor is serviced.

a) *The oxygen sensor has a permanently attached pigtail and electrical connector which should not be removed from the sensor. Damage or removal of the pigtail or electrical connector can adversely affect operation of the sensor.*

b) *Grease, dirt and other contaminants should be kept away from the electrical connector and the louvered end of the sensor.*

c) *Do not use cleaning solvents of any kind on the oxygen sensor.*

d) *Do not drop or roughly handle the sensor.*

e) *The silicone boot must be installed in the correct position to prevent the boot from being melted and to allow the sensor to operate properly.*

15   If the oxygen sensor fails any of these tests, replace it with a new part.

Replacement

Refer to illustration 4.19

Note: *Because it is installed in the exhaust manifold or pipe, which contracts when cool, the oxygen sensor may be very difficult to loosen when the engine is cold. Rather than risk damage to the sensor (assuming you are planning to reuse it in another manifold or pipe), start and run the engine for a minute or two, then shut it off. Be careful not to burn yourself during the following procedure.*

16   Disconnect the cable from the negative terminal of the battery.

17   If necessary, raise the vehicle and place it securely on jackstands.

18   Carefully disconnect the electrical connector from the sensor.

19   Carefully unscrew the sensor from the exhaust manifold **(see illustration)**. **Caution:** *Excessive force may damage the threads.*

4.12a On front heated oxygen sensors, check the oxygen sensor heater voltage on terminal C on the harness side of the connector

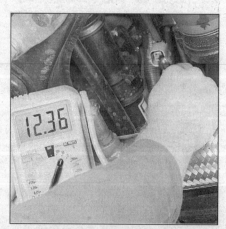

4.12b Checking the rear heated oxygen sensor voltage supply

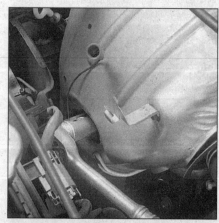

4.19 Remove the oxygen sensor using a special slotted socket, if available

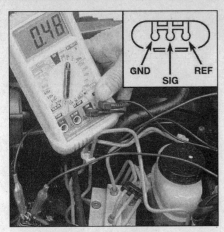

4.25 Backprobe the center terminal signal wire (+) and the GND (-) to check for signal voltage from the TPS and confirm that the signal voltage is approximately 0.5 volts

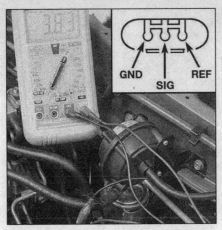

4.26 Open the throttle to wide open and observe that the signal voltage increases to approximately 4.0 volts

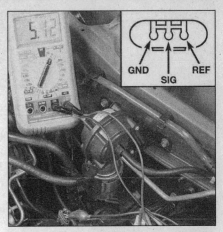

4.27 Check the reference voltage to the TPS with a voltmeter. Probe REF (+) and GND (-) and make sure the reference voltage is approximately 5.0 volts

20   Anti-seize compound must be used on the threads of the sensor to facilitate future removal. The threads of new sensors will already be coated with this compound, but if an old sensor is removed and reinstalled, recoat the threads.
21   Install the sensor and tighten it securely.
22   Reconnect the electrical connector of the pigtail lead to the main engine wiring harness.
23   Lower the vehicle and reconnect the cable to the negative terminal of the battery.

Throttle Position Sensor (TPS)

General description

24   The Throttle Position Sensor (TPS) is located on the end of the throttle shaft on the throttle body. By monitoring the output voltage from the TPS, the ECM can determine fuel delivery based on throttle valve angle (driver demand). A broken or loose TPS can cause intermittent bursts of fuel from the injector and an unstable idle because the ECM thinks the throttle is moving. Any problems in the TPS or circuit will set a code 43 on 1993 and 1994 models and 0403 on 1995 and later models. **Note:** *Vehicles equipped with an automatic transaxle use a wide open and closed throttle position switch as part of the throttle angle detection system. The switch shuts on/off only at these particular points in the throttle angle range. The switch is wired directly to the automatic transaxle control module.*

Check

Refer to illustrations 4.25, 4.26 and 4.27
25   To check the TPS, turn the ignition switch to ON (engine not running) and connect the probes of the voltmeter into the ground wire and signal wire on the backside of the electrical connector. This test checks for the proper signal voltage from the TPS **(see illustration)**. **Note 1:** *Use the TPS harness connector mounted on the bracket for all of the tests. Be careful when backprobing the electrical connector. Do not damage the*

wiring harness or pull on any connectors to make clean contact. **Note 2:** *The ground wire is black (-) and the signal wire is white (+). Refer to the wiring diagrams at the end of Chapter 12 for additional information.* **Note 3:** *Automatics equipped with the wide open and closed throttle position switch can be checked using an ohmmeter. At closed throttle position, the switch should have NO continuity. At wide open throttle, the switch should have continuity.*
26   The sensor should read 0.50 to 1.0-volt at closed throttle. Have an assistant depress the accelerator pedal to simulate full throttle and the sensor should increase voltage to 4.0 to 5.0-volts **(see illustration)**. If the TPS voltage readings are incorrect, replace it with a new unit.
27   Also, check the TPS reference voltage. With the ignition key ON (engine not running), connect the positive (+) probe of the voltmeter **(see illustration)** onto the voltage reference wire (red/yellow wire). There should be approximately 5.0 volts sent from the ECM to the TPS.

Adjustment

28   With the TPS installed, loosen the two mounting screws. Install the voltmeter and check signal voltage at idle (ignition key ON - engine not running).
29   To adjust the TPS:
   Manual transaxle vehicles - rotate the TPS body and observe the signal voltage. It should be 0.3 to 0.5 volts. Rotate the TPS until the correct voltage is attained and tighten the mounting screws
   Automatic transaxle vehicles - with the engine warmed up and idling, check the closed throttle position switch circuit using an ohmmeter. Probe terminal numbers 1 and 2 and observe that at 900 +/- 150 rpm the switch continuity goes OFF to ON. This is the base setting.

Replacement

Refer to illustration 4.30
30   Remove the two retaining screws **(see**

4.30 Remove the two screws (arrows) from the TPS and separate it from the throttle body

illustration) and separate the TPS from the throttle body.
31   Install the new TPS leaving the mounting screws loose. Connect a voltmeter to the signal and ground wires as in Step 25. Rotate the sensor until the voltmeter reading is 0.45 to 0.55-volt.
32   Tighten the screws securely.

Mass Airflow (MAF) sensor

Refer to illustrations 4.34, 4.36 and 4.41

General Information

33   The Mass Airflow Sensor (MAF) is located on the air intake duct. This sensor uses a hot wire sensing element to measure the amount of air entering the engine. The air passing over the hot wire causes it to cool. Consequently, this change in temperature can be converted into an analog voltage signal to the ECM which in turn calculates the required fuel injector pulse width.

Check

34   Check for power to the MAF sensor. Dis-

6

4.34 Disconnect the MAF electrical connector and check for battery voltage to the MAF sensor on the large orange wire

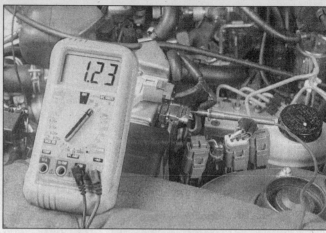

4.36 Backprobe the small orange wire and check the MAF signal voltage

connect the MAF sensor electrical connector and connect a voltmeter to the large orange wire (+) and ground (-) **(see illustration)**. There should be battery voltage.

35   Reconnect the electrical connector and backprobe the MAF sensor small orange wire with the voltmeter and check the voltage **(see illustration 4.36)**. The voltage should be less than 1.0 volt with the ignition switch ON (engine not running).

36   Start the engine and allow it to idle. The signal voltage from the MAF sensor should increase to approximately 0.85 to 1.35 volts on 1993 and 1994 models or 1.0 to 1.7 volts on 1995 and later models **(see illustration)**. If the MAF sensor voltage does not vary, replace the sensor.

37   Raise the engine rpm and observe that the signal voltage increases slightly but not exceeding 4.0 volts.

38   If the voltage readings are correct, check the wiring harness for open circuits or other damage.

Replacement

39   Disconnect the electrical connector from the MAF sensor.

40   Remove the air cleaner assembly (see Chapter 4).

41   Remove the four bolts **(see illustration)** and detach the MAF sensor from the air cleaner housing.

42   Installation is the reverse of removal.

Vehicle Speed Sensor (VSS)

General description

43   The Vehicle Speed Sensor (VSS) is located on the transaxle (see Chapter 7B). This sensor works in conjunction with a reed switch which is installed into the speedometer unit. The reed switch transforms vehicle speed into a pulsing voltage signal that is translated by the ECM and provided as information for other systems for fuel and transmission shift control. Any problems with the VSS will usually set a Code 14 on 1993 and 1994 models or 0104 on 1995 and later models.

Check

Refer to illustration 4.45

44   To check the vehicle speed sensor, unplug the electrical connector in the wiring harness near the sensor. Using an ohmmeter, check the resistance across the two terminals of the Vehicle Speed Sensor wiring harness connector. The resistance of the VSS should be approximately 250 ohms. If the resistance is incorrect, remove the VSS from the transaxle.

45   Place the VSS on a bench and check for a pulsing voltage. Backprobe the electrical connector using two pins and, while slowly spinning the VSS gear, confirm that the voltage signal pulses from 0 to 0.5 volts. Use the A/C scale on the voltmeter **(see illustration)**. If the VSS doesn't respond as described, replace the VSS.

Replacement

46   To replace the VSS, disconnect the electrical connector from the VSS and remove the retaining bolt and lift the VSS

from the transaxle.

47   Installation is the reverse of removal.

Intake air temperature (IAT) sensor (1995 and later models)

General description

48   The air temperature sensor is located inside the air intake duct. This sensor acts as a resistor which changes value according to the temperature of the air entering the engine. Low temperatures produce a high resistance value (for example, at 68-degrees F the resistance is 2.3 to 2.7 K-ohms) while high temperatures produce low resistance values (at 176-degrees F the resistance is 270 to 380 ohms). The ECM supplies approximately 5-volts (reference voltage) to the air temperature sensor. The voltage will change according to the temperature of the incoming air. The voltage will be high when the air temperature is cold and low when the air temperature is warm. Any problems with the air temperature sensor will usually set a code 0401 on OBD-II systems.

4.41 Remove the four bolts (arrows) and separate the MAF sensor from the air cleaner housing

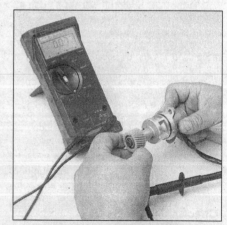

4.45 Using a voltmeter set on the AC scale, backprobe the VSS harness connector and check for voltage pulses that range from 0 to 0.5 volts as the gear is turned slowly

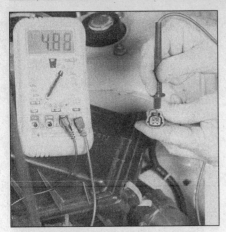

4.50 Disconnect the IAT electrical connector and check the IAT sensor reference voltage

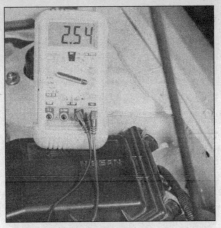

4.52 Working on the IAT sensor, check the resistance of the sensor when the engine temperature and ambient air temperature inside the intake system is cold and then when it is warm

4.57 Location of the EGR gas temperature sensor (arrow)

Check

Refer to illustration 4.50 and 4.52

49 To check the air temperature sensor, disconnect the two prong electrical connector and turn the ignition key ON, but do not start the engine.

50 Measure the voltage (reference voltage). The voltmeter should read approximately five-volts **(see illustration)**.

51 If the voltage signal is not correct, have the ECM diagnosed by a dealer service department or other repair shop.

52 Measure the resistance across the air temperature sensor terminals **(see illustration)**. The resistance should be HIGH when the air temperature is LOW. Next, start the engine and let it idle. Allow the engine to reach normal operating temperature. Turn the ignition OFF, disconnect the air temperature sensor and measure the resistance across the terminals. The resistance should be LOW when the air temperature is HIGH. If the sensor does not exhibit this change in resistance, replace it with a new part.

Power steering pressure switch

General description

53 The power steering pressure switch is attached to the power steering high pressure line. Turning the steering wheel increases power steering fluid pressure and engine load. The pressure switch will close before the load can cause an idle problem. A pressure switch that will not open or an open circuit from the ECM will cause timing to retard at idle, affecting idle quality. A pressure switch that will not close, or an open circuit, may cause the engine to die when the power steering system is used heavily.

Check

54 Disconnect the electrical connector to the pressure switch.

55 Connect an ohmmeter across the terminals of the pressure switch. Start the engine and check for continuity as the steering

wheel is turned. There should be no continuity when the steering wheel is not being turned.

56 If the switch fails this test, replace the switch.

EGR gas temperature sensor

General description

Refer to illustration 4.57

57 Some models are equipped with an EGR gas temperature sensor mounted near the EGR valve, installed into the EGR tube **(see illustration)**. This sensor detects the temperature of the exhaust as it moves through the EGR valve. The information is sent to the ECM and in turn the EGR on/off time is regulated precisely and more efficiently. Any malfunction with the EGR gas temperature sensor will set a code 35 on 1993 and 1994 models and 0305 on 1995 and later models.

Check

Refer to illustration 4.58

58 Disconnect the harness connector for the EGR gas temperature sensor **(see illustration)** and measure the resistance of the sensor. Resistance should decrease as temperature increases and at 212-degrees F should measure 80 to 98 K-ohms.

Removal and installation

59 Disconnect the harness connector for the EGR gas temperature sensor and using and open-end wrench, remove the sensor from the intake manifold.

60 Installation is the reverse of removal.

Knock sensor

Refer to illustration 4.61

General description

61 The knock sensor is located in the engine block under the intake manifold **(see illustration)**. The knock sensor detects abnormal vibration in the engine. The sensor

4.58 Disconnect the electrical connector from the EGR gas temperature sensor and check the resistance cold and then with the engine completely warmed up to operating temperature

produces an AC output voltage which increases with the severity of the knock. The signal is fed into the ECM and the timing is retarded up to 10 degrees to compensate for severe detonation. Any problems with the

4.61 The knock sensor is located near the oil filter

knock sensor will set a code 34 on 1993 and 1994 models and 0304 on 1995 and later models.

Check

62   Disconnect the electrical connector at the knock sensor. Using an ohmmeter set on the high scale, check for resistance between the terminals. It should be 500 to 620 K-ohms (use the 10M scale on ohmmeter).

63   Also check for a loose knock sensor. Tighten with an open end wrench if necessary.

Camshaft position sensor

Note: *A description and replacement procedure of the camshaft sensor can be found in Chapter 5, Section 5.*

Check

64   Disconnect the cable from the negative terminal of the battery, then remove the distributor (see Chapter 5).

65   Remove the rotor retaining screw and pull the rotor off the shaft (see Chapter 5).

66   Remove the sensor dust cover retaining screws and lift off the cover.

67   Inspect the camshaft signal plate for damage and dirt intrusion. Blow any accumulated dust out of the distributor and reinstall the cover and rotor.

68   Inspect the electrical connections for damage and corrosion and correct any defects.

69   With the distributor removed, reconnect the electrical connector to the camshaft sensor, turn the ignition key ON (engine not running) and check the camshaft sensor.

70   Backprobe the electrical connector and check for battery voltage on the orange wire (+). Refer to the wiring diagrams at the end of Chapter 12 for additional information. If battery voltage is not present, check the circuit from the camshaft sensor to the ECM main relay for an open circuit or a defective relay.

71   Check for the 180-degree crankshaft position signal. This will be indicated by a fluctuation on the voltmeter from zero to five volts, two times for every complete rotation of the distributor shaft, which rotates one time for every two rotations of the crankshaft. Backprobe the blue wire (+) and rotate the distributor shaft.

72   Next, check the one-degree timing signal. This will be indicated by a fluctuation on the voltmeter from zero to five volts, 360 times every complete rotation of the distributor. Backprobe the yellow wire (+) and rotate the crankshaft very slowly to detect these voltage changes.

73   If the camshaft sensor signal test results are incorrect, replace the distributor as a complete unit

Crankshaft position sensor (1995 and later models)

General information

Refer to illustrations 4.76 and 4.78

74   1995 and later models are equipped

4.76 Checking the crankshaft sensor resistance

4.78 Remove the crankshaft sensor mounting bolt (arrow) and lift the sensor from the transaxle bellhousing

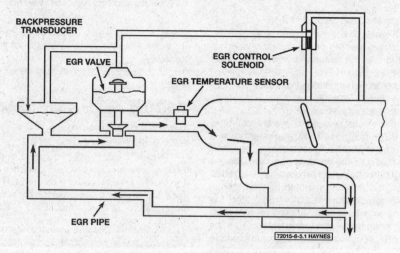

5.1 Schematic of the EGR system

with a crankshaft position sensor that is mounted on the transaxle bellhousing. It detects changes in crankshaft speed using a permanent magnet, core and coil. The changing gap causes the magnetic field near the sensor to change this in turn varies the voltage signal to the ECM. This sensor is used only as the on-board diagnostic device for detecting engine misfire.

75   Working on the harness side of the electrical connector, check for reference voltage to the sensor with the ignition key ON (engine not running). Voltage should be approximately 5.0 volts.

76   .Next, check the resistance of the crankshaft sensor. It should be 432 to 528 ohms **(see illustration)**.

Replacement

77   Disconnect the electrical connector from the crankshaft sensor.

78   Remove the crankshaft sensor bolt **(see illustration)**.

79   Remove the crankshaft sensor from the

engine compartment. Installation is the reverse of removal.

Park/Neutral Back-up light switch

Refer to Chapter 7 for diagnosis and replacement procedures.

---

5   Exhaust Gas Recirculation (EGR) system

General description

Refer to illustrations 5.1, 5.5, 5.6 and 5.12

1   The EGR system is used to lower NOx (oxides of nitrogen) emission levels caused by high combustion temperatures. The EGR recirculates a small amount of exhaust gases into the intake manifold **(see illustration)**. The additional mixture lowers the temperature of combustion thereby reducing the formation of NOx compounds.

2   The EGR systems are equipped with an

EGR and Canister Control Solenoid valve which receives ported and manifold vacuum. The manifold vacuum system utilizes a vacuum tap in the air intake system positioned after the throttle valve. The ported vacuum control system uses a vacuum tap in the throttle body which is exposed to an increasing percentage of manifold vacuum as the throttle valve is opened during acceleration.

3   A backpressure transducer (BPT) valve monitors the exhaust backpressure as the engine rpm increases or decreases to aid in controlling the EGR vacuum signal.

Check

4   Check all hoses for cracks, kinks, broken sections and proper connection. Inspect all system connections for damage, cracks and leaks.

5   To check the EGR system operation, bring the engine up to operating temperature and, with the transmission in Neutral (parking brake set and tires blocked to prevent movement), allow it to idle for 70 seconds. Open the throttle abruptly so the engine speed is between 2,000 and 3,000 rpm and then allow it to close. The EGR valve stem should move if the control system is working properly. The test should be repeated several times. Movement of the stem indicates the control system is functioning correctly (see illustration).

6   If the EGR valve stem does not move, check all of the hose connections to make sure they are not leaking or clogged. Disconnect the vacuum hose and apply ten inches of vacuum with a hand-held vacuum pump (see illustration). If the stem still does not move, replace the EGR valve with a new one. If the valve does open, measure the valve travel to make sure it is approximately 1/8-inch. Also, the engine should run roughly when the valve is open. If it doesn't, the passages are probably clogged.

7   Apply vacuum with the pump and then clamp the hose shut. The valve should stay open for 30 seconds or longer. If it does not, the diaphragm is leaking and the valve should be replaced with a new one.

8   If the EGR valve is not receiving port or ported vacuum, the throttle body unit must be removed to check and clean the slotted port in the throttle bore and the vacuum passages and orifices in the throttle body. Use solvent to remove deposits and check for flow with light air pressure.

9   If the engine idles roughly and it is suspected the EGR valve is not closing, remove the EGR valve and inspect the poppet and seat area for deposits.

10   If the deposits are more than a thin film of carbon, the valve should be cleaned. To clean the valve, apply solvent and allow it to penetrate and soften the deposits, making sure that none gets on the valve diaphragm, as it could be damaged.

11   Use a vacuum pump to hold the valve open and carefully scrape the deposits from the seat and poppet area with a tool. Inspect the poppet and stem for wear and replace

5.5 Be careful not to burn your hand when checking for EGR valve diaphragm movement

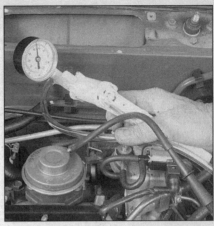

5.6 Use a hand-held vacuum pump to apply vacuum to the EGR valve

5.12 Connect the vacuum pump hose to the bottom line out of the BPT and apply vacuum. The BPT should hold vacuum

the valve with a new one if wear is found.

12   Locate the BPT valve and plug one of the ports with a finger (see illustration). Use a hand-held vacuum pump and apply vacuum to the valve. Check for any sign of leaks.

13   With the engine running, check for a vacuum signal from the intake manifold. This test will determine if manifold vacuum is reaching the various valves and components of the EGR system.

14   Also apply vacuum to the port (vacuum hose) that is routed to the EGR valve. The gauge should hold vacuum. If not, check for broken hoses or ruptured diaphragms in the EGR valve.

15   Check the EGR solenoid (if equipped). Using a hand-held vacuum pump, apply vacuum and then apply battery voltage. The solenoid should activate and allow vacuum to pass through the solenoid.

16   Also remove the vacuum hose(s) from the solenoid and check for vacuum to the EGR valve.

Component replacement

EGR valve

Refer to illustration 5.21, 5.25a, 5.25b and 5.25c

17   When buying a new EGR valve, make sure that you have the right EGR valve. Use

the stamped code located on the top of the EGR valve.

18   Detach the cable from the negative terminal of the battery.

19   Remove the air cleaner housing assembly (see Chapter 4).

20   Detach the vacuum line from the EGR valve.

21   Remove the EGR valve mounting bolts (see illustration).

5.21 Remove the EGR mounting nuts (arrows) and remove the EGR valve from the intake manifold

6

5.25a Remove the EGR pipe flange nut using a large
open-end wrench

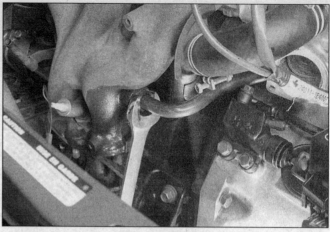

5.25b Disconnect the EGR pipe flange at the exhaust manifold

5.25c Use a flare-nut wrench to remove
the BPT line

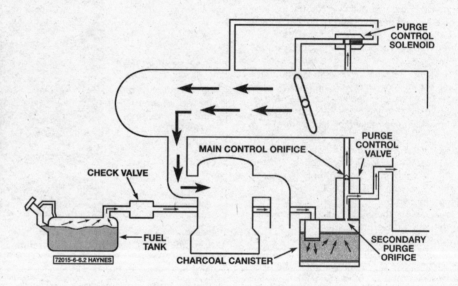

6.2 Schematic of the EVAP system

22   Remove the EGR valve and gasket from the manifold. Discard the gasket.

23   With a wire wheel, buff the exhaust deposits from the EGR valve mounting surface on the manifold and, if you plan to use the same valve, the mounting surface of the valve itself. Look for exhaust deposits in the valve outlet. Remove deposit build-up with a screwdriver. **Caution:** *Never wash the valve in solvents or degreaser - both agents will permanently damage the diaphragm. Sand blasting is also not recommended because it will affect the operation of the valve.*

24   If the EGR passage contains an excessive build-up of deposits, clean it out with a wire wheel. Make sure that all loose particles are completely removed to prevent them from clogging the EGR valve or from being ingested into the engine.

25   If there are large amounts of deposits within the EGR valve, remove the EGR pipe from the exhaust manifold **(see illustrations)** and clean out the deposits inside the tube. **Note:** *Remove the BPT valve and pipe and clean any deposits from the passages.*

26   Installation is the reverse of removal.

EGR vacuum control solenoid

27   Detach the cable from the negative terminal of the battery.

28   Unplug the electrical connector from the solenoid.

29   Clearly label and detach both vacuum hoses.

30   Remove the solenoid mounting screw and remove the solenoid.

31   Installation is the reverse of removal.

6   Evaporative Emissions Control System (EVAP)

Refer to illustrations 6.2, 6.10 and 6.14

General description

1   This system is designed to trap and store fuel vapors that evaporate from the fuel tank, throttle body and intake manifold.

2   The Evaporative Emission Control System (EVAP) consists of a charcoal-filled canister and the lines connecting the canister to the fuel tank, ported vacuum and intake manifold vacuum **(see illustration)**.

3   Fuel vapors are transferred from the fuel tank, throttle body and intake manifold to a canister where they are stored when the engine is not operating. When the engine is running, the fuel vapors are purged from the canister by a purge control solenoid that directs vacuum to the purge control valve (on top of charcoal canister) and consumed in the normal combustion process. Small amounts of fuel vapors flow into the intake manifold through the secondary purge orifice. When engine speed and vacuum increases, the ECM activates the purge control solenoid thereby routing fuel vapors into the intake manifold through the main and secondary purge orifices.

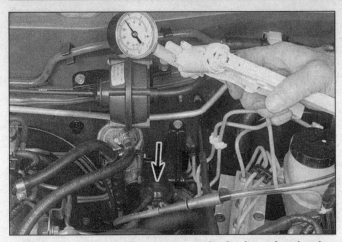

6.10 Apply vacuum to the purge control valve (arrow) and make sure that it holds vacuum for at least 20 seconds

6.14 Location of the charcoal canister mounting bolts (arrows) (1993 through 1997 models)

Check

4   Poor idle, stalling and poor driveability can be caused by an inoperative purge control solenoid, a damaged canister, split or cracked hoses or hoses connected to the wrong tubes.

5   Evidence of fuel loss or fuel odor can be caused by fuel leaking from fuel lines or a cracked or damaged canister, an inoperative bowl vent valve, an inoperative purge valve, disconnected, misrouted, kinked, deteriorated or damaged vapor or control hoses or an improperly seated air cleaner or air cleaner gasket.

6   Inspect each hose attached to the canister for kinks, leaks and breaks along its entire length. Repair or replace as necessary.

7   Inspect the canister. If it is cracked or damaged, replace it.

8   Look for fuel leaking from the bottom of the canister. If fuel is leaking, replace the canister and check the hoses and hose routing.

9   Apply a short length of hose to the lower tube of the purge control valve and attempt to blow through it. Little or no air should pass into the canister (a small amount of air will pass because the canister has a constant purge hole).

10   With a hand-held vacuum pump, apply vacuum to the purge control valve signal tube (upper tube) **(see illustration)**.

11   If the purge control valve does not hold vacuum for at least 20 seconds, the purge control valve is leaking and must be replaced.

12   If the diaphragm holds vacuum, apply battery voltage to the canister control solenoid and observe that vacuum (vapors) are allowed to pass through to the intake system. **Note:** *The purge control solenoid and the EGR control solenoid are the same component. Follow the testing procedure in Section 5 to test the purge control solenoid.*

Component replacement

13   Clearly label, then detach, all vacuum

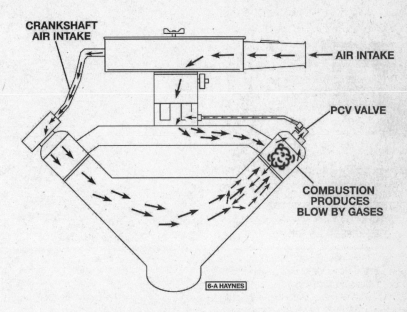

7.1 Gas flow in typical PCV system

lines from the canister.

14   On 1993 through 1997 models, working under the hood, remove the mounting bolts **(see illustration)** and pull the canister assembly out. On 1998 and later models, working under the drivers side rear fender, remove the mounting hardware and pull the canister assembly out.

15   Installation is the reverse of removal.

7 Positive Crankcase Ventilation (PCV) system

Refer to illustration 7.1

1   The Positive Crankcase Ventilation (PCV) system reduces hydrocarbon emissions by scavenging crankcase vapors. It does this by circulating fresh air from the air cleaner through the crankcase, where it mixes with blow-by gases and is then rerouted through a PCV valve to the intake manifold **(see illustration)**.

2   The main components of the PCV system are the PCV valve, a breather separator, a fresh air filtered inlet and the vacuum hoses connecting these two components with the engine.

3   To maintain idle quality, the PCV valve restricts the flow when the intake manifold vacuum is high. If abnormal operating conditions arise, the system is designed to allow excessive amounts of blow-by gases to flow back through the crankcase vent tube into the air cleaner to be consumed by normal combustion.

4   Checking and replacement of the PCV valve and filter is covered in Chapter 1.

8 Catalytic converter

General description

Refer to illustration 8.1

1   The catalytic converter **(see illustration)** is an emission control device added to the exhaust system to reduce pollutants from the exhaust gas stream. A single-bed converter design is used in combination with a three-way (reduction) catalyst. The catalytic coating on the three-way catalyst contains platinum and rhodium, which lowers the levels of oxides of nitrogen (NOx) as well as hydrocarbons (HC) and carbon monoxide (CO).

Check

2   The test equipment for a catalytic converter is expensive and highly sophisticated.

If you suspect that the converter on your vehicle is malfunctioning, take it to a dealer or authorized emissions inspection facility for diagnosis and repair.

3   Whenever the vehicle is raised for servicing of underbody components, check the converter for leaks, corrosion and other damage. If damage is discovered, the converter should be replaced.

Replacement

4   Because the converter part of the exhaust system, converter replacement requires removal of the exhaust pipe assembly (see Chapter 4). Take the vehicle, or the exhaust system, to a dealer or a muffler shop.

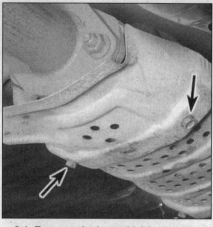

8.1 Remove the heat shield mounting nuts (arrows) to gain access to the catalytic converter for inspection

Chapter 7 Part A
Manual transaxle

Contents

Specifications

Torque specifications

	Ft-lbs
Engine-to-transaxle bolts	
Transaxle-to-engine block bolts	29 to 36
Transaxle-to-oil pan bolts	22 to 30

1 General information

The vehicles covered by this manual are equipped with either a 5-speed manual transaxle or a 4-speed automatic transaxle. Information on the manual transaxle is included in this Part of Chapter 7. Service procedures for the automatic transaxle are contained in Chapter 7, Part B.

The manual transaxle is a compact, two-piece, lightweight aluminum alloy housing containing both the transmission and differential assemblies.

Because of the complexity of the transaxle and the special tools needed to work on it, internal repair procedures for the manual transaxle are beyond the scope of this manual. The information in this Chapter is devoted to removal and installation procedures.

2 Shift linkage - removal and installation

Refer to illustrations 2.2, 2.3, 2.4 and 2.5
1   Raise the vehicle and support it securely on jackstands.
2   Remove the catalytic converter heat shield (see illustration).
3   Disconnect the support and control rod from the transaxle (see illustration).
4   Remove the nut and bolt that attach the control rod to the shift lever (see illustration). Disconnect the return spring, remove the nuts that attach the support rod to the shift lever socket and mass damper and remove the shift linkage assembly. Inspect the condition of all the bushings at

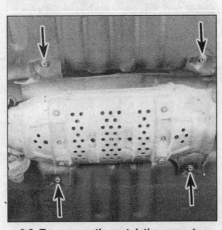

2.2 To remove the catalytic converter heat shield, remove these screws (arrows)

both ends of the support and control rods. If they're cracked or worn, replace them. Inspect the return spring rubber; replace it if

2.3 To disconnect the control rod from the transaxle, remove this nut and bolt (arrow); to disconnect the support rod, remove the two nuts (not visible in this photo) on top of the bracket to which the forward end of the support rod is attached

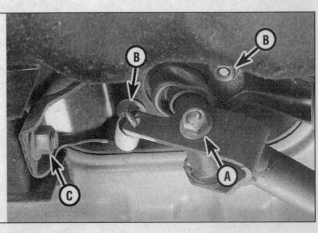

2.4 To disconnect the control rod from the shift lever, remove nut and bolt "A"; to disconnect the support rod from the shift lever socket, remove the two nuts indicated by arrows "B"; to disconnect the support rod from the mass damper, remove nut "C"

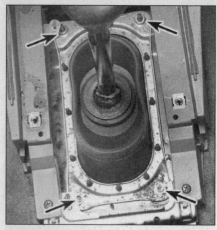

2.5 To replace the large shift lever dust boot, remove the two front nuts (upper arrows) and the two rear nuts from below (not visible in this photo)

3.4 Carefully pry out the driveaxle oil seal with a seal removal tool or a screwdriver; make sure you don't damage the seal bore or the new seal may leak

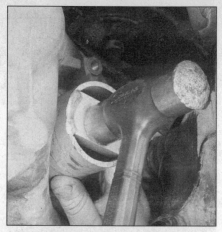

3.5 Use a seal installer, a large socket or a piece of pipe to install the new seal

it's cracked or torn. Also inspect the condition of the rubber insulator portion of the mass damper holder bracket. If the rubber is cracked or torn, replace the holder bracket.

5   Remove the console (see Chapter 11). Detach the shift lever dust boot from the floor **(see illustration)**, pull out the shift lever and slide off the dust boot, seat, insulator and shift lever socket. Inspect these parts for cracks and tears. Replace as necessary. The condition of the shift lever socket and the seat are especially important; if either of these parts is damaged, shifting will be difficult. Also inspect the condition of the large rubber shift lever boot. If it's damaged, replace it.

6   Installation is the reverse of removal. Make sure you lubricate all friction surfaces - the spherical bearing surface of the shift lever, the inside of the shift lever socket and the inside of the seat - with silicone grease. Also be sure to lubricate all support and control rod bushings and collars with silicone grease. Finally, make sure you tighten all fasteners securely.

3   Oil seal replacement

1   Oil leaks can occur as a result of worn seals or O-rings. Replacement of these seals or O-rings is relatively easy, since the repairs can usually be performed without removing the transaxle from the vehicle.

Driveaxle oil seals

Refer to illustrations 3.4 and 3.5

2   The driveaxle oil seals are located on the sides of the transaxle, where the inner ends of the driveaxles are splined into the differential side gears. If you suspect that a driveaxle oil seal is leaking, raise the vehicle and support it securely on jackstands. If the seal is leaking, you'll see lubricant on the side of the transaxle, below the seal.

3   Remove the driveaxle (left side) or

3.7 To remove the vehicle speed sensor, disconnect the electrical connector, remove the hold-down bolt (arrow) and remove the sensor assembly from the transaxle

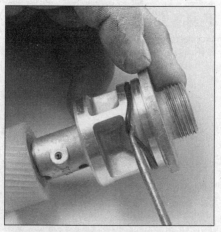

3.9 Remove the oil seal O-ring from the pinion gear assembly with a small hook tool or a small screwdriver; make sure you don't gouge the groove

driveaxle and extension shaft (right side) (see Chapter 8).

4   Using a screwdriver or seal removal tool, carefully pry the seal out of the transaxle bore **(see illustration)**.

5   Using a large section of pipe or a large deep socket as a drift, install the new oil seal. Drive it into the bore squarely and make sure it's fully seated **(see illustration)**. Lubricate the lip of the new seal with multi-purpose grease.

6   Install the driveaxle (or driveaxle and extension shaft). Be careful not to damage the lip(s) of the new seal(s). Check the transaxle lubricant level and add some, if necessary, to bring it up to the required level (see Chapter 1).

Vehicle speed sensor O-ring

Refer to illustrations 3.7 and 3.9

7   The vehicle speed sensor **(see illustration)** is located on the transaxle housing. Look for lubricant around the housing to

determine if the O-ring is leaking.

8   Disconnect the electrical connector, remove the hold-down bolt **(see illustration 3.7)** and remove the pinion assembly and vehicle speed sensor from the transaxle.

9   Using a scribe or a small screwdriver, remove the O-ring seal **(see illustration)**.

10   Install a new O-ring on the pinion gear housing.

11   Installation is the reverse of removal. Check the lubricant level and add some, if necessary, to bring it up to the proper level (see Chapter1).

Control rod seal

Refer to illustrations 3.14a, 3.14b, 3.14c, 3.15, 3.16 and 3.17

12   Raise the vehicle and place it securely on jackstands.

13   Disconnect the transaxle control rod from the yoke (see Section 2).

14   Remove the yoke retaining pin, the yoke and the dust boot **(see illustrations)**.

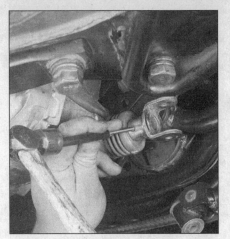

3.14a Drive out the yoke retaining pin with a hammer and punch . . .

3.14b . . . remove the yoke . . .

3.14c . . . and remove the dust boot

3.15 Pry out the control rod seal

3.16 Install the new control rod seal with a large deep socket

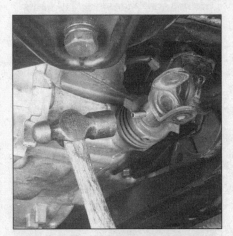

3.17 Install the dust boot and yoke and secure the yoke with a new roll pin

15 Remove the control rod seal **(see illustration)**.
16 Install a new seal **(see illustration)**.
17 Install the dust boot, yoke and the yoke retaining pin **(see illustration)**.
18 Install the control rod (see Section 2).

19 Remove the jackstands and lower the vehicle. Check the lubricant level and add some, if necessary, to bring it up to the proper level (see Chapter 1).

4 Transaxle mount - check and replacement

Refer to illustration 4.2
1 Raise the vehicle and place it securely on jackstands.
2 Insert a large screwdriver or prybar between the mount and the bracket and pry up **(see illustration)**.
3 The transaxle should not move excessively away from the mount. If it does, replace the mount. Also, if the rubber appears to be cracked, torn or otherwise deteriorated, replace the mount.
4 To replace a mount, support the transaxle with a jack, remove the nuts and bolts and remove the mount. It may be necessary to raise the transaxle slightly to provide enough clearance to remove the mount.
5 Installation is the reverse of removal.

4.2 To check a transaxle mount, place a large screwdriver or prybar between the mount and the bracket and try to lever the transaxle up and down; if it moves excessively, replace the mount

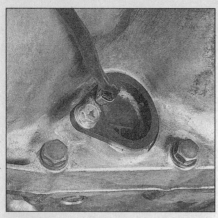

5.1 The back-up light and neutral position switch is located on the lower left side of the transaxle

5.3 To check the back-up light and neutral position switch, disconnect the electrical connector (arrow), probe terminals 2 and 4 with an ohmmeter and verify that there's continuity between these two terminals when the transaxle is in Reverse, but no continuity in any other gear; to check the neutral position switch, probe terminals 1 and 3 with an ohmmeter and verify that there's continuity between these two terminals when the transaxle is in Neutral, but no continuity in any other gear

5 Back-up light and neutral position switch - check and replacement

Check

Refer to illustrations 5.1 and 5.3

1   Raise the vehicle and place it securely on jackstands. Locate the back-up light switch **(see illustration)**.
2   Follow the wiring harness from the switch to the electrical connector. Unplug the connector.
3   Place the transaxle in Reverse and verify that there's continuity between connector terminals 2 and 4 **(see illustration)**. Put the transaxle in Neutral and verify that there's continuity between terminals 1 and 3. Verify that there's no continuity between terminals 2 and 4, and 1 and 3, in any other gear.
4   If continuity isn't as specified, replace the back-up light switch (see below).

Replacement

5   Drain the transaxle lubricant (see Chapter 1).
6   Disconnect the switch electrical connector.

7   Remove the hold-down bolt and pull the switch straight out of the transaxle.
8   Apply a light coat of clean oil to a new O-ring, install the new switch and O-ring, and tighten the hold-down bolt securely.
9   Plug in the electrical connector.
10   Check the switch as described above to ensure it's working properly.
11   Fill the transaxle with the specified lubricant (see Chapter 1).
12   Remove the jackstands and lower the vehicle.

6 Transaxle - removal and installation

Removal

Refer to illustrations 6.3, 6.7, 6.16, 6.17a, 6.17b, 6.18a, 6.18b and 6.20

1   Remove the battery and the battery tray (see Chapter 5).
2   Remove the air cleaner housing, mass airflow sensor and air intake duct (see Chapter 4).
3   Remove the clutch release cylinder from the transaxle (see Chapter 8). Remove the C-

clip **(see illustration)** and disconnect the clutch hydraulic line at the bracket on top of the transaxle.
4   Disconnect the electrical connector for the crankshaft position sensor (see Chapter 6) and the vehicle speed sensor (see Section 3). Disconnect all ground wires.
6   Remove the starter motor from the transaxle (see Chapter 5).
7   Remove the upper transaxle-to-engine bolts **(see illustration)**.
8   Loosen the wheel lug nuts. Raise the vehicle and place it securely on jackstands. Remove the wheels.
9   Disconnect the electrical connector for the back-up light and neutral position switch (see Section 5).
10   Remove the section of exhaust pipe under the engine (see Chapter 4).
11   Drain the transaxle fluid (see Chapter 1).
12   Disconnect the support rod and the shift control rod from the transaxle (see Section 2).
13   Remove the driveaxles (see Chapter 8).
14   Support the engine (see Chapter 2). This can be done from above with an engine hoist or from underneath by placing a jack (with a wood block as an insulator) under the engine oil pan. The engine must be supported at all times while the transaxle is out of the vehicle.

6.3 To detach the clutch hydraulic line from the transaxle, remove this C-clip and pull the line out of the bracket

6.7 Upper transaxle-to-engine bolts (arrows)

6.16 To remove the left transaxle mount, remove these four bolts (arrows)

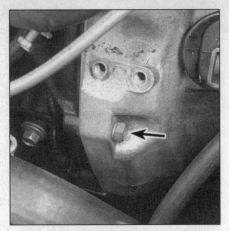

6.17a Front bellhousing-to-engine bolt (arrow)

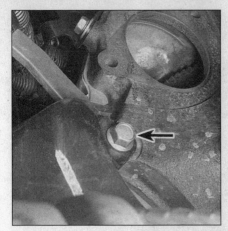

6.17b The rear engine-to-bellhousing bolt (arrow) is located right below the starter (this bolt is also the upper attaching bolt for the large bracket attached to the engine and transaxle)

6.18a Rear bracket-to-transaxle bolt (long bolt)

6.18b Remove the short bolt that attaches the rear bracket to the transaxle, use a long extension and socket inserted through the hole on the right side of the bracket (above the exhaust pipe)

6.20 With the transaxle supported on a transmission jack or floor jack, carefully pull the transaxle away from the engine until the input shaft clears the clutch hub

15 Support the transaxle with a jack (preferably a special jack made for this purpose). If you're using a floor jack, be sure to place a wood block between the lifting pad and the transaxle to protect the cast aluminum housing. Safety chains will help steady the transaxle on the jack.

16 Raise the engine and transaxle slightly and disconnect the left transaxle mount **(see illustration)** and the oil pan-to-bellhousing nuts and bolts (see Chapter 2).

17 Remove the front bellhousing-to-engine bolt and the rear engine-to-bellhousing bolt **(see illustrations)** and remove the bolts that attach the large bracket to the transaxle.

18 Remove the bolts which attach the engine/transaxle mounting bracket to the rear of the transaxle **(see illustrations)**.

19 Make a final inspection of the transaxle for any wires and hoses that have been overlooked.

20 Lower the left (driver's) end of the engine, then roll the transaxle and jack toward the side of the vehicle **(see illustration)**. Once the input shaft is clear of the splines in the clutch hub, lower the transaxle and remove it from under the vehicle. Try to keep the transaxle as level as possible.

21 While the transaxle is removed, be sure to inspect the clutch components (see Chapter 8). In most cases, new clutch components should be routinely installed whenever the transaxle is removed.

Installation

22 If removed, install the clutch components (see Chapter 8).

23 With the transaxle secured to the jack as on removal, raise it into position and then carefully engage the input shaft with the splines in the clutch hub. Do not use excessive force to install the transaxle - if the input shaft does not slide into place, readjust the angle of the transaxle so it is in the same plane as the engine. If the engine and transaxle are in the same plane, but the input shaft still won't engage the clutch hub, turn the input shaft slightly and the splines on the shaft will engage properly with the splines in the clutch hub.

24 Install the transaxle-to-engine bolts. Tighten all engine-to-transaxle bolts to the torque listed in this Chapter's Specifications.

25 Install the transaxle mount nuts and bolts. Tighten all nuts and bolts securely.

26 Remove all transaxle and engine supports.

27 Install the various items removed previously. Refer to Chapter 8 for driveaxle installation, Chapter 4 for exhaust pipe installation, and Chapter 5 for starter motor installation.

28 Make a final check that all electrical wiring has been connected and that the transaxle has been filled with the specified lubricant to the proper level (see Chapter 1). Lower the vehicle.

29 Connect the negative battery cable. Road test the vehicle to check for proper transaxle operation and check for leakage.

7 Transaxle overhaul - general information

1 Overhauling a manual transaxle is a difficult job for the do-it-yourselfer. It involves the disassembly and reassembly of many small

7A

parts. Numerous clearances must be precisely measured and, if necessary, changed with select fit spacers and snap-rings. As a result, if transaxle problems arise, it can be removed and installed by a competent do-it-yourselfer, but overhaul should be left to a transmission repair shop. Rebuilt transaxles may be available - check with your dealer parts department and auto parts stores. At any rate, the time and money involved in an overhaul is almost sure to exceed the cost of a rebuilt unit.

2   Nevertheless, it's not impossible for an inexperienced mechanic to rebuild a transaxle if the special tools are available and the job is done in a deliberate step-by-step manner so nothing is overlooked.

3   The tools necessary for an overhaul include internal and external snap-ring pliers, a bearing puller, a slide hammer, a set of pin punches, a dial indicator and possibly a hydraulic press. In addition, a large, sturdy workbench and a vise or transaxle stand will be required.

4   During disassembly of the transaxle, make careful notes of how each piece comes off, where it fits in relation to other pieces and what holds it in place. Your notes plus the manufacturer's shop manual, which contains exploded views, will make it easlier to get the transaxle back together.

5   Before taking the transaxle apart for repair, it will help if you have some idea what area of the transaxle is malfunctioning. Certain problems can be closely tied to specific areas in the transaxle, which can make component examination and replacement easier. Refer to the *Troubleshooting* section at the front of this manual for information regarding possible sources of trouble.

Chapter 7 Part B
Automatic transaxle

Contents

Specifications

Torque specifications

	Ft-lbs (unless otherwise indicated)
Driveplate-to-torque converter bolts	33 to 43
Transaxle-to-engine bolts	
Transaxle-to-engine block bolts	29 to 36
Starter motor bolts	22 to 27
Rear engine mount bracket-to-transaxle bolt	54 to 61
Oil pan-to-transaxle bolts	22 to 27

1 General information

All models covered by this manual are equipped with either a 5-speed manual transaxle or a 4-speed automatic transaxle. All information on the automatic transaxle is included in this Part of Chapter 7. Information for the manual transaxle can be found in Part A of this Chapter.

Because of the complexity of the automatic transaxle and the specialized equipment needed to service it, this Chapter contains only those procedures related to general diagnosis, routine maintenance, adjustment, and removal and installation.

If the transaxle requires major repair work, it should be taken to a dealer service department or an automotive or transmission repair shop. You can, however, save money by removing and installing the transaxle yourself, even if the repair work is done by a shop.

2 Diagnosis - general

Note: *Automatic transaxle malfunctions may be caused by five general conditions: poor engine performance, improper adjustments, hydraulic malfunctions, mechanical malfunctions or malfunctions in the computer or its signal network. Diagnosis of these problems should always begin with a check of the easily repaired items: fluid level and condition (see Chapter 1), shift linkage adjustment and throttle linkage adjustment. Next, perform a road test to determine if the problem has* been corrected or if more diagnosis is necessary. If the problem persists after the preliminary tests and corrections are completed, additional diagnosis should be done by a dealer service department or transmission repair shop. Refer to the Troubleshooting section at the front of this manual for information on symptoms of transaxle problems.

Preliminary checks

1 Drive the vehicle to warm the transaxle to normal operating temperature.
2 Check the fluid level as described in Chapter 1:
  a) If the fluid level is unusually low, add enough fluid to bring the level within the designated area of the dipstick, then check for external leaks (see below).
  b) If the fluid level is abnormally high, drain off the excess, then check the drained fluid for contamination by coolant. The presence of engine coolant in the automatic transmission fluid indicates that a failure has occurred in the internal radiator walls that separate the coolant from the transmission fluid (see Chapter 3).
  c) If the fluid is foaming, drain it and refill the transaxle, then check for coolant in the fluid, or a high fluid level.
3 Make sure the engine idle speed is correct (see Chapter 4). If the idle speed is incorrect, have it adjusted by a dealer service department before proceeding.
4 Inspect the shift cable (see Section 4). Make sure that it's properly adjusted and operates smoothly.

Fluid leak diagnosis

5 Most fluid leaks are easy to locate visually. Repair usually consists of replacing a seal or gasket. If a leak is difficult to find, the following procedure may help.
6 Identify the fluid. Make sure it's transmission fluid and not engine oil or brake fluid (automatic transmission fluid is a deep red color).
7 Try to pinpoint the source of the leak. Drive the vehicle several miles, then park it over a large sheet of cardboard. After a minute or two, you should be able to locate the leak by determining the source of the fluid dripping onto the cardboard.
8 Make a careful visual inspection of the suspected component and the area immediately around it. Pay particular attention to gasket mating surfaces. A mirror is often helpful for finding leaks in areas that are hard to see.
9 If the leak still cannot be found, clean the suspected area thoroughly with a degreaser or solvent, then dry it.
10 Drive the vehicle for several miles at normal operating temperature and varying speeds. After driving the vehicle, visually inspect the suspected component again.
11 Once the leak has been located, the cause must be determined before it can be properly repaired. If a gasket is replaced but the sealing flange is bent, the new gasket will not stop the leak. The bent flange must be straightened.
12 Before attempting to repair a leak, check to make sure that the following conditions are corrected or they may cause another leak. **Note:** *Some of the following*

conditions cannot be fixed without highly specialized tools and expertise. Such problems must be referred to a transmission shop or a dealer service department.

Gasket leaks

13 Check the pan periodically. Make sure the bolts are tight, no bolts are missing, the gasket is in good condition and the pan is flat (dents in the pan may indicate damage to the valve body inside).

14 If the pan gasket is leaking, the fluid level or the fluid pressure may be too high, the vent may be plugged, the pan bolts may be too tight, the pan sealing flange may be warped, the sealing surface of the transaxle housing may be damaged, the gasket may be damaged or the transaxle casting may be cracked or porous. If sealant instead of gasket material has been used to form a seal between the pan and the transaxle housing, it may be the wrong sealant.

Seal leaks

15 If a transaxle seal is leaking, the fluid level or pressure may be too high, the vent may be plugged, the seal bore may be damaged, the seal itself may be damaged or improperly installed, the surface of the shaft protruding through the seal may be damaged or a loose bearing may be causing excessive shaft movement.

16 Make sure the dipstick tube seal is in good condition and the tube is properly seated. Periodically check the area around the speedometer gear or sensor for leakage. If transmission fluid is evident, check the O-ring for damage.

Case leaks

17 If the case itself appears to be leaking, the casting is porous and will have to be repaired or replaced.

18 Make sure the oil cooler hose fittings are tight and in good condition.

Fluid comes out vent pipe or fill tube

19 If this condition occurs, the transaxle is overfilled, there is coolant in the fluid, the case is porous, the dipstick is incorrect, the vent is plugged or the drain-back holes are plugged.

Diagnostic trouble codes

20 The computer for the automatic transaxle has a self-diagnostic capability; it continually monitors important information sensor and output actuator circuits for malfunctions. When a monitored circuit is damaged, shorted or disconnected, a diagnostic trouble code is stored in the computer's memory. At a dealer service department, stored trouble codes are extracted from computer memory with a proprietary diagnostic instrument known as CONSULT. Codes can also be extracted with some generic scanners. However, neither the CONSULT nor scanners are generally available to the home mechanic, so the following procedure is provided to enable you to extract any stored codes by using the OD OFF indicator light on the instrument cluster (above the fuel gauge).

21 Start the engine and warm it up to its normal operating temperature.

22 Turn the ignition switch to the OFF position.

23 Turn the ignition switch to the ACC position.

24 Put the overdrive switch in the ON position.

25 Move the selector lever to the P position.

26 Turn the ignition switch to the ON position, but don't start the engine.

27 The OD OFF indicator light should come on for about two seconds.

a) *If the OD OFF indicator light doesn't come on, there's something wrong with the transaxle computer or with the indicator light circuit. We don't recommend testing the resistance or voltage of the computer terminals; drawing too much current or putting too much voltage through the terminals can destroy the computer. Have the OD OFF indicator light circuit tested and repaired by a dealer service department or other qualified repair shop.*

b) *If the OD OFF indicator light comes on, proceed to the next Step.*

28 On 1993 and 1994 models, turn the ignition key to ACC. On 1995 and later models, turn the ignition key to OFF.

29 Move the shift lever to the D position.

30 Put the overdrive switch in the OFF position.

31 Turn the ignition switch to the ON position, but do not start the engine.

32 Wait for at least two seconds after the ignition switch is turned to ON, then move the shift lever to the 2 position.

33 Put the overdrive switch in the ON position.

34 Move the shift lever to the 1 position.

35 Put the overdrive switch in the OFF position.

36 Depress the accelerator pedal all the way to the floor, then release it.

37 The computer is now in its output mode. It will begin displaying any stored trouble code(s) by flashing the OD OFF indicator light in a sequence of long and short flashes specific to each stored code.

38 If all monitored circuits are operating correctly, you will see one long flash, followed by one long pause, followed by ten short flashes of equal duration with short pauses of equal duration between them.

39 Each code is represented by the position of a long flash in the ten-consecutive-flash sequence:

a) *If the first flash of the ten-flash sequence is long, the revolution sensor circuit is shorted or disconnected.*

b) *If the second of the ten flashes is long, the vehicle speed sensor circuit is shorted or disconnected.*

c) *If the third flash is long, the throttle position sensor sensor circuit is shorted or disconnected.*

d) *If the fourth flash is long, the shift solenoid valve A circuit is shorted or disconnected.*

e) *If the fifth flash is long, the shift solenoid valve B circuit is shorted or disconnected.*

f) *If the sixth flash is long, the overrun clutch solenoid circuit is shorted or disconnected.*

g) *If the seventh flash is long, the torque converter clutch solenoid valve circuit is shorted or disconnected.*

h) *If the eighth flash is long, the fluid temperature sensor is disconnected or the computer power source circuit is damaged.*

i) *If the ninth flash is long, the engine speed signal circuit is shorted or disconnected.*

j) *If the tenth flash is long, the line pressure solenoid valve circuit is shorted or disconnected.*

k) *If the OD OFF indicator light flashes on and off, alternating back and forth between long flashes and long pauses, all of equal length, either the battery power is low, the battery has been disconnected for a long time, or the battery has been connected incorrectly.*

l) *If the OD OFF indicator light doesn't come on, either the neutral start/back-up light switch, or the overdrive switch or the throttle position switch circuit is disconnected, or the computer is damaged.*

40 Except for the vehicle speed sensor and throttle position sensor, both of which are covered in Chapter 6, repairing the rest of the malfunctions listed above is beyond the scope of the home mechanic. If one of more of these codes is displayed, have the transaxle repaired by a dealer service department or automatic transmission shop.

Erasing a diagnostic trouble code

41 If you have the transaxle repaired at a dealer or transmission shop, they will erase the trouble code when they're done making the repair. If you make a repair yourself, here's how to erase the code when you're done:

42 If the ignition switch remains on after a repair, turn it off once, wait three seconds, then turn it on again.

43 Perform the self-diagnostic procedure described above.

44 Change the diagnostic test mode from Mode II to Mode I by turning the mode selector on the ECM (see Chapter 6).

---

3 Shift cable - check, adjustment and replacement

Check

1 Move the shift lever from the "P" position to the "1" position. You should be able to feel the detents in each range. If you can't feel the detents, or if the pointer indicating

3.4 Loosen the shift cable locknut

3.5 Pull down on the shift cable and, holding the cable and manual shaft lever in this position, tighten the locknut (it's not really necessary to remove the locknut; it's removed in this photo so you can see the slotted adjustment hole in the end of the cable)

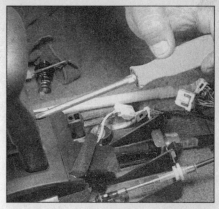

3.9a Remove this Phillips screw in the front of the shift lever handle . . .

3.9b . . . unplug the electrical connector for the overdrive switch leads (the white wires) . . .

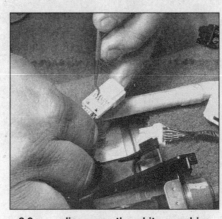

3.9c . . . disengage the white overdrive switch leads from the connector with an awl . . .

3.9d . . . and carefully pull the shift lever handle straight up to remove it

the ranges is incorrectly aligned, adjust the shift cable.

Adjustment

Refer to illustrations 3.4 and 3.5

2   Place the shift lever in the "P" position.
3   Raise the vehicle and support it securely on jackstands.
4   Loosen the shift cable-to-manual lever locknut **(see illustration)** and place the manual shaft lever in the "P" position.
5   Pull down on the cable **(see illustration)**

and, holding the cable and manual shaft lever in this position, install and tighten the locknut.
6   Move the shift lever from "P" to "1" again. Make sure that it moves smoothly and quietly.

7   Remove the jackstands and lower the vehicle.

Replacement

Refer to illustrations 3.9a, 3.9b, 3.9c, 3.9d, 3.10a, 3.10b, 3.11, 3.12 and 3.15

8   Remove the center console and the center dash trim bezel (see Chapter 11).
9   Remove the shift lever handle **(see illustrations)**.
10   Remove the shift lever indicator panel **(see illustrations)**.

7B

3.10a   To remove the shift lever indicator panel, remove these two screws (arrows) and this small console bracket . . .

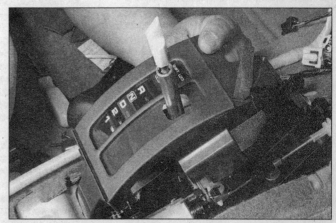

3.10b   . . . then pop the plastic retainers loose and lift the indicator panel straight up

3.11 To disconnect the shift cable from the shift lever, remove this nut

3.12 To detach the shift cable from the shift lever base, pry off this retaining clip

3.15 To detach the shift cable from its bracket on the transaxle, remove this C-clip (arrow)

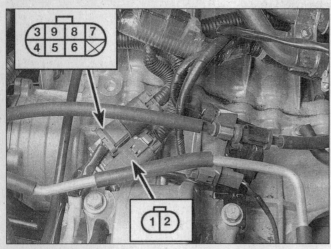

4.2 Unplug the electrical connectors for the neutral start/back-up light switch and check continuity at the indicated terminals

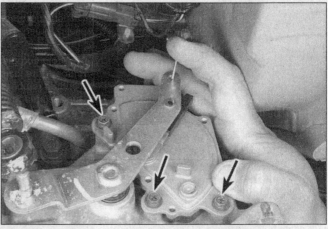

4.4 Disconnect the shift cable, loosen the switch retaining screws (arrows), and insert a 5/32-inch drill bit through the adjusting holes in the manual shaft lever and the switch

11   Disconnect the shift cable from the shift lever **(see illustration)**.
12   Disconnect the shift cable from the shift lever base **(see illustration)**.
13   Raise the vehicle and place it securely on jackstands.
14   Disconnect the shift cable from the manual shaft lever **(see illustration 3.4)**.
15   Remove the C-clip from the shift cable bracket **(see illustration)** and disengage the cable from the bracket.
16   Installation is the reverse of removal. Be sure to adjust the cable when you're done.

---

4   Neutral start/back-up light switch - check, adjustment and replacement

Check

Refer to illustration 4.2

1   If the engine will start with the shift lever in any position other than Park or Neutral, check and adjust the neutral start switch.
2   Unplug the switch electrical connectors

and check continuity as follows **(see illustration)**:

a) *With the shift lever in Park, there should be continuity between terminals 1 and 2, and between terminals 3 and 4.*
b) *With the shift lever in Reverse, there should be continuity between terminals 3 and 5.*
c) *With the shift lever in Neutral, there should be continuity between terminals 1 and 2, and between terminals 3 and 6.*
d) *With the shift lever in Drive, there should be continuity between terminals 3 and 7.*
e) *With the shift lever in 2, there should be continuity between terminals 3 and 8.*
f) *With the shift lever in 1, there should be continuity between terminals 3 and 9.*

3   If the switch fails any of these continuity checks, disconnect the shift cable from the manual lever and retest the switch.

a) *If the switch passes all the continuity checks this time, reconnect the shift cable and adjust it, then retest the switch.*
b) *If the switch still fails any of the continuity tests, remove it from the transaxle*

and try testing it again on the bench.
c) *If the switch passes all the continuity tests on the bench, install it and adjust it (see below).*
d) *If the switch still fails any of the continuity tests, replace it (see below).*

Adjustment

Refer to illustration 4.4

4   Disconnect the shift cable from the manual lever (see Section 3), loosen the switch retaining screws, set the manual lever at the Neutral position and insert a 0.16-inch pin (or a 5/32-inch drill bit) through the adjustment holes in both the manual shaft lever and the switch **(see illustration)**. Make sure the pin is perpendicular to the switch and the lever. Tighten the switch retaining screws securely.
5   Recheck the switch continuity as described above. If switch continuity is still not as specified, replace the switch.

Replacement

6   Disconnect the negative cable from the battery.

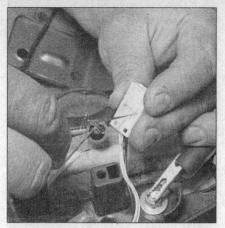

5.3 To test the shift lock solenoid, apply battery voltage to the solenoid terminals and verify that there's an audible "click"

5.8 Remove the retaining screw (arrow), unplug the electrical connector and remove the solenoid

5.11a To disconnect the upper end of the key interlock cable from the key lock cylinder, remove the lock plate from the cylinder . . .

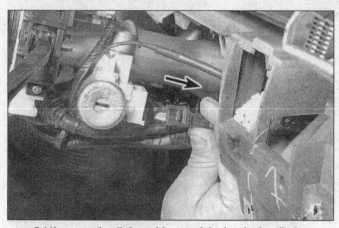

5.11b . . . and pull the cable out of the key lock cylinder

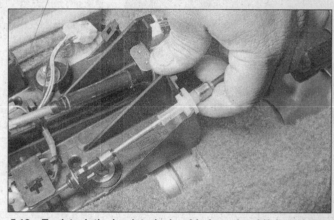

5.12a To detach the key interlock cable from the shift lever base, pull it straight up . . .

7 Shift the transaxle into Neutral.
8 Unplug the switch electrical connectors **(see illustration 4.2)**.
9 Remove the switch retaining screws **(see illustration 4.4)** and remove the switch.
10 Installation is the reverse of removal. Don't tighten the retaining screws until you have adjusted the switch as described in Step 4.

5 Shift lock system - description, check and component replacement

Description

1 The shift lock system prevents the shift lever from being shifted out of Park or Neutral until the brake pedal is applied. Other than the following simple component checks, diagnosis of the shift lock system should be left to a dealer service department.

Check

Refer to illustration 5.3

2 Remove the center console (see Chapter 11).
3 Unplug the electrical connector for the shift lock solenoid. Apply battery voltage to the solenoid terminals **(see illustration)** and verify that there's an audible "click." On 1993 through 1995 models, use terminals 8 (blue wire with black stripe) and 9 (black wire); on 1996 and 1997 models, use terminal 5 (red wire) of the A/T device electrical connector and terminal 1 (blue wire) of the shift lock solenoid connector.
4 If the shift lock solenoid doesn't click when energized by the battery, replace it.

Component replacement

Shift lock solenoid

Refer to illustration 5.8

5 Remove the center console (see Chapter 11).
6 Remove the gear position indicator.
7 Unplug the electrical connector for the shift lock solenoid **(see illustration 5.3)**.
8 Remove the solenoid **(see illustration)**.
9 Installation is the reverse of removal.

Key interlock cable

Refer to illustrations 5.11a, 5.11b, 5.12a, 5.12b, 5.12c and 5.13

10 If the key interlock cable breaks, you'll have to remove the steering column cover and the center console to replace it (see

5.12b . . . then squeeze the two plastic tangs on the shift lever cam together with a pair of pliers . . .

Chapter 11).
11 Up at the key lock cylinder, remove the lock plate from the cylinder and disconnect the upper end of the shift lock cable **(see illustrations)**.
12 Down at the shift lever, detach the key interlock cable from the shift lever base **(see illustration)**, then disconnect it from the plas-

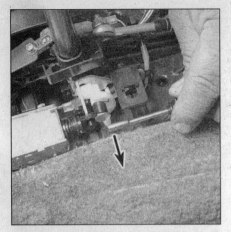

5.12c . . . and simultaneously pull the key interlock rod off the tangs

6.3b Unplug the electrical connector for the revolution sensor

tic cam on the shift lever **(see illustrations)**.
13   Unlock the slider from the adjuster holder and remove the key interlock rod; to insert the key interlock rod into the adjuster holder on the new cable, push it in until it locks into place **(see illustration)**.

5.13 Unlock the slider from the adjuster holder by squeezing the two tabs (arrows) on the slider and pull out the key interlock rod; to attach the key interlock rod to the new cable, insert it into the adjuster holder and push it in until it locks into place

14   Installation is otherwise the reverse of removal.

---

6   Automatic transaxle - removal and installation

Removal

Refer to illustrations 6.3a, 6.3b, 6.3c, 6.5, 6.6, 6.10a, 6.10b, 6.13, 6.14, 6.18a, 6.18b, 6.18c, 6.18d, 6.19a and 6.19b

1   Remove the battery and the battery tray (see Chapter 5).
2   Remove the air cleaner and resonator (see Chapter 4).
3   Unplug the electrical connectors for the automatic transaxle solenoid, the revolution sensor, the neutral start/back-up light switch and, on 1995 and later models, the crankshaft position sensor **(see illustrations)**. Unplug the connector for the vehicle speed sensor

6.3a Unplug all electrical connectors (arrows) on top of the transaxle bellhousing; also detach the vent hose (arrow on far right)

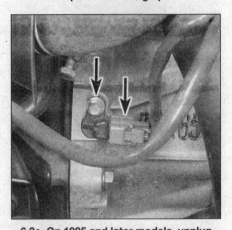

6.3c On 1995 and later models, unplug the electrical connector for the crankshaft position sensor; then unbolt and remove the sensor so that it won't be damaged during removal of the transaxle

(see Chapter 6). It's a good idea to remove the crankshaft position sensor so that it won't

6.5 To remove the left transaxle mount, remove the indicated bolts and nut (arrows)

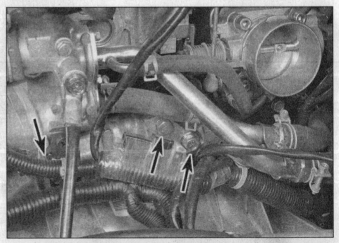

6.6 Remove the indicated upper transaxle-to-engine bolts (arrows) (the bolt on the far right is the upper starter motor bolt; the other starter bolt is to the right, behind the big wiring harness)

6.10a Disconnect this oil cooler line (arrow) . . .

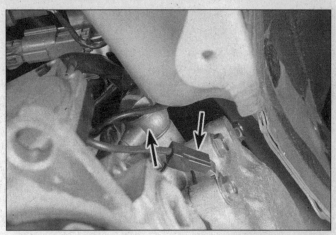

6.10b . . . and this one (arrow); detach this ground strap (arrow) from the left side of the transaxle housing

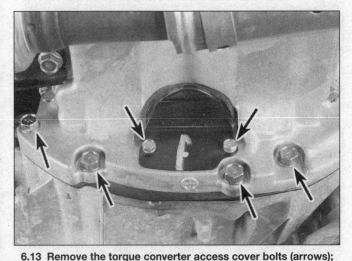

6.13 Remove the torque converter access cover bolts (arrows); the four big bolts (arrows) secure the engine oil pan casting to the transaxle - all four must be removed

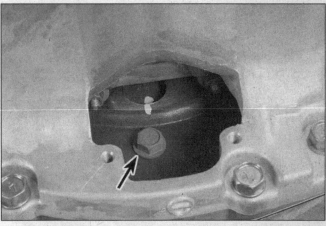

6.14 Mark the relationship of the torque converter to the driveplate, then remove the torque converter-to-driveplate bolts (arrow) by rotating the crankshaft to bring each bolt into the access cover window

be damaged during transaxle removal.

4   Disconnect the vent hose (see illustration 6.3a).

5   Remove the left transaxle mount (see illustration).

6   Remove the upper transaxle-to-engine bolts (see illustration).

7   Remove the starter motor (see Chapter 5).

8   Loosen the wheel lug nuts. Raise the vehicle and support it securely on jackstands. Remove the wheels.

9   Drain the transaxle/differential fluid (see Chapter 1).

10   Disconnect the oil cooler lines (see illustrations). Plug the hoses to keep out dirt and moisture. Disconnect the ground strap next to the left cooler line fitting.

11   Disconnect the shift cable from the transaxle (see Section 3).

12   Remove the driveaxles (see Chapter 8).

13   Remove the torque converter cover (see illustration).

14   Mark the relationship of the torque converter to the driveplate (see illustration) so they can be installed in the same position.

6.18a To remove the center member, unbolt it from the front mount . . .

15   Remove the torque converter mounting bolts. Turn the crankshaft for access to each one in turn.

16   Support the engine using a hoist from above, or a jack and a wood block under the

6.18b . . . and from the rear mount . . .

oil pan to spread the load.

17   Support the transaxle with a jack - preferably a special jack made for this purpose. Safety chains will help steady the transaxle on the jack.

18   Remove the center member (see illustrations).

6.18c . . . then remove these two bolts from the front end
of the member . . .

6.18d . . . and these three (arrows) from the rear end

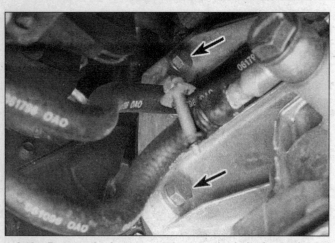

6.19a Remove the front transaxle-to-engine bolts (arrows) . . .

6.19b . . . and the rear engine-to-transaxle bolt (arrow) (this bolt is
located right below the starter motor)

19   Remove the lower engine oil pan-to-transaxle bolts (see illustration 6.13). Remove the front transaxle-to-engine bolt and the rear engine-to-transaxle bolt (see illustrations)
20   Lower the transaxle slightly and make a final check for any wiring harnesses or lines that may still be connected.
21   Move the transaxle to the side to disengage it from the engine block dowel pins and make sure the torque converter is detached from the driveplate. Secure the torque converter to the transaxle so that it will not fall out during removal. Lower the transaxle from the vehicle.

Installation

22   Make sure that the torque converter is securely engaged in the transaxle prior to installation.

23   With the transaxle secured to the jack, raise it into position. Be sure to keep it level so the torque converter does not slide forward. Connect the cooler lines.
24   Move the transaxle carefully into place until the dowel pins are engaged and the torque converter is engaged.
25   Turn the torque converter to line up the bolt holes with the holes in the driveplate. The match marks on the torque converter and driveplate, made during step 14, must line up.
26   Install the engine oil pan-to-transaxle, engine-to-transaxle and transaxle-to-engine bolts. Tighten them to the torque values listed in this Chapter's Specifications.
27   Install the torque converter-to-driveplate bolts. Tighten them to the torque listed in this Chapter's Specifications. Install the torque converter cover and tighten the bolts securely.

28   Install the center member and tighten the bolts securely.
29   Install the left transaxle mount. Tighten all bolts and nuts securely.
30   Remove the jacks supporting the transaxle and the engine.
31   Install the dipstick tube.
32   Install the starter motor (see Chapter 5).
33   Connect the shift cable (see Section 3).
34   Plug in the electrical connectors for the automatic transaxle solenoid, revolution sensor, vehicle speed sensor, neutral start/back-up light switch and, on 1995 and later models, the crankshaft position sensor
35   Connect the driveaxles to the transaxle (see Chapter 8).
36   Remove the jackstands and lower the vehicle.
37   Fill the transaxle with the proper type and amount of fluid (see Chapter 1). Run the vehicle and check for fluid leaks.

Chapter 8
Clutch and driveaxles

Contents

Specifications

General

Inner CV joint boot length	3-53/64 to 3-29/32 inches
Outer CV joint boot length	3-21/64 to 3-11/32 inches

Torque specifications

	Ft-lbs (unless otherwise indicated)
Clutch master cylinder retaining nuts	72 to 96 in-lbs
Clutch release cylinder retaining bolts	22 to 30
Clutch release cylinder bleeder screw	51 to 86 in-lbs
Clutch pressure plate bolts	16 to 22
Driveaxle/hub nut	174 to 231
Driveaxle support bearing retainer bolts (right-side only)	108 to 168 in-lbs
Driveaxle support bearing bracket-to-engine block bolts	19 to 26

1 General information

The information in this Chapter deals with the components from the rear of the engine to the front wheels, except for the transaxle, which is dealt with in Chapter 7A and 7B. For the purposes of this Chapter, these components are grouped into two categories: clutch and driveaxles. Separate Sections within this Chapter offer general descriptions and checking procedures for both groups.

Since nearly all the procedures covered in this Chapter involve working under the vehicle, make sure it's securely supported on sturdy jackstands or a hoist where the vehicle can be easily raised and lowered.

2 Clutch - description and check

1 All vehicles with a manual transaxle use a single dry plate, diaphragm spring type clutch. The clutch disc has a splined hub which allows it to slide along the splines of the transaxle input shaft. The clutch and pressure plate are held in contact by spring pressure exerted by the diaphragm in the pressure plate.

2 The clutch release system is hydraulically operated. The release system consists of the clutch pedal, the clutch master cylinder, the clutch release cylinder, the hydraulic line between the master cylinder and release cylinder, and the clutch release bearing.

3 When pressure is applied to the clutch pedal to release the clutch, the clutch master cylinder transmits this movement to the clutch release cylinder, which moves the clutch release lever. As the lever pivots, the shaft fingers push against the release bearing. The bearing pushes against the fingers of the diaphragm spring of the pressure plate assembly, which in turn releases the clutch plate.

4 Terminology can be a problem regarding the clutch components because common names have in some cases changed from that used by the manufacturer. For example, the clutch release cylinder is sometimes referred to as a slave cylinder, the driven plate is also called the clutch plate or disc, the pressure plate assembly is also known as the clutch cover, and the clutch release bearing is sometimes called a throw-out bearing.

5 Other than replacing components that have obvious damage, some preliminary checks should be performed to diagnose a clutch system failure.

a) Before proceeding, check and, if necessary, adjust clutch pedal freeplay and height (see Chapter 1).
b) To check "clutch spin down time," run the engine at normal idle speed with the transaxle in Neutral (clutch pedal up - engaged). Disengage the clutch (pedal down), wait several seconds and shift the transaxle into Reverse. No grinding noise should be heard. A grinding noise would most likely indicate a problem in the pressure plate or the clutch disc.
c) To check for complete clutch release, run the engine (with the parking brake applied to prevent movement) and hold the clutch pedal approximately 1/2-inch from the floor. Shift the transaxle between 1st gear and Reverse several times. If the shift is not smooth, component failure is indicated.
d) Visually inspect the clutch pedal bushing at the top of the clutch pedal to make sure there is no sticking or excessive wear.

8

3.2 To disconnect the clutch master cylinder pushrod from the clutch pedal, remove this retaining clip and clevis pin (arrow)

e) *Under the vehicle, verify that the clutch release lever is solidly mounted on the ball stud.*

f) *Make sure that the hydraulic lines aren't leaking at either the master cylinder or the release cylinder (see Sections 3 and 4). Bleed the system if necessary (see Section 5).*

3.3c Unscrew the hydraulic line fitting from the clutch master cylinder with a flare-nut wrench to protect the fitting nut

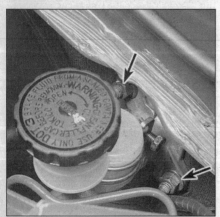

3.4 To detach the clutch master cylinder from the firewall, remove these two nuts (arrows)

3.3a To gain easier access to the hydraulic line and mounting nuts for the clutch master cylinder, remove these screws (arrows) and set aside the small relay box and the Automatic Speed Control Device (ASCD) pump

3   Clutch master cylinder - removal, overhaul and installation

Note: *Before beginning this procedure, contact local parts stores and dealer service departments concerning the purchase of a rebuild kit or a new master cylinder. Availability and cost of the necessary parts may dictate whether the cylinder is rebuilt or replaced with a new one. If you decide to rebuild the cylinder, inspect the bore as described in Step 9 before purchasing parts.*

Removal

Refer to illustrations 3.2, 3.3a, 3.3b, 3.3c and 3.4

1   Disconnect the negative cable from the battery.
2   Under the dashboard, disconnect the pushrod from the top of the clutch pedal. It's held in place with a clevis pin. To remove the clevis pin, remove the clip **(see illustration)**.
3   Remove the small relay panel and the

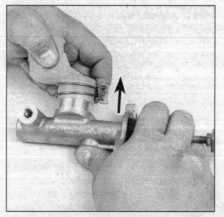

3.5 To detach the reservoir from the master cylinder, loosen the reservoir clamp screw and pull off the reservoir

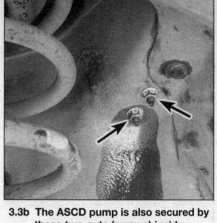

3.3b The ASCD pump is also secured by these two nuts (arrows) inside the wheelhousing

ASCD (cruise control) pump **(see illustrations)**. Disconnect the hydraulic line at the clutch master cylinder **(see illustration)**. Use a flare-nut wrench on the fitting to protect it from being rounded off. Have rags handy - some fluid will be spilled when the line is disconnected. **Caution:** *Don't allow brake fluid to spill onto the paint; it will damage the finish.*
4   Remove the nuts **(see illustration)** which attach the master cylinder to the firewall. As you remove the master cylinder, make sure you don't spill any fluid.

Overhaul

Refer to illustrations 3.5, 3.6a, 3.6b, 3.6c, 3.7a, 3.7b, 3.8, 3.14 and 3.16
5   Remove the reservoir cap and drain all fluid from the master cylinder. Loosen the reservoir clamp screw **(see illustration)**, then pull off the reservoir, prying gently if necessary.
6   Remove the dust boot, depress the pushrod and pry out the snap-ring with a small screwdriver **(see illustrations)**, then pull out the pushrod and washer.
7   Remove the stopper bolt and pull out

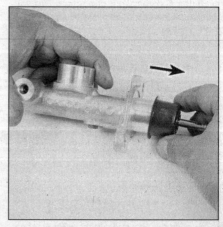

3.6a Remove the dust boot

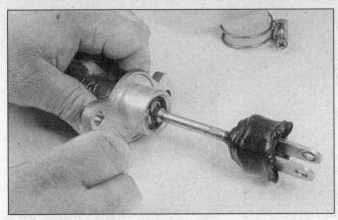

3.6b Remove the snap-ring with a small screwdriver . . .

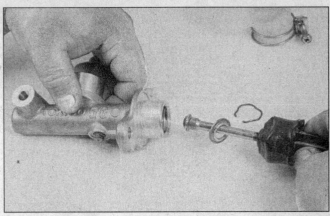

3.6c . . . then pull out the pushrod and washer

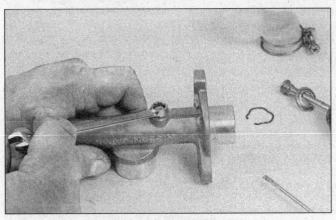

3.7a Remove the stopper bolt . . .

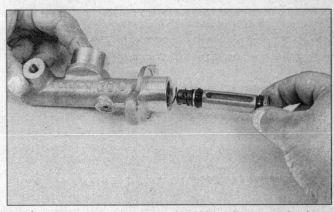

3.7b . . . then remove the piston assembly and return spring (note the groove in the piston for the stopper bolt)

the piston assembly and return spring **(see illustrations)**.

8   Wash all the parts with brake cleaner, then lay them out neatly for inspection **(see illustration)**.

9   Inspect the bore of the master cylinder for deep scratches, score marks and ridges. The surface must be smooth to the touch. If the bore isn't perfectly smooth, replace the master cylinder with a new or rebuilt unit.

10   Inspect the piston/cup assembly for scratches, score marks and ridges. The surface must be smooth to the touch. Inspect the piston cups for tears and excessive wear. If the piston/cup assembly isn't perfectly smooth, or either cup is damaged or excessively worn, replace the piston/cup assembly.

11   If you're rebuilding the master cylinder, use the new parts contained in the rebuild kit. Follow any specific instructions included with the kit. Wash those parts which must be reused with brake cleaner, denatured alcohol or clean brake fluid. DO NOT use petroleum-based solvents.

12   Some rebuild kits will include a complete piston/cup assembly; others will only supply the cups. If you're using a kit that supplies the cups only, remove the old cups from the piston and install new cups. Make sure the new cups face the same direction, toward the front of the master cylinder **(see illustration 3.8)**.

13   Attach the spring to the front end of the piston.

14   Lubricate the bore of the cylinder and the cups with plenty of clean brake fluid or brake assembly lube **(see illustration)**.

15   Carefully guide the piston/cup assembly

8

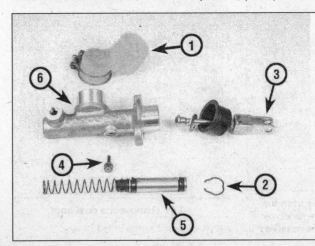

3.8 Wash all the parts thoroughly with brake cleaner, then lay them out for inspection:

1   *Reservoir*
2   *Snap-ring*
3   *Pushrod, washer, dust boot and clevis*
4   *Stopper bolt*
5   *Piston/cup assembly and return spring*
6   *Master cylinder body*

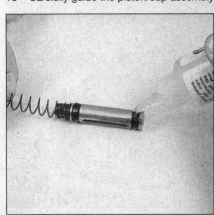

3.14 Lubricate the internal parts with brake assembly lube or clean brake fluid

3.16   Lubricate the contact surface of the pushrod with anti-seize compound or multi-purpose grease

4.3a   Loosen the banjo fitting bolt . . .

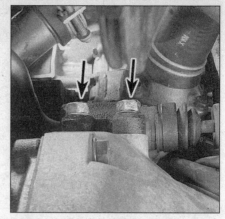

4.3b   . . . then detach the clutch release cylinder from the transaxle by removing these two bolts (arrows)

4.4   Pull the release cylinder down and unscrew the hydraulic line fitting

into the bore, with the slot in the piston aligned with the threaded hole in the master cylinder for the stopper bolt; make sure the spring end is installed first, with the pushrod end of the piston closest to the opening, and don't damage the cups. Compress the spring and piston assembly and install the stopper bolt.

16   Lubricate the piston end of the pushrod **(see illustration)**, then install the pushrod and washer, depress the pushrod and install

a new snap-ring. Make sure the snap-ring seats completely in its groove. Install the dust boot.

17   Install the fluid reservoir and tighten the reservoir clamp screw securely.

Installation

18   Position the master cylinder on the firewall and install the mounting nuts, but don't tighten them yet.

19   Connect the hydraulic line to the master cylinder, moving the cylinder slightly as necessary to thread the fitting properly into the bore. Don't cross-thread the fitting as it's installed.

20   Tighten the master cylinder mounting nuts and the hydraulic line fitting securely.

21   Connect the pushrod to the clutch pedal.

22   Fill the clutch master cylinder reservoir with brake fluid conforming to DOT 3 specifications and bleed the clutch system (see Section 5).

23   Check the clutch pedal height and freeplay (see Chapter 1).

---

4   Clutch release cylinder - removal, overhaul and installation

Note: *Before beginning this procedure, contact a local parts store and dealer service*

department concerning the purchase of a rebuild kit or a new release cylinder. Availability and cost of the necessary parts may dictate whether the cylinder is rebuilt or replaced with a new one. If you decide to rebuild the cylinder, inspect the bore as described in Step 9 before purchasing parts.

Removal

Refer to illustrations 4.3a, 4.3b and 4.4

1   Disconnect the negative cable from the battery.

2   Raise the vehicle and support it securely on jackstands.

3   Loosen the clutch hydraulic line banjo bolt **(see illustration)**. Remove the release cylinder mounting bolts **(see illustration)**.

4   Disconnect the hydraulic line at the release cylinder **(see illustration)**. Have a small can and rags handy - some fluid will be spilled when the line is disconnected.

5   Remove the release cylinder.

Overhaul

Refer to illustrations 4.6, 4.7a, 4.7b, 4.8, 4.9 and 4.10

6   Remove the boot and the pushrod **(see illustration)**.

7   Tap the cylinder on a block of wood to eject the piston and piston cup **(see illustrations)**.

4.6   Remove the dust boot and pushrod from the release cylinder

4.7a   Tap the cylinder on a block of wood . . .

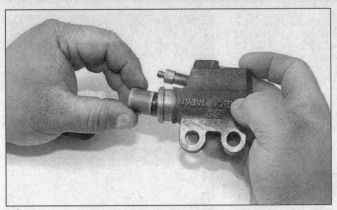

4.7b ... to free the piston and piston cup from the release cylinder

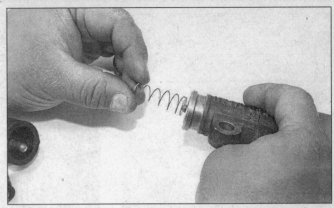

4.8 Remove the return spring from the release cylinder

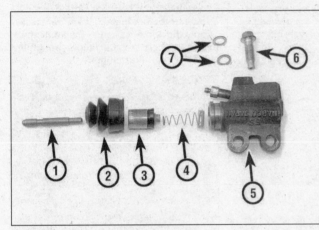

4.9 Thoroughly wash all the parts with brake cleaner and lay them out for inspection

1 Pushrod
2 Dust boot
3 Piston and piston cup
4 Return spring
5 Release cylinder
6 Hydraulic line banjo bolt
7 Sealing washers (always replace)

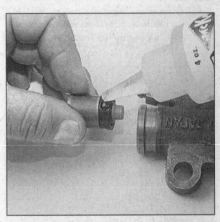

4.10 Lubricate the piston and piston cup, then install the piston assembly; make sure the piston cup faces the right direction (toward the bore)

8   Remove the spring from inside the cylinder **(see illustration)**.

9   Wash all the parts in brake cleaner, lay them out **(see illustration)** and inspect as follows:

a) *Inspect the dust boot for cracks and tears. Replace it if damaged.*

b) *Replace the piston and piston cup regardless of their condition (these two parts must be replaced every time the release cylinder is disassembled).*

c) *Inspect the spring for corrosion and deformation. Replace it if damaged.*

d) *Carefully inspect the bore of the cylinder. Check for deep scratches, score marks and ridges. The bore must be smooth to the touch. If any imperfections are found, the release cylinder must be replaced.*

10   Reassembly is the reverse of disassembly. Be sure to lubricate all parts with clean brake fluid or brake assembly lube. Make sure the piston cup is installed facing the right direction (*toward* the inside of the cylinder) **(see illustration)**.

Installation

11   Connect the hydraulic line to the release cylinder before bolting the release cylinder in place. Use new sealing washers. Tighten the banjo bolt finger tight - don't torque it until the release cylinder is securely bolted in place.

12   Install the release cylinder on the clutch housing. Make sure the pushrod is seated in the release lever pocket. Install the retaining bolts and tighten them to the torque listed in this Chapter's Specifications.

13   Tighten the hydraulic line banjo bolt to the torque listed in this Chapter's Specifications.

14   Fill the clutch master cylinder with brake fluid (conforming to DOT 3 specifications).

15   Bleed the system (see Section 5).

16   Lower the vehicle and connect the negative battery cable.

5   Clutch hydraulic system - bleeding

Refer to illustration 5.4

1   The hydraulic system should be bled of all air whenever any part of the system has been removed or if the fluid level has been allowed to fall so low that air has been drawn into the master cylinder. The procedure is similar to bleeding a brake system.

2   Fill the master cylinder with new brake fluid conforming to DOT 3 specifications. **Caution:** *Do not re-use any of the fluid coming from the system during the bleeding operation or use fluid which has been inside an open container for an extended period of time.*

3   Raise the vehicle and place it securely on jackstands to gain access to the release cylinder.

4   Locate the bleeder valve on the clutch release cylinder **(see illustration)**. Remove the dust cap which fits over the bleeder valve

5.4 To bleed the clutch hydraulic release system, push one end of a clear plastic hose over the bleeder screw on the release cylinder (arrow) and submerge the other end of the hose in a bottle of clean brake fluid

8

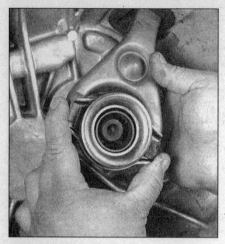

6.3 Disengage the clutch release lever from the ball stud, then remove the release bearing and release lever

6.4 To check the bearing, hold it by the outer race and rotate the inner race while applying pressure; if the bearing doesn't turn smoothly or if it's noisy, replace the bearing.

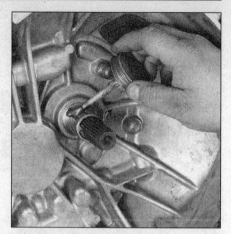

6.5 Apply a light coat of high-temperature grease to the bearing surface of the retainer (before installing the transaxle, apply the same grease to the input shaft splines to help the shaft slide through the clutch hub)

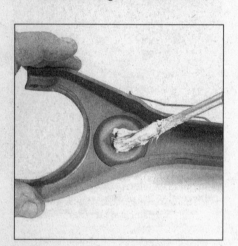

6.6a Using high-temperature grease, lubricate the ball stud socket in the back of the release lever . . .

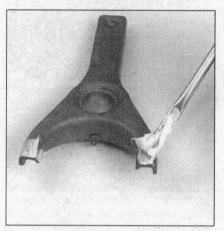

6.6b . . . the lever ends, the depression for the release cylinder pushrod . . .

6.6c . . . and the ball stud

and push a length of plastic hose over the valve. Place the other end of the hose into a clear container with about two inches of brake fluid in it. The hose end must be submerged in the fluid.

5   Have an assistant depress the clutch pedal and hold it. Open the bleeder valve on the release cylinder, allowing fluid to flow through the hose. Close the bleeder valve when fluid stops flowing from the hose. Once closed, have your assistant release the pedal.

6   Continue this process until all air is evacuated from the system, indicated by a full, solid stream of fluid being ejected from the bleeder valve each time and no air bubbles in the hose or container. Keep a close watch on the fluid level inside the clutch master cylinder reservoir; if the level drops too low, air will be sucked back into the system and the process will have to be started over again.

7   Install the dust cap and lower the vehicle. Check carefully for proper operation before placing the vehicle in normal service.

6   Clutch release bearing and lever, removal, inspection and installation

Warning: *Dust produced by clutch wear and deposited on clutch components may contain asbestos, which is hazardous to your health. DO NOT blow it out with compressed air and DO NOT inhale it. DO NOT use gasoline or petroleum-based solvents to remove the dust. Brake system cleaner should be used to flush it into a drain pan. After the clutch components are wiped clean with a rag, dispose of the contaminated rags and cleaner in a labeled, covered container.*

Removal

Refer to illustration 6.3

1   Disconnect the negative cable from the battery.

2   Remove the transaxle (see Chapter 7).

3   Disengage the clutch release lever from the ball stud, then remove the bearing and lever **(see illustration)**.

Inspection

Refer to illustration 6.4

4   Hold the bearing by the outer race and rotate the inner race while applying pressure **(see illustration)**. If the bearing doesn't turn smoothly or if it's noisy, replace the bearing/hub assembly with a new one. Wipe the bearing with a clean rag and inspect it for damage, wear and cracks. Don't immerse the bearing in solvent; it's sealed for life, to do so would ruin it. Also check the release lever for cracks and bends.

Installation

Refer to illustrations 6.5, 6.6a, 6.6b, 6.6c, 6.7a and 6.7b

5   Lubricate the bearing surface of the retainer with high-temperature grease **(see illustration)**.

6   Lubricate the release lever ball socket, lever ends and release cylinder pushrod socket with high-temperature grease **(see illustrations)**.

7   Attach the retainer spring and the release bearing to the release lever **(see illustrations)**.

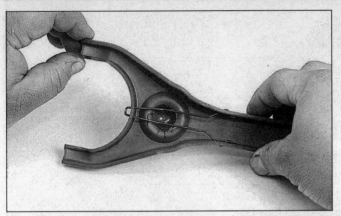

6.7a Install the retainer spring onto the release lever

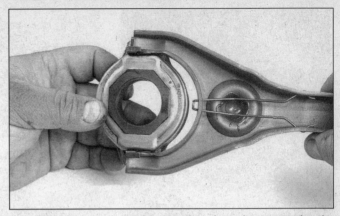

6.7b Slide the release bearing onto the release lever engaging the clips on the ends

7.5 Mark the relationship of the pressure plate to the flywheel (in case you're going to re-use the same pressure plate)

7.7a Remove the pressure plate and the clutch disc

7.7b Remove the pilot bearing with a small slide hammer

8   Slide the release bearing onto the transaxle input shaft bearing retainer while passing the end of the release lever through the opening in the clutch housing. Push the clutch release lever onto the ball stud until it's firmly seated.

9   Apply a light coat of high-temperature grease to the face of the release bearing where it contacts the pressure plate diaphragm fingers.

10   The remainder of installation is the reverse of removal.

7   Clutch components - removal, inspection and installation

Warning: *Dust produced by clutch wear and deposited on clutch components may contain asbestos, which is hazardous to your health. DO NOT blow it out with compressed air and DO NOT inhale it. DO NOT use gasoline or petroleum-based solvents to remove the dust. Brake system cleaner should be used to flush the dust into a drain pan. After the clutch components are wiped clean with a rag, dispose of the contaminated rags and cleaner in a labeled, covered container.*

Removal
Refer to illustrations 7.5, 7.7a and 7.7b

1   Access to the clutch components is normally accomplished by removing the transaxle, leaving the engine in the vehicle. If, of course, the engine is being removed for major overhaul, then the opportunity should always be taken to check the clutch for wear and replace worn components as necessary. However, the relatively low cost of the clutch components compared to the time and labor involved in gaining access to them warrants their replacement any time the engine or transaxle is removed, unless they are new or in near-perfect condition. The following procedures assume that the engine will stay in place.

2   Remove the transaxle from the vehicle (see Chapter 7A). Support the engine while the transaxle is out. Preferably, an engine hoist should be used to support it from above. However, if a jack is used underneath the engine, make sure a piece of wood is used between the jack and oil pan to spread the load. **Caution:** *The pick-up for the oil pump is very close to the bottom of the oil pan. If the pan is bent or distorted in any way, engine oil starvation could occur.*

3   The release fork and release bearing can

remain attached to the transaxle; however, you should inspect them (see Section 6) while the transaxle is removed.

4   To support the clutch disc during removal, install a clutch alignment tool through the clutch disc hub.

5   Carefully inspect the flywheel and pressure plate for indexing marks. The marks are usually an X, an O or a white letter. If they cannot be found, scribe marks yourself so the pressure plate and the flywheel will be in the same alignment during installation **(see illustration)**.

6   Slowly loosen the pressure plate-to-flywheel bolts. Work in a diagonal pattern and loosen each bolt a little at a time until all spring pressure is relieved.

7   Hold the pressure plate securely and completely remove the bolts, followed by the pressure plate and clutch disc **(see illustration)**. Use a small slide hammer to remove the pilot bearing **(see illustration)**.

Inspection
Refer to illustrations 7.10, 7.12a and 7.12b

8   Ordinarily, when a problem occurs in the clutch, it can be attributed to wear of the clutch driven plate assembly (clutch disc). However, all components should be inspected at this time.

8

7.10 Examine the clutch disc for evidence of excessive wear, such as burned friction material, loose rivets, worn hub splines and distorted damper cushions or springs

7.12b Examine the pressure plate friction surface for score marks, cracks and evidence of overheating (blue spots)

9   Inspect the flywheel for cracks, heat checking, score marks and other damage. If the imperfections are slight, a machine shop can resurface it to make it flat and smooth. Refer to Chapter 2 for the flywheel removal procedure.
10   Inspect the lining on the clutch disc.

NORMAL FINGER WEAR

EXCESSIVE WEAR

EXCESSIVE FINGER WEAR

BROKEN OR BENT FINGERS

7.12a Replace the pressure plate if any of these conditions are noted

There should be at least 1/16-inch of lining above the rivet heads. Check for loose rivets, distortion, cracks, broken springs and other obvious damage (see illustration). As mentioned above, ordinarily the clutch disc is replaced as a matter of course, so if in doubt about the condition, replace it with a new one.
11   The release bearing should be replaced along with the clutch disc (see Section 6).
12   Check the machined surface and the diaphragm spring fingers of the pressure plate (see illustrations). If the surface is grooved or otherwise damaged, replace the pressure plate assembly. Also check for obvious damage, distortion, cracking, etc. Light glazing can be removed with emery cloth or sandpaper. If a new pressure plate is indicated, new or factory rebuilt units are available.

Installation

Refer to illustrations 7.13 and 7.14
13   Install a new pilot bearing (see illustration).
14   Carefully wipe the flywheel and pressure plate machined surfaces clean. It's important that no oil or grease is on these surfaces or the lining of the clutch disc. Handle these parts only with clean hands. Position the clutch disc and pressure plate with the clutch held in place with an alignment tool (see illustration). Make sure it's installed properly (most replacement clutch plates will be marked "flywheel side" or something similar - if not marked, install the clutch disc with the damper springs or cushion toward the transaxle).
15   Tighten the pressure plate-to-flywheel

7.13 Tap a new pilot bearing into place with a hammer and punch; make sure it's fully seated, but don't pound too hard on it or you could damage it

7.14 Center the clutch disc in the pressure plate with a clutch alignment tool, then tighten the pressure plate-to-flywheel bolts

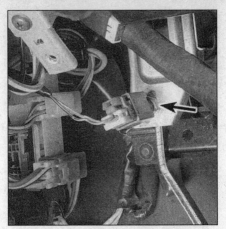

8.1 The clutch start switch (arrow) is located at the top of the clutch pedal

bolts only finger tight, working around the pressure plate.

16   Center the clutch disc by ensuring the alignment tool is through the splined hub and into the recess in the crankshaft. Wiggle the tool up, down or side-to-side as needed to bottom the tool. Tighten the pressure plate-to-flywheel bolts a little at a time, working in a crisscross pattern to prevent distortion of the cover. After all of the bolts are snug, tighten them to the torque listed in this Chapter's Specifications. Remove the alignment tool.

17   Using high-temperature grease, lubricate the inner groove of the release bearing (see Section 6). Also place grease on the release lever contact areas and the transaxle input shaft.

18   Install the clutch release bearing (see Section 6).

19   Install the transaxle, clutch release cable and all components removed previously, tightening all fasteners to the proper torque specifications.

8 Clutch start switch - check and replacement

Refer to illustrations 8.1 and 8.5

1   The clutch start switch (**see illustration**) is located at the top of the clutch pedal.

2   Verify that the engine will not start when the clutch pedal is released.

3   Verify that the engine will start when the clutch pedal is depressed all the way.

4   If the clutch start switch doesn't perform as described above, adjust the clutch pedal (see Chapter 1). The switch should now operate properly. If it doesn't, check switch continuity.

5   Verify that there is continuity between the clutch start switch terminals (**see illustration**) when the pedal is depressed.

6   Verify that no continuity exists between the switch terminals when the pedal is released.

7   If the switch fails either of these continuity tests, replace it. Loosen the nut near the body of the switch, then unscrew the switch.

8.5 Verify that there is continuity between the clutch start switch terminals when the pedal is depressed and no continuity when the pedal is released

Unplug the electrical connector. Installation is the reverse of removal.

8   Adjust the clutch pedal (see Chapter 1).

9   Verify that the engine doesn't start when the clutch pedal is released.

9 Driveaxles - general information and inspection

1   Power is transmitted from the transaxle to the wheels through a pair of driveaxles. The inner end of each driveaxle is splined into the differential side gears. The outer ends of the driveaxles are splined to the axle hubs and locked in place by a large nut.

2   The inner ends of the driveaxles are equipped with sliding constant velocity joints, which are capable of both angular and axial motion. The inner joints are the "double-offset" type, which consists of ball bearings running between an inner race and an outer cage. They can also be disassembled, cleaned and inspected, but they must be replaced as a single unit if defective.

3   The outer CV joints are the "Rzeppa" (pronounced "sheppa") or "Birfield" type, which also consists of ball bearings running

10.4a Remove the cotter pin . . .

between an inner race and an outer cage. However, Rzeppa/Birfield joints are capable of angular - but not axial - movement. These outer joints should be cleaned, inspected and repacked whenever replacing an outer CV joint boot, but they cannot be disassembled. If an outer joint is damaged, it must be replaced.

4   The boots should be inspected periodically for damage and leaking lubricant. Torn CV joint boots must be replaced immediately or the joints can be damaged. Boot replacement involves removal of the driveaxle (see Section 10). **Note:** *Some auto parts stores carry "split" type replacement boots, which can be installed without removing the driveaxle from the vehicle. This is a convenient alternative; however, the driveaxle should be removed and the CV joint disassembled and cleaned to ensure the joint is free from contaminants such as moisture and dirt which will accelerate CV joint wear.* The most common symptom of worn or damaged CV joints, besides lubricant leaks, is a clicking noise in turns, a clunk when accelerating after coasting and vibration at highway speeds. To check for wear in the CV joints and driveaxle shafts, grasp each axle (one at a time) and rotate it in both directions while holding the CV joint housings, feeling for play indicating worn splines or sloppy CV joints. Also check the driveaxle shafts for cracks, dents and distortion.

10 Driveaxles - removal and installation

Removal

Refer to illustrations 10.4a, 10.4b, 10.4c, 10.5, 10.6, 10.9, 10.10a, 10.10b and 10.10c

1   Disconnect the cable from the negative terminal of the battery.

2   Loosen the front wheel lug nuts, raise the vehicle and support it securely on jackstands.

3   Remove the wheel.

4   Remove the cotter pin, the nut lock and the felt washer from the driveaxle/hub nut (**see illustrations**).

8

10.4b . . . the bearing nut lock and the felt washer from the driveaxle/hub nut

10.5 Remove the driveaxle/hub nut and washer. To prevent the hub from turning, wedge a prybar between two of the wheel studs and allow the prybar to rest against the ground or the floorpan of the vehicle

10.10a If you're removing the left driveaxle, carefully pry the inner CV joint out of the transaxle and remove the driveaxle assembly

10.6 If the driveaxle splines are "frozen," knock the driveaxle loose with a hammer and brass punch

10.10b If you're removing the right driveaxle, remove these three bolts (arrows) (third bolt not visible in this photo) . . .

10.9 Pull out on the steering knuckle and detach the driveaxle from the hub

10.10c . . . mark the relationship of the bearing retainer to the support bracket and remove the driveaxle assembly

5   Remove the driveaxle/hub nut and washer. To prevent the hub from turning, wedge a prybar between two of the wheel studs and allow the prybar to rest against the ground or the floorpan of the vehicle **(see illustration)**.

6   If the driveaxle splines are "frozen," free them by tapping the end of the driveaxle with a soft-faced hammer or a hammer and a brass punch **(see illustration)**.

7   Remove the engine splash shields (see Chapter 1). Place a drain pan underneath the transaxle to catch the lubricant that may spill out when the driveaxles are removed.

8   Disconnect the control arm from the steering knuckle (see Chapter 10).

9   Pull out on the steering knuckle and detach the driveaxle from the hub **(see illustration)**.

10   If you're removing the left driveaxle, carefully pry the inner CV joint out of the transaxle **(see illustration)** and remove the driveaxle assembly. The inner CV joint housing on the right (passenger side) driveaxle terminates at a support bracket. To detach the right driveaxle assembly from the bracket,

remove the three retainer-to-bracket bolts, mark the relationship of the bearing to the support bracket and pull out the driveaxle assembly **(see illustrations)**. Do not try to separate the bearing from the inner CV joint until you have the entire assembly on the bench (see Section 12). **Note:** *When removing the left (driver's) side driveaxle on models with an automatic transaxle, it may not be possible to pry the inner CV joint out of the transaxle. In this case, it will be necessary to remove the right side driveaxle and insert a screwdriver through the differential side gears and knock the left side shaft free.*

11   Install a new driveaxle oil seal (see Chapter 7).

Installation

12   Installation is the reverse of the removal procedure, but with the following additional points:

a) *When installing the left driveaxle, push the driveaxle in sharply to seat the retaining ring on the inner CV joint in its groove in the differential side gear (the right driveaxle assembly has no retaining ring - it's secured by the three bolts which attach it to the support bracket). When installing the right driveaxle,*

tighten the retainer-to-support bearing bracket bolts to the torque listed in this Chapter's Specifications.

b) *Tighten the control arm balljoint-to-steering knuckle nut to the torque listed in the Chapter 10 Specifications.*

c) *Tighten the driveaxle/hub nut to the torque listed in this Chapter's Specifications, then install the nut lock and a new cotter pin.*

d) *Install the wheel and lug nuts, lower the vehicle and tighten the lug nuts to the torque listed in the Chapter 1 Specifications.*

e) *Check the transaxle or differential lubricant and add, if necessary, to bring it to the proper level (see Chapter 1).*

11 Support bearing assembly - check and replacement

Note: *This procedure applies to the support bearing for the right driveaxle assembly.*

1   Remove the right driveaxle assembly (see Section 10).

2   Rotate the extension shaft and listen to the bearing. It should operate smoothly and quietly.

12.2a Cut off the old boot clamps with a pair of diagonal cutters

12.2b Slide the boot back and wipe off as much of the old CV grease as possible

12.3 Mark the shaft, inner race, cage and outer race (housing) so they can be reassembled in the original relationship to each other

3   If the bearing is rough or noisy, unbolt the support bearing bracket from the engine block and remove the extension shaft, bearing and bracket. Take the assembly, a new bearing and three new dust shields (small one on the end of the extension shaft and two larger ones, on either side of the bearing) to an automotive machine shop. They will have the right tools to press off the old bearing and press on the new one.

4   Install the extension shaft assembly, tightening the bearing support-to-engine block bolts to the torque listed in this Chapter's Specifications.

5   Install the driveaxle (see Section 10).

12 Driveaxle boot replacement and CV joint inspection

Note: *If the CV joints must be overhauled (usually due to torn boots), explore all options before beginning the job. Complete rebuilt driveaxles are available on an exchange basis, which eliminates much time and work. Whichever route you choose to take, check on the cost and availability of parts before disassembling the vehicle.*

All units

1   Remove the driveaxle (see Section 10).

Inner CV joint

Disassembly

Refer to illustrations 12.2a, 12.2b, 12.3, 12.4, 12.5 and 12.6

2   Remove both boot clamps and discard them, then slide the boot out of the way **(see illustrations)**.

3   Mark the shaft, the inner race, the cage and the outer race (housing) so they can be reassembled in the original position **(see illustration)**.

4   Pry the wire ring bearing retainer from the housing **(see illustration)**.

5   Pull the housing off the inner bearing assembly **(see illustration)**.

6   Remove the snap-ring from the groove in the axleshaft with a pair of snap-ring pliers **(see illustration)**.

7   Slide the inner race off the axleshaft. If the splines are stuck, apply some penetrant and give it a few careful taps with a hammer and a brass punch.

8   Using a screwdriver or piece of wood, pry the ball bearings from the cage. Be care-

12.4 Pry the retainer from the housing with a small screwdriver

ful not to scratch the inner race, the ball bearings or the cage. Remove the cage.

9   Remove the stop ring for the inner race **(see illustration 12.24c)**.

Inspection

Refer to illustrations 12.10a and 12.10b

10   Clean the components with solvent to

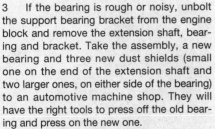

12.5 Slide the housing off the bearing assembly; some of the bearings may fall out when the race is removed, so be ready to catch them

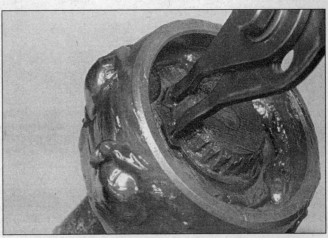

12.6 Remove the snap-ring from the groove in the axleshaft with a pair of snap-ring pliers

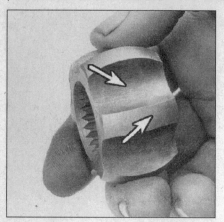

12.10a Inspect the inner race lands and grooves for pitting, score marks, cracks and other signs of wear and damage

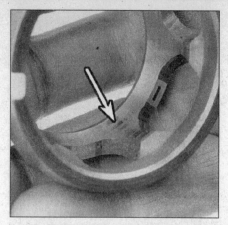

12.10b Inspect the cage for cracks, pitting and score marks (shiny, polished spots are normal and will not adversely affect CV joint performance

12.11 Wrap the axleshaft splines with tape to protect the boot, then slide the small boot clamp and boot onto the axleshaft and remove the tape

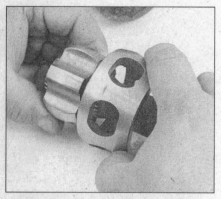

12.13 Install the stop ring, then slide the cage and the inner race onto the shaft splines until the inner race butts against the stop ring (make sure you have the cage on first - it can't be installed after the inner race is installed because it won't fit over the inner race)

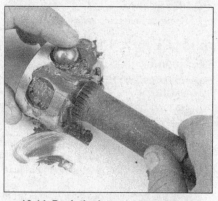

12.14 Pack the inner race and cage assembly with grease, by hand, until grease is worked completely into the assembly

12.17a Slide the housing over the assembled inner CV joint - make sure none of the balls fall out

remove all traces of grease. Inspect the cage and races for pitting, score marks, cracks and other signs of wear and damage. Shiny, polished spots are normal and will not adversely affect CV joint performance **(see illustrations)**.

Reassembly

Refer to illustrations 12.11, 12.13, 12.14, 12.17a, 12.17b, 12.18, 12.19, 12.20a, 12.20b, 12.20c, 12.20d and 12.20e

11   Wrap the axleshaft splines with tape to avoid damaging the boot. Slide the small boot clamp and boot onto the axleshaft **(see illustration)**, then remove the tape. Slide the large boot clamp over the boot.

12   Install the cage on the axleshaft with the smaller diameter side of the cage facing toward the center of the shaft.

13   Install a new stop ring for the inner race, then install the inner race onto the axleshaft **(see illustration)**, with the matchmark on the race (or the larger diameter side) aligned with the mark on the end of the axleshaft.

14   Install the snap-ring that retains the

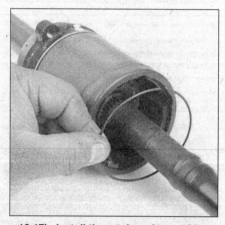

12.17b Install the retainer ring, making sure it seats in its groove

inner race. Make sure it's completely seated in its groove by trying to push the inner race off the shaft.

15   Move the cage up over the inner race, aligning the match marks. Press the ball bearings into the cage windows with your thumbs. If they won't stay in place, apply CV joint grease to hold them.

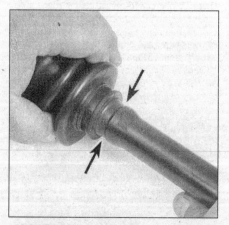

12.18 Make sure the small diameter end of the boot seats in this groove (arrows)

16   Fill the outer race and boot with CV joint grease (normally included with the new boot kit). Pack the inner race and cage assembly with grease, by hand, until grease is worked completely into the assembly **(see illustration)**.

17   Slide the inner race, balls and cage into the CV joint housing and install the wire ring

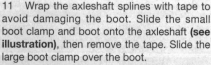

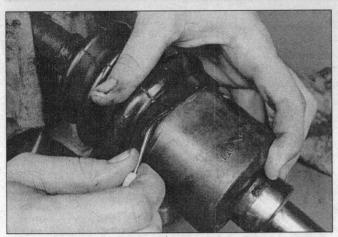

12.19 Equalize the pressure inside the boot with atmospheric pressure by inserting a *dull* screwdriver between the boot and the outer race; make sure you don't damage the boot

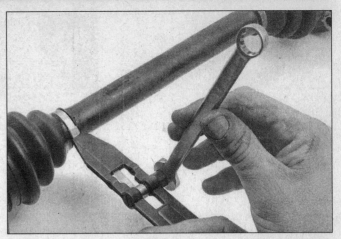

12.20a You'll need a special tightening tool to install "band" type boot clamps: Install the band with its end pointing in the direction of axle rotation and tighten it securely . . .

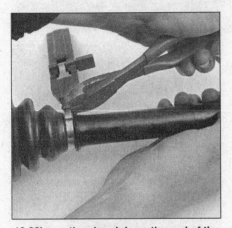

12.20b . . . then bend down the end of the clamp back and cut off the excess

12.20c If you're installing crimp-type boot clamps, you'll need a pair of special crimping pliers (available at most auto parts stores)

12.20d To install fold-over type boot clamps, bend the tang down . . .

12.20e . . . then tap the tabs over to hold it in place

bearing retainer **(see illustrations)**.

18   Wipe any excess grease from the axle boot groove on the outer race. Seat the small diameter of the boot in the recessed area on the axleshaft **(see illustration)**. Push the other end of the boot onto the CV joint housing and move the race in or out until there's no deformation (distortion or dents) in the boot.

19   Adjust the length of the joint (measured from one end of the boot to the other) **(see illustration 12.31)**. Equalize the pressure in the boot by inserting a dull screwdriver between the boot and the outer race **(see illustration)**. Don't damage the boot with the tool.

20   Install the boot clamps. There are three types of clamps you're likely to encounter: the band type, which requires a special tightening tool, the crimp type (which also requires a special tool), or the fold-over type **(see illustrations)**.

21   Install a new circlip on the inner CV joint stub axle.

22   Install the driveaxle (see Section 10).

Outer CV joint

Disassembly

Refer to illustrations 12.24a, 12.24b and 12.24c

23   Remove the boot clamps and separate the boot from the outer CV joint **(see illustration 12.2)**.

24   Clamp the axleshaft in a bench vise

(equipped with protective jaws) and drive off the outer CV joint with a brass hammer **(see illustration)**, then remove the retainer ring and the stop ring for the inner race **(see illustrations)**. Slide off the old boot.

12.24a To remove the outer CV joint, clamp the axleshaft in a bench vise and tap the joint off with a brass hammer - if the joint doesn't come off fairly easily, use a hammer and brass punch positioned on the inner race

8

12.24b Remove the retainer ring for the outer CV joint assembly . . .

12.24c . . . then remove the stop ring for the inner race

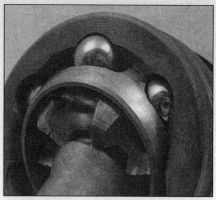

12.26 After the old grease has been rinsed away and the cleaning solvent has been blown out with compressed air, move the inner race through its full range of motion and inspect the bearing surfaces for wear or damage - if any of the balls, the race or the cage look damaged, replace the outer joint assembly

12.28 Install a new retainer ring for the outer CV joint assembly

25   Thoroughly wash the outer CV joint in clean solvent and blow it dry with compressed air, if available. The outer joint shouldn't be disassembled, so it's difficult to wash away all the old grease, and to rid the bearing of solvent once it's clean. But it's imperative that the job be done thoroughly, so take your time and do it right.

Inspection

Refer to illustration 12.26
26   Move the inner race through its full range of motion to expose the bearings, cage and bearing surfaces of the inner race (see illustration). Inspect the bearing surfaces for signs of wear. If the CV joint is worn, replace it.

Reassembly

Refer to illustrations 12.28, 12.29, 12.30 and 12.31
27   Slide the new outer boot onto the driveaxle. It's a good idea to wrap vinyl tape around the shaft splines to prevent damage to the boot (see illustration 12.11).
28   Remove the tape and install a new inner race stop ring and a new retainer ring (see illustration 12.24c and the accompanying illustration).
29   Pack the CV joint with CV joint grease through the splined hole in the inner race. Force the grease into the joint by inserting a wooden dowel through the splined hole and pushing it to the bottom of the joint. Repeat

12.29 Apply CV joint grease through the splined hole, then insert a wooden dowel (slightly smaller in diameter than the hole) into the hole and push down - the dowel will force the grease into the joint

this procedure until the bearing is completely packed (see illustration).
30   Tap the outer CV joint into place with a hammer and a wood block (see illustration).
31   Slide the boot into position (see illustration 12.18). When the boot is in position, add the remainder of the grease in the boot replacement kit to the CV joint boot. Slide the boot on and adjust the length (see illustration), equalize the pressure inside the boot (see illustration 12.19) and install the new clamps (see illustrations 12.20a through 12.20e).

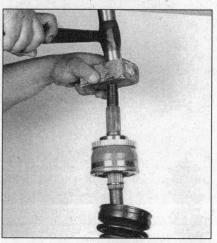

12.30 To install the outer CV joint, put the axleshaft in a bench vise (equipped with protective jaws) and tap on the CV joint with a hammer and a block of wood; drive the joint onto the axleshaft splines until the retainer ring on the shaft seats in the groove in the inner race of the joint

All units

32   Install the driveaxle (see Section 10).

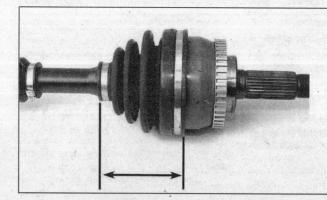

12.31 Adjust the boot to the length indicated in this Chapter's Specifications and tighten the clamps

Chapter 9   Brakes

Contents

Specifications

General

Brake fluid type	See Chapter 1

Disc brakes

Minimum pad thickness	See Chapter 1
Brake disc minimum thickness	
Front	0.787 inch*
Rear	0.315 inch*
Maximum disc runout (front and rear)	0.0028 inch
Maximum disc thickness variation	
Front	0.0004 inch
Rear	0.0008 inch

Rear drum brakes

Shoe friction material minimum thickness	See Chapter 1
Maximum inside diameter	0.906 inch*
Maximum out-of-round	0.0012 inch

If different specifications are cast into the disc or drum, they supersede information printed here

Power brake booster

Output rod length	13/32-inch
Booster-to-clevis dimension	4-59/64 inches

Brake pedal adjustments

Free height	
Manual transaxle	6-21/32 to 7-3/64 inches
Automatic transaxle	6-31/32 to 7-23/64 inches
Freeplay	3/64 to 7/64 inch
Depressed height	3-1/2 inches

Brake light switch

Plunger-to-pedal stopper clearance	0.012 to 0.039 inch

Torque specifications

	Ft-lbs (unless otherwise indicated)
Brake booster-to-body mounting nuts	108 to 144 in-lbs
Brake caliper	
Caliper mounting bolts (front and rear)	16 to 23
Torque plate bolts	
Front	53 to 72
Rear	28 to 38
Brake hose-to-caliper banjo bolt	144 to 168 in-lbs
Caliper and wheel cylinder bleeder screws	61 to 78 in-lbs
Master cylinder-to-brake booster retaining nuts	
1993 through 1995	70 to 96 in-lbs
1996 and later	108 to 132 in-lbs
Wheel cylinder retaining bolts	52 to 95 in-lbs

9

1   General information

The vehicles covered by this manual are equipped with hydraulically operated front and rear brake systems. The front brakes are disc type and the rear brakes are drum or disc type. Both the front and rear brakes are self adjusting. The disc brakes automatically compensate for pad wear, while the drum brakes incorporate an adjustment mechanism which is activated as the parking brake is applied.

Hydraulic system

The hydraulic system consists of two separate circuits. The master cylinder has separate reservoirs for the two circuits, and, in the event of a leak or failure in one hydraulic circuit, the other circuit will remain operative. A dual proportioning valve on the firewall provides brake balance between the front and rear brakes.

Power brake booster

The power brake booster, utilizing engine manifold vacuum and atmospheric pressure to provide assistance to the hydraulically operated brakes, is mounted on the firewall in the engine compartment.

Parking brake

The parking brake operates the rear brakes only, through cable actuation. It's activated by a lever mounted in the center console.

Service

After completing any operation involving disassembly of any part of the brake system, always test drive the vehicle to check for proper braking performance before resuming normal driving. When testing the brakes, perform the tests on a clean, dry, flat surface. Conditions other than these can lead to inaccurate test results.

Test the brakes at various speeds with both light and heavy pedal pressure. The vehicle should stop evenly without pulling to one side or the other. Avoid locking the brakes, because this slides the tires and diminishes braking efficiency and control of the vehicle.

Tires, vehicle load and wheel alignment are factors which also affect braking performance.

2   Anti-lock Brake System (ABS) - general information

1   The Anti-lock Brake System (ABS) is designed to maintain vehicle steerability, directional stability and optimum deceleration under severe braking conditions and on most road surfaces. It does so by monitoring the rotational speed of each wheel and controlling the brake line pressure to each wheel during braking. This prevents the wheels from locking up.

Components

Actuator assembly
Refer to illustration 2.2

2   The actuator assembly **(see illustration)** consists of an electric hydraulic pump and three solenoid valves: front left, front right and rear. The electric pump provides hydraulic pressure to charge the reservoirs in the actuator, which supplies pressure to the braking system. The solenoid valves modulate brake line pressure during ABS operation. The body contains four valves - one for each wheel. The pump, the reservoirs and the solenoid valves are all housed in the actuator assembly.

Speed sensors

3   The speed sensors, which are located at each wheel, generate a small sine wave current when the toothed sensor rotors are turning. This analog voltage signal is monitored by the ABS control unit, which converts it to a digital signal from which it can determine wheel rotational speed.

4   The front speed sensors are mounted on the steering knuckles in close relationship to the toothed sensor rotors, which are integral with the outer constant velocity (CV) joints.

5   The rear wheel sensors are bolted to the axle carriers (rear knuckles). The toothed sensor rotors are integral with the rear wheel hub/bearing assemblies.

ABS computer

6   The ABS control unit, which is mounted behind the left (driver's side) kick panel, is the "brain" of the ABS system. The function of the control unit is to monitor and process information received from the wheel speed sensors to control the hydraulic line pressure, avoiding wheel lock up. The control unit also monitors the system for malfunctions, even when the ABS system is inactive during normal driving conditions.

7   Each time you start the engine, the system turns on the ABS warning light (to the left of the speedometer) for about a second. As soon as the engine is running, the light should go off. The system then performs a self-test the first time the vehicle speed exceeds four mph. You may hear a mechanical noise during the test; this is normal. If the system detects a problem, the ABS light will come on and remain on. A diagnostic code will also be stored in the control unit, which indicates the problem area or component.

Diagnosis and repair
Refer to illustrations 2.8a and 2.8b

8   If the ABS warning light on the dash comes on and stays on while the vehicle is in operation, the ABS system requires attention. Diagnosis is quite complex, involving a number of lengthy diagnostic procedures, so we don't recommend attempting to fix the ABS system at home. However, if you're willing to do a little work, you can obtain a diagnostic trouble code - which will indicate the general area of the problem - as follows:

a) *Drive the vehicle above 20 mph for at least one minute.*
b) *Stop the vehicle and turn off the engine.*

2.2 The ABS actuator assembly houses an electric hydraulic pump, a reservoir and three solenoid valves

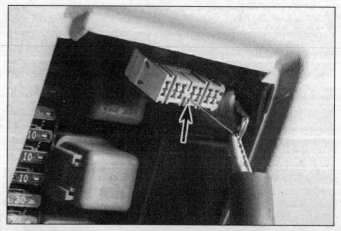

2.8a To put the control unit into output mode, ground the L terminal (arrow) on the Data Link Connector

c) Ground terminal L on the Data Link Connector (see illustration).
d) With terminal L grounded, turn the ignition switch to On.
e) After 3.6 seconds, the ABS warning light will begin flashing any stored trouble codes. The code is determined by counting the number of on-and-off flashes (see illustration). The sequence always begins with a 3.6-second "off"

period, followed by a flash, then a 1.6-second off period, then two flashes. This "Code 12" (the "start" code, not a trouble code) is followed by another 3.6-second off period, then the trouble codes are displayed, in the order in which they were stored, starting with the latest stored code. All codes are two-digit codes, so the first flash(es) indicate the tens place, followed by a longer

delay, followed by the single-digit flash(es). For example, a sequence of four flashes, then a pause, followed by a sequence of five flashes, would indicate a Code 45 (front left actuator solenoid valve).
f) Count the number of flashes of the ABS light, then refer to the accompanying table.

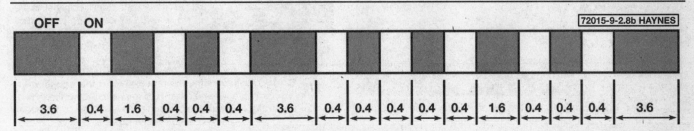

2.8b The ABS light on the dash indicates a problem in a particular circuit by the number of flashes. For example, three 0.4-second flashes, followed by a 1.6-second interval, followed by two more 0.4-second flashes indicates a code 32, which means there's a short circuit in the rear right sensor circuit

Code	Malfunctioning part	Code	Malfunctioning part
45	Front left actuator solenoid valve	31	Rear right sensor (open circuit)
41	Front right actuator solenoid valve	32	Rear right sensor (short circuit)
55	Rear actuator solenoid valve	18	Sensor rotor
25	Front left sensor (open circuit)	61	Actuator motor or motor relay
26	Front left sensor (short circuit)	63	Solenoid valve relay circuit (except power supply for relay coil)
21	Front right sensor (open circuit)		
22	Front right sensor (short circuit)	57	Power supply (low voltage)
35	Rear left sensor (open circuit)	16	Brake light switch circuit
36	Rear left sensor (short circuit)	71	Control module

Symptom	Malfunctioning part or circuit
Warning light says on when ignition switch is turned on	Control module power supply circuit Warning light bulb circuit Control module or control module connector Solenoid valve relay stuck Power supply for solenoid valve relay coil
Warning light stays on during self-diagnosis	Control module
Warning light does not come on when ignition switch is turned on	Fuse, warning light bulb or warning light circuit Control module
Warning light does not come on during self-diagnosis	Control module

9   A trouble code can be set by a simple malfunction. Although you can't troubleshoot the types of malfunctions listed in the accompanying trouble code table, you can check the following things:
a) Check the brake fluid level in the reservoir.
b) Verify that all electrical connectors are securely connected.
c) Check the fuses.
d) Check the brake system (see Chapter 1).

e) Check the brake pads (see Section 3).
f) Check the brake pedal (see Section 15).
10   After verifying that all of the above are okay, try to erase the stored trouble code(s) as follows: Unground terminal L; the ABS warning light should remain on. Within 12.5 seconds, ground the L terminal three successive times; each ground must last more than one second. The ABS light should now go out.

a) If the light stays on, take the vehicle to a dealer service department or other qualified repair shop and have the ABS system repaired.
b) If the light doesn't stay on, drive the vehicle above 20 mph for at least one minute and verify that the warning light on the dash doesn't come on again. If it does, take the vehicle to a dealer service department or other qualified repair shop and have the ABS system repaired.

9

3.5 Use a C-clamp to depress the piston into the caliper before removing the caliper and pads

3.6a Before disassembling the brake, wash it thoroughly with brake system cleaner and allow it to dry - position a drain pan under the brake to catch the residue - DO NOT use compressed air to blow off brake dust!

3.6b Remove these two bolts (upper and lower arrows) to detach the caliper from the torque plate; the middle arrow points to the brake hose banjo bolt (which shouldn't be removed unless the caliper requires service)

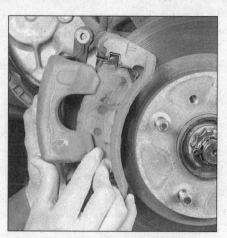

3.6c Lift the caliper off the pads

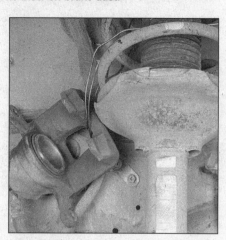

3.6d Hang the caliper out of the way with a piece of coat hanger or wire

3.6e Remove the outer brake pad

3.6f Remove the inner brake pad

3 Disc brake pads - replacement

Refer to illustrations 3.5 and 3.6a through 3.6uu

Warning: *Disc brake pads must be replaced on both front wheels at the same time - never replace the pads on only one wheel. Also, the dust created by the brake system may contain asbestos, which is harmful to your health. Never blow it out with compressed air and don't inhale any of it. An approved filtering mask should be worn when working on the brakes. Do not, under any circumstances, use petroleum-based solvents to clean brake parts. Use brake system cleaner only!*

1   Remove the cap from the brake fluid reservoir.
2   Loosen the front or rear wheel lug nuts, raise the front or rear of the vehicle and support it securely on jackstands. Block the wheels at the opposite end.
3   Remove the wheels. Work on one brake assembly at a time, using the assembled brake for reference if necessary.
4   Inspect the brake disc carefully as outlined in Section 5. If machining is necessary, follow the information in that Section to remove the disc, at which time the pads can be removed as well.
5   Push the piston back into its bore to provide room for the new brake pads. A C-clamp can be used to accomplish this (**see illustration**). As the piston is depressed to the bottom of the caliper bore, the fluid in the master cylinder will rise. Make sure that it doesn't overflow. If necessary, siphon off some of the fluid. **Warning:** *This procedure does not apply to rear caliper pistons; rear pistons use a ratcheting mechanism inside*

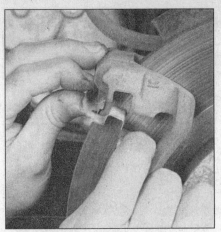

3.6g Remove the upper pad retainer

3.6h Remove the lower pad retainer

3.6i Remove the shim from the outer brake pad

3.6j Remove the shim cover from the inner brake pad

3.6k Remove the shim from the inner brake pad

3.6l Remove the caliper pins and the dust boots

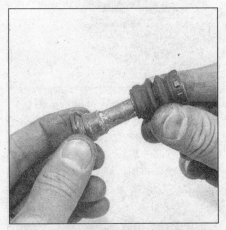

3.6m Inspect the boots for cracks and tears and replace them if they're damaged

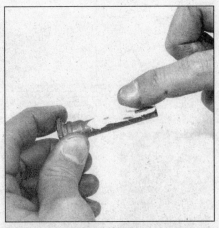

3.6n Clean the caliper pins, inspect them for scoring and corrosion, and replace them if necessary; coat the pins with high-temperature grease . . .

3.6o . . . and install them in the torque plate

the caliper which will be damaged if you try to depress the piston with a C-clamp. Rear pistons are retracted with a pair of needle-nose pliers, as shown in the photo sequence for the rear pad replacement procedure.

6   Follow the accompanying photos **(see illustrations 3.6a through 3.6y),** for the actual front pad replacement procedure. Be sure to stay in order and read the caption under each illustration. If you're replacing rear brake pads, follow the second photo sequence **(see illustrations 3.6z through 3.6uu).**

7   When reinstalling the caliper, be sure to tighten the mounting bolts to the torque listed in this Chapter's Specifications.

8   After the job has been completed, firmly depress the brake pedal a few times to bring the pads into contact with the disc. Check the level of the brake fluid, adding some if necessary. Check the operation of the brakes carefully before placing the vehicle into normal service.

9

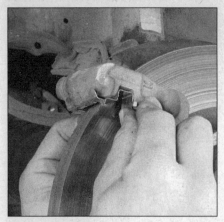

3.6p Install the upper pad retainer

3.6q Install the lower pad retainer

3.6r Make sure the upper and lower pad retainers are properly seated as shown

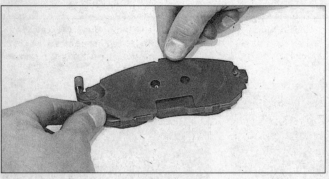

3.6s Install the shim . . .

3.6t . . . and the shim cover on the inner pad

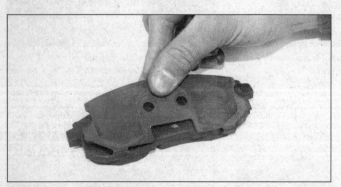

3.6u Install the shim on the outer pad

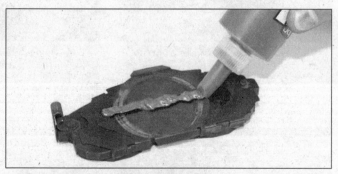

3.6v Apply anti-squeal compound to the back of both pads (let the compound "set up" a few minutes before installing them)

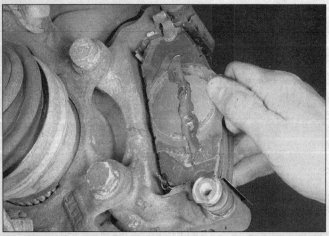

3.6w Install the inner brake pad

3.6x Install the outer brake pad

3.6y Install the caliper and the caliper bolts, tightening the caliper bolts to the torque listed in this Chapter's Specifications. **Note:** *If the caliper won't fit over the pads, use a C-clamp to push the piston into the caliper a little further*

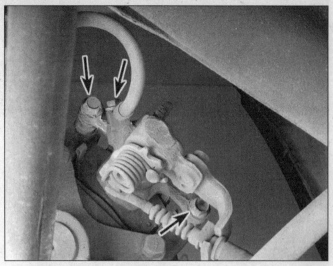

3.6z Remove the rear caliper pins (left and right arrows); the center arrow points to the banjo bolt for the brake hose - it isn't necessary to remove this bolt unless you're planning to remove the caliper for overhaul

3.6aa Remove the caliper . . .

3.6bb . . . and hang it out of the way with a piece of wire

3.6cc Remove the outer brake pad

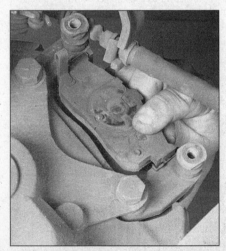

3.6dd Remove the inner brake pad

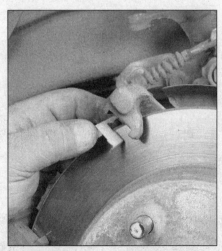

3.6ee Remove the upper pad retainer

3.6ff Remove the lower pad retainer

9

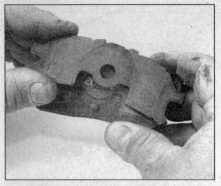

3.6gg Remove the shim from the outer pad

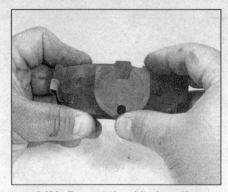

3.6hh Remove the shim from the inner pad

3.6ii Remove the caliper pins and boots

3.6jj Inspect the pins for scoring and the boots for cracks and tears

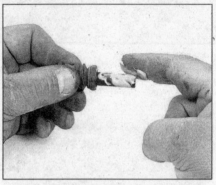

3.6kk Wipe off the pins and lubricate them with high-temperature grease before installing them

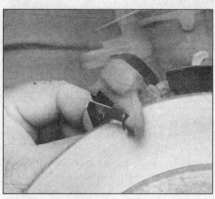

3.6ll Install the upper pad retainer

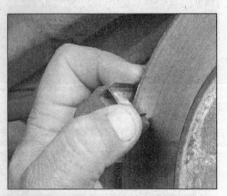

3.6mm Install the lower pad retainer

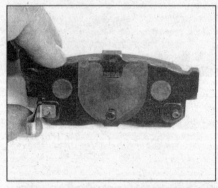

3.6nn Install the shim on the inner pad

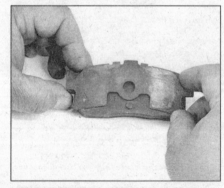

3.6oo Install the shim on the outer pad

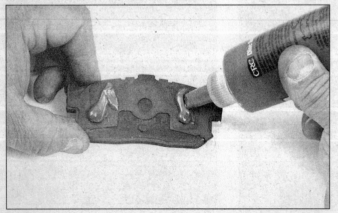

3.6pp Apply anti-squeal compound to the backs of both pads (let the compound "set up" a few minutes before installing the pads)

3.6qq Install the inner pad - notice the small projection on the pad backing plate (arrow); this must fit into a notch in the face of the caliper piston

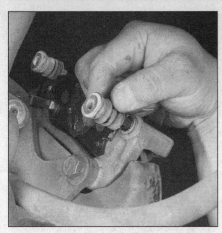

3.6rr Install the caliper pins and boots

3.6ss Install the outer pad

3.6tt Using a pair of needle-nose pliers, turn the piston clockwise to retract it into the caliper - make sure the piston is positioned so that one of the notches in the piston face will engage with the projection on the pad backing plate

---

4 Disc brake caliper - removal, overhaul and installation

Warning: *The dust created by the brake system may contain asbestos, which is harmful to your health. Never blow it out with compressed air and don't inhale any of it. An approved filtering mask should be worn when working on the brakes. Do not, under any circumstances, use petroleum-based solvents to clean brake parts. Use brake system cleaner only!*

Removal

Refer to illustrations 4.2a, 4.2b and 4.3

1   Loosen the front or rear wheel lug nuts, raise the front or rear of the vehicle and place it securely on jackstands. Block the wheels at the opposite end. Remove the front or rear wheel.

2   To disconnect the parking brake cable from the rear caliper, unbolt the cable bracket from the caliper and disengage the cable from the toggle lever **(see illustrations)**.

3   Remove the banjo bolt and discard the old copper washers **(see illustration 3.6a or 3.6z)**. Disconnect the brake hose from the caliper. Plug the brake hose to keep contaminants out of the brake system and to prevent losing any more brake fluid than is necessary **(see illustration)**. See Section 3 for the rest of the caliper removal procedure (it's part of brake pad replacement).

Overhaul

Front caliper

Refer to illustrations 4.4, 4.6 and 4.9

Note: *If an overhaul is indicated (usually because of fluid leakage), explore all options before beginning the job. New and factory rebuilt calipers are available on an exchange basis, which makes this job quite easy. If you decide to rebuild the calipers, make sure a rebuild kit is available before proceeding. Always rebuild the calipers in pairs - never rebuild just one of them. We have also included a step-by-step sequence for overhauling a rear caliper. Nevertheless, please note that rebuilding a rear caliper is more difficult than rebuilding a front caliper. If a rear caliper is leaking or otherwise defective, we*

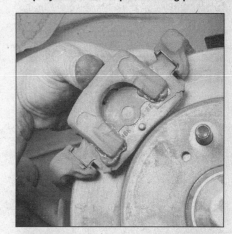

3.6uu Install the caliper and caliper bolts and tighten the bolts to the torque listed in this Chapter's Specifications

recommend buying a new or rebuilt unit.

4   Place a wood block between the piston and the caliper frame to prevent damage as it

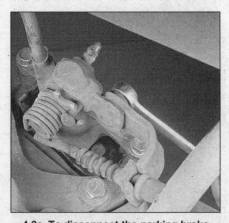

4.2a To disconnect the parking brake cable assembly from the rear brake caliper, remove the this bolt and detach the cable bracket . . .

4.2b . . . then disengage the cable end from the toggle lever

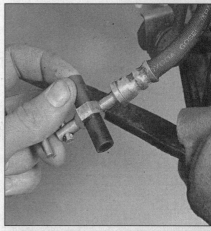

4.3 Using a piece of rubber hose of the appropriate size, plug the brake line

9

4.4 With the caliper padded to catch the piston, use compressed air to force the piston out of its bore - make sure your hands or fingers are not between the piston and caliper

4.6 To remove the seal from the caliper bore, use a plastic or wooden tool, such as a pencil

4.9 Install the dust boot flange in the upper groove of the cylinder, then insert the piston into the boot (NOT the bore) at an angle and, using a rotating motion, work the piston completely into the dust boot

is ejected. To remove the piston from the caliper, apply compressed air to the brake fluid hose connection on the caliper body **(see illustration)**. Use only enough pressure to ease the piston out of its bore. **Warning:** *Be careful not to place your fingers between the piston and the caliper, as the piston may* come out with some force.

5   Inspect the mating surfaces of the piston and caliper bore wall. If there is any scoring, rust, pitting or bright areas, replace the complete caliper unit with a new one.
6   If these components are in good condition, remove the piston seal from the caliper bore using a wooden or plastic tool **(see illustration)**. Metal tools may damage the

cylinder bore.
7   Wash all the components with brake system cleaner.
8   Submerge the new piston seal in brake fluid and install it into the lower groove in the caliper bore, making sure it isn't twisted.
9   Install the boot in the upper groove in the caliper bore. Make sure the flange on the boot seats in the groove completely. Lubricate the piston with clean brake fluid, carefully slide it into the new boot **(see illustration)**, then position it squarely in the caliper bore and apply firm (but not excessive) pressure to install it. Make sure the piston boot seats in the groove in the piston.

Rear caliper

Refer to illustrations 4.10a through 4.10zz
10   Follow the accompanying photo sequence **(see illustrations)**. Stay in order, read the captions carefully and pay attention to the way the parts fit together in the photos.
11   After you have fully disassembled the rear caliper, wash all parts with brake system cleaner.

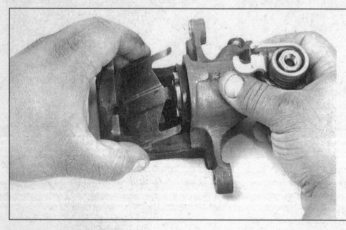

4.10a Remove the pad retainer from the caliper

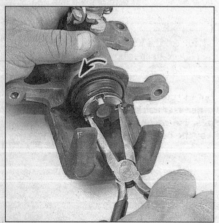

4.10b Using a pair of needle-nose pliers, withdraw the piston from the caliper bore by turning it counterclockwise

4.10c Once the piston is far enough out of the caliper bore, remove the outer circumference of the dust boot from its groove in the bore

4.10d Remove the piston from the caliper

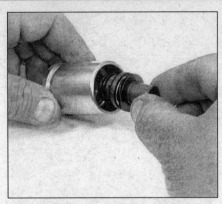

4.10e Remove the snap-ring from the piston

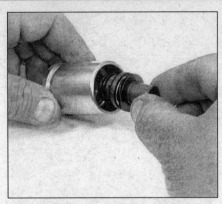

4.10f Remove the adjuster nut and four washers from the piston

4.10g Remove the snap-ring from the caliper bore . . .

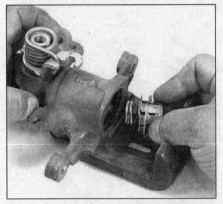

4.10h . . . then remove the spring cover, the spring, . . .

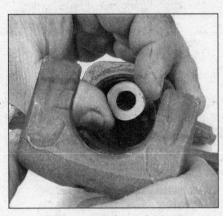

4.10i . . . and the spring seat

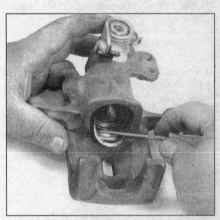

4.10j Remove the old piston seal with a pencil or an awl (if you use an awl, make sure you don't scratch the piston bore or groove)

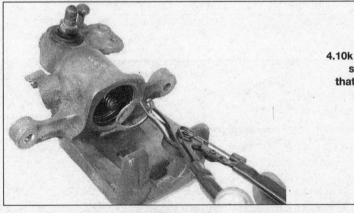

4.10k Remove the snap-ring that retains . . .

4.10l . . . the key plate (removed from the caliper bore for clarity)

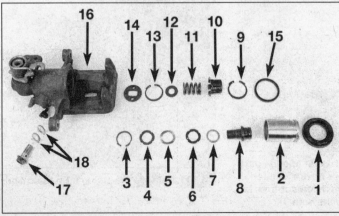

4.10m Carefully lay out the pieces you've disassembled so far, in the exact order in which they were removed; pay attention to details, like which side of each spacer faces up, which way the bearing is installed, etc.

1	Dust boot	10	Spring cover
2	Piston	11	Spring
3	Snap-ring	12	Spring seat
4	Spacer	13	Snap-ring
5	Wave washer	14	Key plate
6	Spacer	15	O-ring
7	Bearing	16	Caliper
8	Adjuster nut	17	Banjo bolt
9	Snap-ring	18	Sealing washers

9

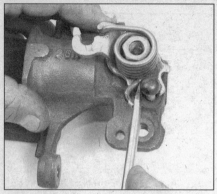

4.10n Pry off the spring from the cam nut (wear safety goggles - the spring can fly off with considerable force if it gets away from you)

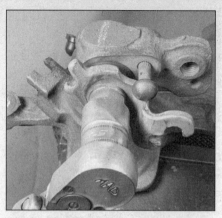

4.10o Place the caliper in a bench vise and remove the cam nut and washer

4.10p Remove the toggle lever from the cam

4.10q Pry off the cam boot

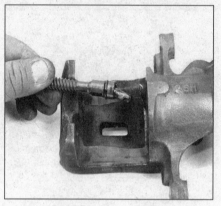

4.10r Pull out the pushrod and the strut (discard the old O-ring)

4.10s Remove the cam

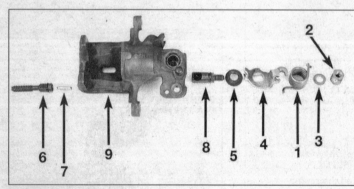

4.10t Lay out these last parts in the order in which they were removed

1	Spring	6	Pushrod and O-ring
2	Cam nut	7	Strut
3	Washer	8	Cam
4	Toggle lever	9	Caliper
5	Cam boot		

4.10u Install a new cam boot

4.10v Lubricate the cam boot . . .

4.10w . . . the cam . . .

4.10x . . . and the pushrod and new O-ring with brake assembly lube or clean brake fluid

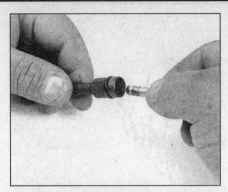

4.10y Insert the strut into the pushrod, then install them in the caliper; make sure they're fully seated into the hole in the cam

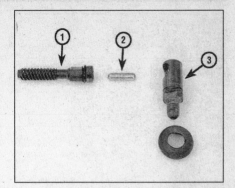

4.10z This is how the pushrod (1), strut (2) and cam (3) must fit together when installed in the caliper (removed for clarity)

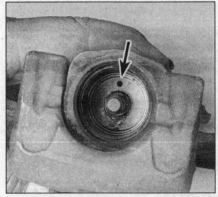

4.10aa In the bottom of the caliper bore, there's a positioning hole (arrow)

4.10bb Make sure the pin (arrow) on the key plate fits into the hole in the bottom of the caliper bore (in other words, the key plate is upside down as shown here; when you install it in the caliper, make sure this pin is facing down, into the bore and engaged with the hole)

4.10cc After installing the key plate - and making sure the pin on the key plate is in the hole in the caliper bore - install the snap-ring that retains the key plate

4.10dd Install the spring seat

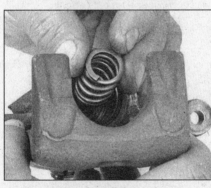

4.10ee Install the spring

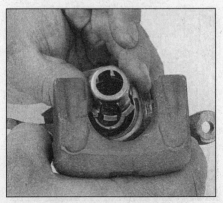

4.10ff Install the spring cover

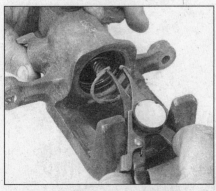

4.10gg To install the spring cover snap-ring . . .

4.10hh . . . place the caliper in a bench vise as shown, slip a deep socket over the spring and spring cover, and carefully compress the spring with a pair of large water pump pliers until the flange of the spring cover is slightly below the snap-ring groove in the caliper bore wall, then install the snap-ring

9

4.10ii Lubricate the adjuster nut with brake lube or clean brake fluid . . .

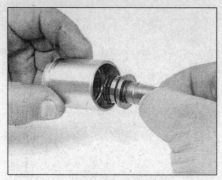

4.10jj . . . then insert the adjuster nut into the piston

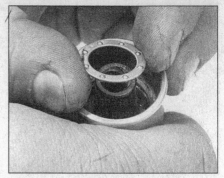

4.10kk Install the bearing . . .

4.10ll . . . spacer . . .

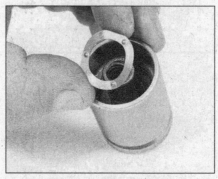

4.10mm . . . wave washer . . .

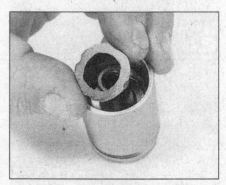

4.10nn . . . and spacer on the adjuster nut . . .

4.10oo . . . then install the snap-ring that retains them

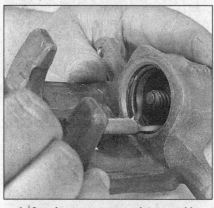

4.10pp Immerse a new piston seal in clean brake fluid or brake assembly lube, then install it into the seal groove in the caliper bore . . .

4.10qq . . . and lubricate the seal and bore with brake lube or clean brake fluid

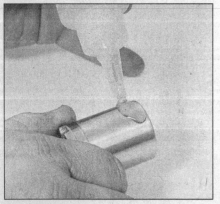

4.10rr Also lubricate the piston

4.10ss Slide the new boot onto the end of the piston and install the flange of the boot into its groove, then rotate (clockwise) the piston into the bore far enough to seat the inner edge of the boot into the groove (arrow) on the piston

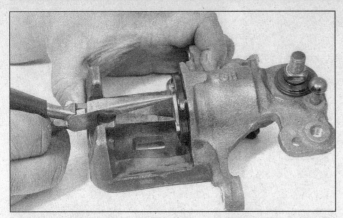

4.10tt To depress the piston, rotate it clockwise with a pair of needle-nose pliers

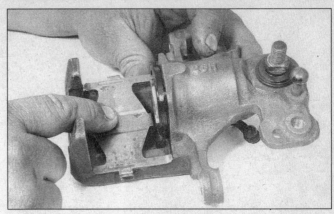

4.10uu Install a new pad retainer

4.10vv Install the toggle lever onto the cam; make sure the lever is oriented exactly as shown

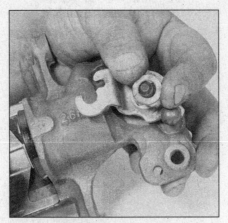

4.10ww Install the washer, . . .

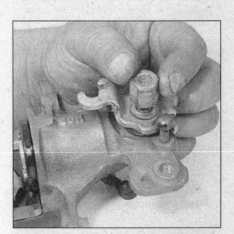

4.10xx . . . and the nut, and tighten the nut securely

4.10yy Install the spring

4.10zz Make sure the spring is installed exactly as shown

5.3 The brake pads on this vehicle were obviously neglected, as they wore down to the rivets and cut deep grooves into the disc - wear this severe means the disc must be replaced

Installation

12   Installation is the reverse of removal. Don't forget to use new copper sealing washers for the brake hose-to-caliper banjo bolt (the rebuild kit should include new washers).

13   Bleed the brake system (see Section 11). Make sure there are no leaks from the hose connections. Test the brakes carefully before returning the vehicle to normal service.

5   Brake disc - inspection, removal and installation

Inspection

Refer to illustrations 5.3, 5.4a, 5.4b and 5.5

1   Loosen the wheel lug nuts, raise the vehicle and support it securely on jackstands. Remove the wheel and install two lug nuts to hold the disc in place. **Note:** *If the lug nuts don't contact the disc when screwed on all the way, install washers under them.*

2   Remove the brake caliper (see Section 4). It isn't necessary to disconnect the brake hose. After removing the caliper bolts, suspend the caliper out of the way with a piece of wire.

3   Visually inspect the disc surface for score marks and other damage. Light scratches and shallow grooves are normal after use and may not always be detrimental to brake operation, but deep scoring - over

9

5.4a To check disc runout, mount a dial indicator as shown and rotate the disc

5.4b Using a swirling motion, remove the glaze from the disc surface with sandpaper or emery cloth

5.5 Use a micrometer to measure disc thickness

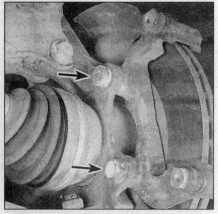

5.6a To remove the front caliper torque plate, remove these bolts (arrows)

5.6b To remove the rear caliper torque plate, remove these bolts (arrows)

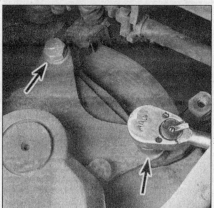

5.7 Remove the brake disc

0.039-inch (1.0 mm) - requires disc removal and refinishing by an automotive machine shop. Be sure to check both sides of the disc **(see illustration)**. If pulsating has been noticed during application of the brakes, suspect disc runout.

4   Tko check disc runout, place a dial indicator at a point about 1/2-inch from the outer edge of the disc **(see illustration)**. Set the indicator to zero and turn the disc. The indicator reading should not exceed the specified allowable runout limit. If it does, the disc should be refinished by an automotive machine shop. **Note:** *When replacing the brake pads, it's a good idea to resurface the discs regardless of the dial indicator reading, as this will impart a smooth finish and ensure a perfectly flat surface, eliminating any brake pedal pulsation or other undesirable symptoms related to questionable discs. At the very least, if you elect not to have the discs resurfaced, remove the glaze from the surface with emery cloth using a swirling motion* **(see illustration)**.

5   It's absolutely critical that the disc not be machined to a thickness under the specified minimum allowable disc refinish thickness. The minimum wear (or discard) thickness is cast into the disc. The disc thickness can be checked with a micrometer **(see illustration)**.

Removal

Refer to illustrations 5.6a, 5.6b and 5.7

6   Remove the two torque plate-to-steering knuckle bolts **(see illustrations)** and detach the torque plate.

7   Remove the lug nuts which you installed to hold the disc in place and remove the disc from the hub **(see illustration)**. If the disc is stuck to the hub and won't come off, install bolts of the proper diameter and thread pitch into the threaded holes between the wheel studs and tighten them, which will force the disc off the hub.

Installation

8   Place the disc in position over the threaded studs.

9   Install the torque plate and caliper assembly over the disc and position it on the steering knuckle. Tighten the torque plate bolts to the torque listed in this Chapter's Specifications.

10   Install the wheel, then lower the vehicle to the ground. Tighten the lug nuts to the torque listed in the Chapter 1 Specifications. Depress the brake pedal a few times to bring the brake pads into contact with the disc. Bleeding won't be necessary unless the brake hose was disconnected from the caliper. Check the operation of the brakes carefully before driving the vehicle.

6   Drum brake shoes - replacement

Refer to illustrations 6.4a through 6.4ll for 1993 through 1997 models or illustrations 6.4mm through 6.4ag for 1998 and later models
Warning: *Drum brake shoes must be replaced on both wheels at the same time - never replace the shoes on only one wheel. Also, the dust created by the brake system may contain asbestos, which is harmful to your health. Never blow it out with compressed air and don't inhale any of it. An approved filtering mask should be worn when working on the brakes. Do not, under any circumstances, use petroleum-based solvents to clean brake parts. Use brake system cleaner only!*
Caution: *Whenever the brake shoes are replaced, the return and hold-down springs should also be replaced. Due to the continuous heating/cooling cycle the springs are subjected to, they lose tension over a period of time and may allow the shoes to drag on the drum and wear at a much faster rate than normal.*

6.2 If a brake drum is "frozen" to the wheel studs, insert two bolts in the holes provided and tighten them until they push against the hub flange, which will force the drum off - notice the maximum allowable diameter cast into the drum surface

6.4a Before disassembling the brake, wash it thoroughly with brake system cleaner and allow it to dry - position a drain pan under the brake to catch the residue - DO NOT use compressed air to blow the brake dust off!

6.4b Remove the hold-down spring from the rear brake shoe; to release a hold-down spring with a brake spring tool, push it in and rotate it 90-degrees, then release pressure and remove the retainer, spring and pin

6.4c Remove the hold-down spring from the front shoe (put the pins, retainers and springs in a plastic bag so you don't lose them)

6.4d Pull the front shoe toward the front of the vehicle far enough to disengage it from the wheel cylinder and the anchor plate

1   Loosen the wheel lug nuts, raise the rear of the vehicle and support it securely on jackstands. Block the front wheels to keep the vehicle from rolling. Remove the wheels.
2   Release the parking brake and remove the brake drums. If the brake drum is difficult to remove, install a two bolts of the proper size and thread pitch into the threaded holes provided **(see illustration)** and turn them in. As the bolts are tightened, they will contact the surface of the wheel bearing flange and push off the drum, which should now come off.
3   **Note:** *All four rear brake shoes must be replaced at the same time, but to avoid mixing up parts, work on only one brake assembly at a time.*

1993 through 1997 models

4   On 1993 through 1997 models, before disassembling anything, clean off the brake assembly with brake system cleaner **(see illustration)**. Next, remove the hub and wheel bearing assembly (see Chapter 10). Then follow the accompanying illustrations for the brake shoe replacement procedures **(see**

6.4e Disengage the lower return spring from the front shoe

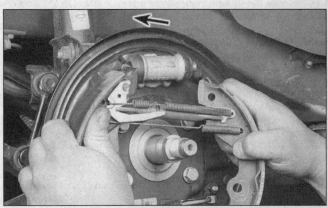

6.4f Pull the rear shoe toward the rear of the vehicle and disengage it from the wheel cylinder

9

6.4g　Remove the adjuster assembly

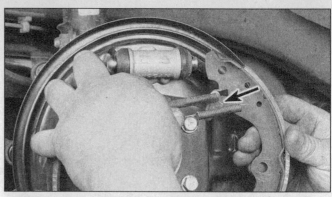

6.4h　Unhook the adjuster spring and the upper return spring from the front shoe and remove the front shoe; don't lose the spacer (arrow) for the adjuster spring

6.4i　Unhook the adjuster spring (1) from the rear shoe; do not unhook the upper return spring (2) yet

6.4j　Remove the adjuster lever

6.4k　Unhook the upper return spring from the rear shoe

6.4m　Clean, then lubricate the contact areas (arrows) on the brake backing plate with high-temperature brake grease

6.4l　Disengage the parking brake cable from the parking brake lever and remove the rear shoe

6.4n　Pop off the retainer ring from the parking brake lever pivot pin on the back of the rear shoe

6.4o　Remove the washer from the parking brake lever pivot pin

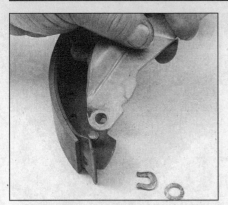

6.4p Remove the parking brake lever from the pin

6.4q If there's no pin in the new rear shoe, knock the parking brake lever pin out of the old shoe and install it in the new shoe

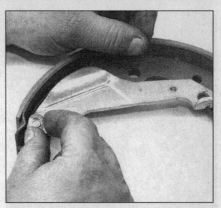

6.4r Install the parking brake lever on the new shoe, then install the washer . . .

6.4s . . . and the retainer ring

6.4t Install the spring on the parking brake cable

6.4u Pull back the spring and insert the plug on the end of the parking brake cable into the slot in the parking brake lever

6.4v Insert the hold-down spring pin through the backing plate and install the rear brake shoe

6.4w Install the hold-down spring and retainer, compress the spring with a spring tool, then rotate the retainer 90-degrees to lock it onto the pin

6.4x Make sure the rear shoe is properly engaged with the wheel cylinder at the top and properly seated against the anchor plate at the bottom

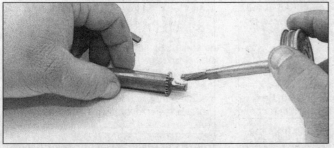

6.4y Lubricate the adjuster screw pivot end . . .

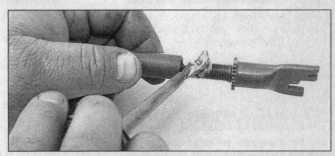

6.4z . . . and the threads with high-temperature brake grease

9

6.4aa Install the adjuster lever
as shown . . .

6.4bb . . . followed by the adjuster
screw assembly

6.4cc Install the front shoe; make sure it's
properly engaged with the wheel cylinder
and the adjuster assembly like this

6.4dd Install the front shoe hold-down
pin, spring and retainer and lock them
into place

6.4ee Hook the upper return spring into
its hole (arrows) in the rear shoe . . .

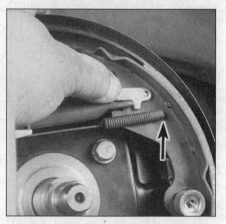

6.4ff . . . and the front shoe; install the
spacer for the adjuster spring . . .

6.4gg . . . and hook the adjuster spring
into the front shoe, through the hole in
the spacer

6.4hh . . . and hook the other end of the
adjuster spring into its notch (arrow) in
the adjuster lever

6.4ii Hook the lower return spring into its
hole in the rear shoe, . . .

illustrations 6.4b through 6.4ll). Be sure to stay in order and read the caption under each illustration.

1998 and later models

5   On 1998 and later models, before disassembling anything, clean off the brake assembly with brake system cleaner **(see illustration)**. Next, remove the hub and wheel bearing assembly (see Chapter 10). Then follow the accompanying illustrations for the brake shoe replacement procedures **(see illustrations 6.4mm through 6.4ag)**. Be sure to stay in order and read the caption under each illustration.

6   Before reinstalling the drum, it should be checked for cracks, score marks, deep scratches and hard spots, which will appear as small discolored areas. If the hard spots cannot be removed with fine emery cloth or if any of the other conditions listed above exist, the drum must be taken to an automotive machine shop to have it resurfaced. **Note:** *Professionals recommend resurfacing the drums each time a brake job is done. Resurfacing will eliminate the possibility of out-of-round drums. If the drums are worn so much that they can't be resurfaced without exceed-*

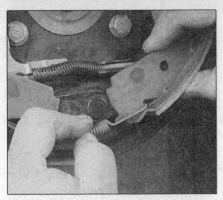

6.4jj . . . and its hole in the front shoe

6.4kk Pull the lower ends of both shoes apart and seat them against the anchor plate as shown

6.4ll This is how your completed brake assembly should look! Now go do the other rear brake

6.4mm Remove the brake hold-down springs and retainers, then pull the hold-down pins through the rear of the backing plate

6.4nn Lift the brake assembly off the backing plate as a single assembly, then remove the upper and lower return springs

6.4oo Disengage the adjuster and the adjuster spring (the one *behind* the adjuster) and remove the leading shoe

6.4pp Disengage and remove the adjuster from the trailing shoe

6.4qq Disengage and remove the adjuster lever from the trailing shoe

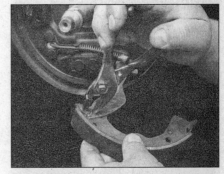

6.4rr Remove the C-clip retainer from the parking brake lever pivot pin and disconnect the parking brake lever from the trailing shoe

6.4ss Disengage the parking brake cable from the parking brake lever

6.4tt Apply a thin film of high-temperature grease to the friction surfaces on the brake backing plate

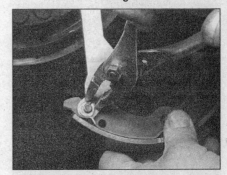

6.4uu Install the parking brake lever on the new trailing shoe and secure it with a new C-clip retainer

9

6.4vv Connect the parking brake cable and install the new trailing shoe; make sure it's properly engaged with the wheel cylinder

6.4ww Insert the hold-down pin through the rear of the backing plate and install the trailing shoe hold-down spring and retainer

6.4xx Attach the adjuster spring to the trailing shoe as shown; note that the spring is hooked into the hole from the *rear*

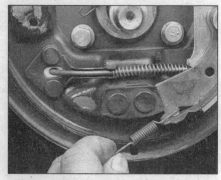

6.4yy Attach the lower return spring to the trailing shoe as shown; note that the spring is hooked into the hole from the *front*

6.4zz Attach the adjuster spring to the post on the *backside* of the leading shoe

6.4ab Insert the adjuster between the two shoes

6.4ac Make sure the adjuster is properly engaged with both shoes

6.4ad Insert the hold-down pin through the rear of the backing plate and install the hold-down spring and retainer for the leading shoe

6.4ae Install the adjuster lever - make sure it's properly engaged with the pivot pin and the adjuster

6.4af Hook the rear end of the return spring to the adjuster lever and the front end to the post on the leading shoe

6.4ag This is how your completed brake assembly should look. Now go do the other rear brake

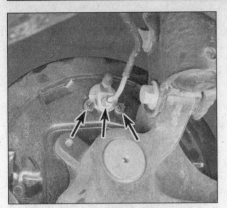

7.4 To remove the wheel cylinder, disconnect the brake line fitting (middle arrow), then remove the two wheel cylinder bolts (outer arrows)

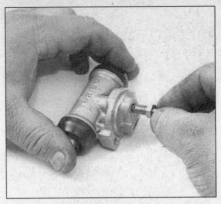

7.7a To disassemble the wheel cylinder, remove the bleeder screw . . .

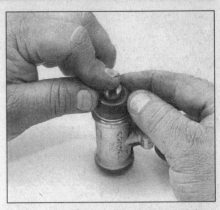

7.7b . . . push in on the piston, peel the outer lip of the dust boot out of its groove in the wheel cylinder and pull off the boot (repeat at other end) . . .

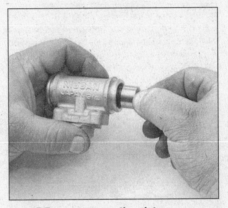

7.7c . . . remove the pistons . . .

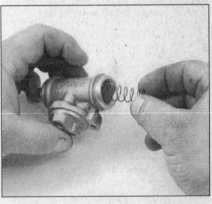

7.7d . . . and remove the spring

ing the maximum allowable diameter, which is stamped into the drum **(see illustration 6.2)**, then new ones will be required. At the very least, if you elect not to have the drums resurfaced, remove the glaze from the surface with emery cloth using a swirling motion.

7   To make a preliminary adjustment of the brake, turn the adjuster star wheel until the brakes just begin to drag on the drum as it is turned, then remove the drum and back-off the star wheel a few clicks. Repeat this procedure until no dragging can be heard when you rotate the drum. Depress the brake pedal firmly several times, then rotate the drum to ensure that the brakes are not dragging. If they are, remove the drum and back off the star wheel a little more.

8   Mount the wheel, install the lug nuts, then lower the vehicle. Tighten the wheel lug nuts to the torque listed in the Chapter 1 Specifications.

9   Make a number of forward and reverse stops and operate the parking brake to adjust the brakes until satisfactory pedal action is obtained.

10   Check the operation of the brakes carefully before driving the vehicle.

7   Wheel cylinder - removal, overhaul and installation

Note: *If an overhaul is indicated (usually because of fluid leaks or sticky operation), explore all options before beginning the job. New wheel cylinders are available, which makes this job quite easy. If you decide to rebuild the wheel cylinder, make sure a rebuild kit is available before proceeding. Never overhaul only one wheel cylinder - always rebuild both of them at the same time.*

Removal

Refer to illustration 7.4

1   Raise the rear of the vehicle and support it securely on jackstands. Block the front wheels to keep the vehicle from rolling.

2   Remove the brake shoe assembly (see Section 6).

3   Remove all dirt and foreign material from around the wheel cylinder.

4   Disconnect the brake line **(see illustration)** with a flare-nut wrench, if available. Don't pull the brake line away from the wheel cylinder.

5   Remove the wheel cylinder mounting bolts.

6   Detach the wheel cylinder from the brake backing plate and place it on a clean workbench. Immediately plug the brake line to prevent fluid loss and contamination.

Overhaul

Refer to illustrations 7.7a, 7.7b, 7.7c, 7.7d, 7.11, 7.12a and 7.12b

7   Remove the bleeder screw, boots, pistons and spring assembly from the wheel cylinder body **(see illustrations).**

8   Clean the parts with brake system cleaner. **Warning:** *Do not, under any circumstances, use petroleum-based solvents to clean brake parts!*

9   Use filtered, unlubricated compressed air to dry the wheel cylinder and blow out the passages.

10   Check the bore for corrosion and score marks. If the bore feels even slightly rough, the cylinder must be replaced with a new one.

11   Reassembly is the reverse of disassembly **(see illustration).**

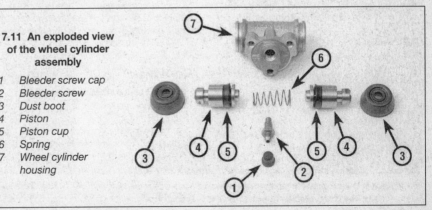

7.11 An exploded view of the wheel cylinder assembly

1   Bleeder screw cap
2   Bleeder screw
3   Dust boot
4   Piston
5   Piston cup
6   Spring
7   Wheel cylinder housing

9

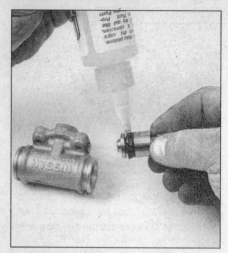

7.12a Lubricate the bore and piston cups
(and make sure the piston cups face in,
toward the wheel cylinder)

7.12b Make sure the outer lips of the dust
boots are seated in their grooves in the
wheel cylinder - slide the boot down over
the piston and push on the lower edge
until it pops into place

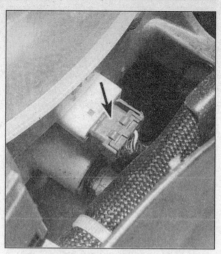

8.2 Unplug the electrical connector
(arrow) for the fluid level warning switch

8.4 Loosen the brake line fittings with a flare-nut wrench (the
other fitting is on the left side, below the fluid level
warning switch)

8.6 Remove the master cylinder mounting nuts (arrows)

12   Be sure to lubricate the bore and the new pistons and cups with clean brake fluid or brake assembly lube **(see illustration)**. Make sure the cup lips face in (toward the inside of the wheel cylinder), and make sure the dust boots are fully seated **(see illustration)**.

Installation

13   Place the wheel cylinder in position and install the bolts finger tight. Connect the brake line to the cylinder, being careful not to cross-thread the fitting. Tighten the wheel cylinder bolts to the torque listed in this Chapter's Specifications.
14   Tighten the brake line securely and install the brake shoe assembly (see Section 6).
15   Bleed the brakes (see Section 11).
16   Check the operation of the brakes carefully before driving the vehicle.

8   Master cylinder - removal, overhaul and installation

Note: *Before deciding to overhaul the master cylinder, check on the availability and cost of a new or factory rebuilt unit and also the availability of a rebuild kit.*

Removal

Refer to illustrations 8.2, 8.4 and 8.6

1   Disconnect the cable from the negative battery terminal.
2   Unplug the electrical connector for the fluid level warning switch **(see illustration)**.
3   Remove as much fluid as possible from the reservoir with a syringe.
4   Place rags under the fittings and prepare caps or plastic bags to cover the ends

of the lines once they're disconnected. **Caution:** *Brake fluid will damage paint. Cover all body parts and be careful not to spill fluid during this procedure.* Loosen the fittings at the ends of the brake lines where they enter the master cylinder **(see illustration)**. To prevent rounding off the flats, use a flare-nut wrench, which wraps around the fitting hex.
5   Pull the brake lines away from the master cylinder and plug the ends to prevent contamination.
6   Remove the nuts attaching the master cylinder to the power booster **(see illustration)**. Pull the master cylinder off the studs to remove it. Again, be careful not to spill the fluid as this is done. Remove and discard the old gasket between the master cylinder and the power brake booster.

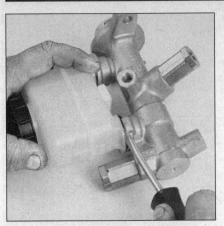

8.8a Pry the reservoir loose from the grommets . . .

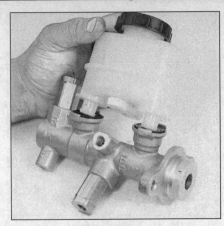

8.8b . . . pull off the reservoir . . .

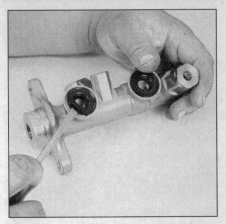

8.8c . . . and pry out the grommets

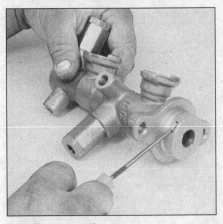

8.9a To remove the stopper cap, pry open the two small tangs on the side with a small screwdriver . . .

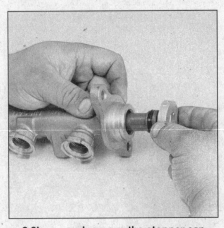

8.9b . . . and remove the stopper cap

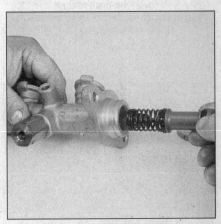

8.10a Remove the primary piston assembly . . .

8.10b . . . tap the master cylinder on a block of wood and remove the secondary piston assembly

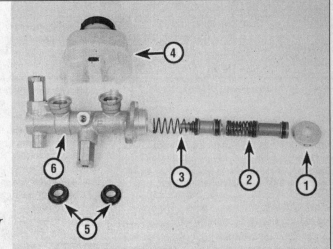

8.11 Lay out the parts like this to ensure proper reassembly - note the direction that the piston cups face

1 Stopper cap
2 Primary piston assembly
3 Secondary piston assembly
4 Reservoir
5 Grommets
6 Master cylinder body

Overhaul

Refer to illustrations 8.8a, 8.8b, 8.8c, 8.9a, 8.9b, 8.10a, 8.10b, 8.11, 8.14a, 8.14b, 8.14c, 8.15a, 8.15b, 8.16a, 8.16b, 8.16c, 8.16d and 8.17

7   Before attempting the overhaul of the master cylinder, obtain the proper rebuild kit.

8   Pull off the reservoir and remove the grommets **(see illustrations)**.

9   Remove the stopper cap **(see illustrations)**.

10   The internal components can now be removed from the bore **(see illustrations)**.

11   Note the sequence in which the parts were disassembled so they can be returned to their original locations. Laying all the parts out in order is one way to ensure everything is reassembled correctly **(see illustration)**.

12   Carefully inspect the bore of the master cylinder. Any deep score marks or other damage will mean a new master cylinder is required. DO NOT attempt to hone the bore.

13   Replace all parts included in the rebuild

9

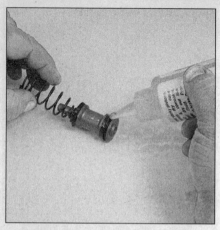

8.14a Lubricate the primary and secondary cups with brake assembly lube or clean brake fluid

8.14b Insert the secondary piston assembly . . .

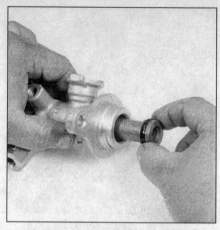

8.14c . . . and the primary piston assembly into the master cylinder

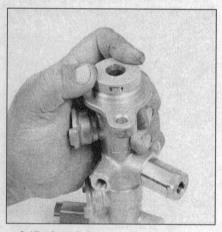

8.15a Install the new stopper cap . . .

8.15b . . . and bend the tangs inward

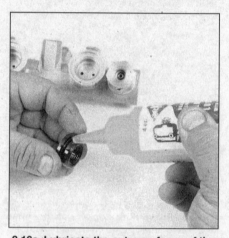

8.16a Lubricate the outer surfaces of the new grommets with brake assembly lube or clean brake fluid to make them easier to push into the master cylinder . . .

kit, following any instructions in the kit. Clean all re-used parts with brake system cleaner. **Warning:** *Do not use any petroleum-based solvents. During reassembly, lubricate all parts liberally with clean brake fluid.*
14   Lubricate the assembled secondary and primary piston assemblies with clean brake fluid or brake assembly lube and insert them into the bore, bottoming them against the end of the master cylinder **(see illustrations)**.
15   Place the cylinder in a bench vise and install the stopper bolt **(see illustrations)**.

16   Install the reservoir grommets and reservoir **(see illustrations)**.
17   If you'd like to make the fluid level in the reservoir easier to read, use a laundry marker to highlight the raised "MAX" and "MIN"

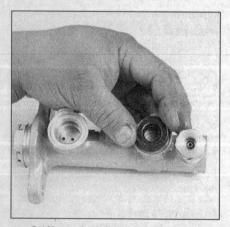

8.16b . . . install the grommets . . .

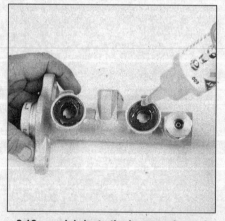

8.16c . . . lubricate the inner surface of the grommets to make the reservoir easier to push into the grommets . . .

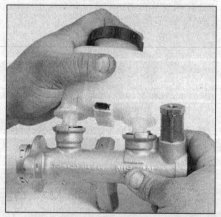

8.16d . . . and install the reservoir by pushing it firmly into the grommets

8.17 While you've got the master cylinder out of the vehicle, this is an excellent time to highlight the "MIN" and "MAX" levels with a laundry marker; this will make them much easier to read

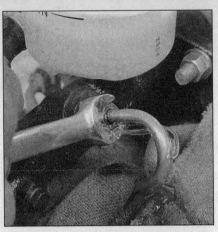

8.28 Have an assistant depress the brake pedal and hold it down, then loosen the fitting nut, allowing the air and fluid to escape; repeat this procedure on both fittings until the fluid is clear of air bubbles

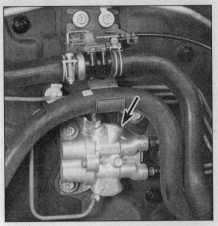

9.1 The proportioning valve (arrow) is mounted on the firewall; to replace it, simply unscrew all the threaded fittings and disconnect and plug the brake lines, then remove the mounting bolt

marks on the side of the reservoir (see illustration).

18   Bench bleed the master cylinder before installing it. You'll have to apply pressure to the master cylinder piston and, at the same time, control flow from the brake line outlets, so put the master cylinder in a vise, with the jaws of the vise clamping on the mounting flange.

19   If you can find them, insert threaded plugs into the brake line outlet holes and snug them down so no air will leak past them, but not so tight that they can't be easily loosened. If you can't find plugs that will fit, you can use your fingers to block the holes (see Step 23).

20   Fill the reservoir with brake fluid of the recommended type (see Chapter 1).

21   Remove one plug and push the piston assembly into the bore to expel the air from the master cylinder. A large Phillips screwdriver can be used to push on the piston assembly.

22   To prevent air from being drawn back into the master cylinder, the plug must be replaced and snugged down before releasing the pressure on the piston.

23   Repeat this procedure until brake fluid, free of air bubbles, is expelled from the brake line outlet hole. When only brake fluid is expelled, repeat the procedure at the other outlet hole and plug. Be sure to keep the master cylinder reservoir filled with brake fluid to prevent the introduction of air into the system.

24   Since high pressure isn't involved in the bench bleeding procedure, an alternative to the removal and replacement of the plugs with each stroke of the piston assembly is available. Before pushing in on the piston assembly, remove the plug as described in Step 20. Before releasing the piston, however, instead of replacing the plug, simply put your finger tightly over the hole to keep air from being drawn back into the master cylin-

der. Wait several seconds for brake fluid to be drawn from the reservoir into the bore, then depress the piston again, removing your finger as brake fluid is expelled. Be sure to put your finger back over the hole each time before releasing the piston, and when the bleeding procedure is complete for that outlet, replace the plug and tighten it before going on to the other port.

Installation

Refer to illustration 8.28

25   Install the master cylinder over the studs on the power brake booster and tighten the nuts only finger-tight at this time. Don't forget to use a new gasket.

26   Thread the brake line fittings into the master cylinder. Since the master cylinder is still a bit loose, it can be moved slightly so the fittings thread in easily. Don't strip the threads as the fittings are tightened.

27   Tighten the mounting nuts to the torque listed in this Chapter's Specifications. Tighten the brake line fittings securely.

28   Fill the master cylinder reservoir with fluid, then bleed the the lines at the master cylinder, followed by bleeding the remainder of the brake system (see Section 11). To bleed the lines at the master cylinder, have an assistant depress the brake pedal and hold it down. Loosen the fitting to allow air and fluid to escape (see illustration). Tighten the fitting, then allow your assistant to return the pedal to its rest position. Repeat this procedure on both fittings until the fluid is free of air bubbles, then bleed the rest of the system. Check the operation of the brake system carefully before driving the vehicle. **Warning:** *If you do not have a firm brake pedal at the end of the bleeding procedure, or have any doubts as to the effectiveness of the brake system, DO NOT drive the vehicle. Have it towed to a dealer service department or other qualified repair shop for diagnosis.*

9   Proportioning valve - replacement

Refer to illustration 9.1

1   The proportioning valve (see illustration) is mounted on the firewall. Its purpose is to limit hydraulic pressure to the rear brakes under heavy braking conditions to prevent rear wheel lockup.

2   The valve is not serviceable; if you suspect it's malfunctioning, have it checked by a dealer service department or other repair shop equipped with the necessary pressure gauges.

3   If the valve is leaking or has been determined to be defective, replace it by unscrewing the brake lines with a flare-nut wrench and unbolting the valve from its mounting bracket. After the new valve is installed, bleed the brake system (see Section 11).

10   Brake hoses and lines - inspection and replacement

Inspection

1   About every six months, with the vehicle raised and supported securely on jackstands, the rubber hoses which connect the steel brake lines with the front and rear brake assemblies should be inspected for cracks, chafing of the outer cover, leaks, blisters and other damage. These are important and vulnerable parts of the brake system and inspection should be complete. A light and mirror will be helpful for a thorough check. If a hose exhibits any of the above conditions, replace it with a new one.

Replacement

Front brake hose

Refer to illustrations 10.3 and 10.4

2   Loosen the wheel lug nuts, raise the

9

10.3 Loosen the threaded fitting on the brake line; use a flare-nut wrench to protect the corners of the nut

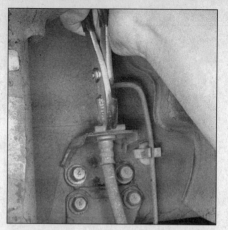

10.4 Pull off the U-clip with a pair of pliers

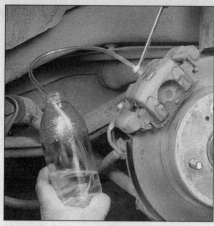

11.8 When bleeding the brakes, a hose is connected to the bleed screw at the caliper or wheel cylinder and then submerged in brake fluid - air will be seen as bubbles in the tube and container (all air must be expelled before moving to the next wheel)

vehicle and support it securely on jackstands. Remove the wheel.

3   At the bracket, unscrew the brake line fitting from the hose **(see illustration)**. Use a flare-nut wrench to prevent rounding off the corners.

4   Remove the U-clip from the female fitting at the bracket with a pair of pliers **(see illustration)**, then pass the hose through the bracket.

5   At the caliper end of the hose, remove the banjo bolt, then separate the hose from the caliper. Note that there are two copper sealing washers on either side of the banjo fitting - they should be replaced with new ones during installation.

6   Remove the U-clip from the strut bracket, then detach the hose from the bracket.

7   To install the hose, pass the caliper fitting end through the strut bracket, then connect the fitting to the caliper with the banjo bolt and new copper washers.

8   Make sure the hose isn't twisted between the caliper and the strut bracket.

9   Route the hose into the frame bracket, again making sure it isn't twisted, then connect the brake line fitting, starting the threads by hand. Install the U-clip, then tighten the fitting securely.

10   Bleed the caliper (see Section 11).

11   Install the wheel and lug nuts, lower the vehicle and tighten the lug nuts to the torque listed in the Chapter 1 Specifications.

Rear brake hose

12   The rear brake hose serves as the flexible connection between two rigid metal lines, one on the body and the other on the axle. Both ends of the hose are attached to these metal lines with threaded fittings and U-clips. Refer to Steps 2, 3 and 4. Be sure to bleed the wheel cylinders when you're done (see Section 11).

Metal brake lines

13   When replacing brake lines, be sure to

use the correct parts. Don't use copper tubing for any brake system components. Purchase steel brake lines from a dealer or auto parts store.

14   Prefabricated brake line, with the tube ends already flared and fittings installed, is available at auto parts stores and dealer parts departments.

15   When installing the new line, make sure it's securely supported in the brackets and has plenty of clearance between moving or hot components.

16   After installation, check the master cylinder fluid level and add fluid as necessary. Bleed the brake system (see Section 11) and test the brakes carefully before driving the vehicle in traffic.

11 Brake hydraulic system - bleeding

Refer to illustration 11.8

Warning: *Wear eye protection when bleeding the brake system. If the fluid comes in contact with your eyes, immediately rinse them with water and seek medical attention.*

Note: *Bleeding the hydraulic system is necessary to remove any air that manages to find its way into the system when it's been opened during removal and installation of a hose, line, caliper or master cylinder.*

1   You'll probably have to bleed the system at all four brakes if air has entered it due to low fluid level, or if the brake lines have been disconnected at the master cylinder.

2   If a brake line was disconnected only at a wheel, then only that caliper or wheel cylinder must be bled.

3   If a brake line is disconnected at a fitting located between the master cylinder and any of the brakes, that part of the system served by the disconnected line must be bled.

4   Remove any residual vacuum from the brake power booster by applying the brake several times with the engine off.

5   Remove the master cylinder reservoir cover and fill the reservoir with brake fluid. Reinstall the cover. **Note 1:** *Check the fluid*

level often during the bleeding operation and add fluid as necessary to prevent the fluid level from falling low enough to allow air bubbles into the master cylinder. **Note 2:** *If you're working on a model equipped with ABS, turn the ignition switch off and disconnect the electrical connectors for the ABS actuator or detach the battery ground cable.*

6   Have an assistant on hand, as well as a supply of new brake fluid, a clear plastic container partially filled with clean brake fluid, a length of 3/16-inch plastic, rubber or vinyl tubing to fit over the bleeder valve and a wrench to open and close the bleeder valve.

7   Beginning at the right rear wheel, loosen the bleeder valve slightly, then tighten it to a point where it's snug but can still be loosened quickly and easily.

8   Place one end of the tubing over the bleeder valve and submerge the other end in brake fluid in the container **(see illustration)**.

9   Have the assistant pump the brakes slowly a few times to get pressure in the system, then hold the pedal down firmly.

10   While the pedal is held down, open the bleeder valve just enough to allow a flow of fluid to leave the valve. Watch for air bubbles to exit the submerged end of the tube. When the fluid flow slows after a couple of seconds, close the valve and have your assistant release the pedal.

11   Repeat Steps 9 and 10 until no more air is seen leaving the tube, then tighten the bleeder valve and proceed to the left front wheel, the left rear wheel and the right front wheel, in that order, and perform the same procedure. Be sure to check the fluid in the master cylinder reservoir frequently.

12   Never use old brake fluid. It contains moisture which will deteriorate the brake system components and could cause the fluid to boil, which could render the brake system inoperative.

13   Refill the master cylinder with fluid at the

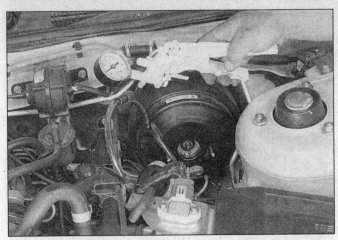

12.5a Apply about 20 in-Hg of vacuum to the power brake booster before measuring the length of the output rod

12.5b Measure the length of the booster output rod from the top of the rod to the master cylinder mounting surface and compare your measurement to the specified length listed in this Chapter's Specifications

end of the operation. If you're working on a model with ABS, be sure to reconnect the electrical connectors to the ABS actuator or reconnect the battery.

14   Check the operation of the brakes. The pedal should feel solid when depressed, with no sponginess. If necessary, repeat the entire process. **Warning:** *Do not operate the vehicle if you're in doubt about the effectiveness of the brake system.*

---

12   Power brake booster - check, replacement and adjustment

Check

Operating check

1   Depress the brake pedal several times with the engine off and make sure there's no change in the pedal reserve distance.
2   Depress the pedal and start the engine. If the pedal goes down slightly, operation is normal.

Airtightness check

3   Start the engine and turn it off after one or two minutes. Depress the brake pedal slowly several times. If the pedal depresses less each time, the booster is airtight.
4   Depress the brake pedal while the engine is running, then stop the engine with the pedal depressed. If there's no change in the pedal reserve travel after holding the pedal for 30 seconds, the booster is airtight.

Output rod length check

Refer to illustrations 12.5a and 12.5b

5   Remove the master cylinder (see Section 8). It isn't necessary to disconnect the lines from the master cylinder, as long as you can move it forward far enough to provide clearance for the following measurement. But make sure you don't kink the metal lines. Apply about 20 in-Hg of vacuum to the brake booster with a hand-operated vacuum pump **(see illustration)**, measure the length of the output rod

(see illustration) and compare your measurement to the dimensions listed in this Chapter's Specifications. If the rod length is outside specifications, replace the booster.

Replacement

Refer to illustration 12.8

Note: *Power brake booster units shouldn't be disassembled. They require special tools not normally found in most automotive repair stations or shops. They're fairly complex and, because of their critical relationship to brake performance, should be replaced with a new or rebuilt one.*

6   Remove the brake master cylinder, if you haven't already done so (see Section 8).
7   Disconnect the vacuum hose leading from the engine to the booster. Be careful not to damage the hose when removing it from the booster fitting.
8   Remove the steering column lower finish panel. Locate the pushrod clevis connecting

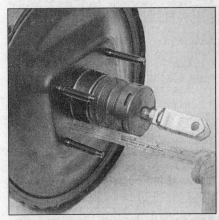

12.8 Remove this retaining clip (right arrow), pull out the clevis pin and detach the pushrod from the brake pedal; the two left power brake booster mounting nuts (arrows) are visible in this photo (the two right nuts, not visible in this photo, are to the right of the booster mounting bracket)

the booster to the brake pedal **(see illustration)**.
9   Remove the clevis pin retaining clip with pliers and pull out the clevis pin.
10   Remove the four nuts and washers holding the brake booster to the firewall **(see illustration 12.8)**.
11   Slide the booster straight out from the firewall until the studs clear the holes.
12   Installation is the reverse of removal. But be sure to measure the following dimension before installing the power brake booster assembly.

Adjustment

Refer to illustration 12.13

13   Measure the distance between the power brake booster and the hole in the clevis **(see illustration)** and compare it to the booster-to-clevis dimension listed in this Chapter's Specifications. If it isn't the same, loosen the adjusting nut and turn the clevis in or out to the specified length, then tighten the nut.

12.13 Measure the distance between the power brake booster and the hole in the clevis and compare your measurement to the dimension listed in this Chapter's Specifications; if they're not the same, adjust the clevis before installing the power brake booster

9

13.4 To adjust parking brake lever travel, turn this adjusting nut until the specified number of clicks is obtained

14.3a On models with rear drum brakes, remove the spring from the parking brake cable . . .

14.3b . . . then remove these two bolts (arrows) to detach the cable housing from the backing plate, and pull the cable assembly out of the plate

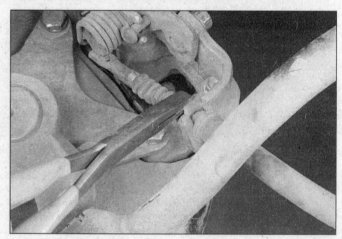

14.4a On models with rear disc brakes, remove the lock plate from the bracket

13   Parking brake - check and adjustment

Check

1   The parking brake lever, when properly adjusted, should travel four to five clicks when a moderate pulling force is applied.
2   If the parking brake lever travels less than the specified minimum number of clicks, it might not be releasing completely and the shoes or pads could even be dragging against the drum or disc. If the lever can be pulled up more than the specified maximum number of clicks, the parking brake may not hold adequately on an incline, allowing the car to roll.

Adjustment

Refer to illustration 13.4
3   To gain access to the parking brake cable adjuster, remove the center console (see Chapter 11).
4   Loosen or tighten the adjusting nut (see illustration) until the desired travel is attained. Turn the nut clockwise to tighten

the cable and decrease the number of clicks at the parking brake lever, or turn it counter-clockwise to loosen the cable and increase the number of clicks at the lever.
5   Install the console (see Chapter 11).

14   Parking brake cables - replacement

Rear cables

Refer to illustrations 14.3a 14.3b, 14.4a, 14.4b, 14.5a, 14.5b, 14.6a, 14.6b and 14.7
1   Make sure the parking brake is completely released.
2   Loosen the rear wheel lug nuts, raise the rear of the vehicle and support it securely on jackstands. Block the front wheels. Remove the wheel.
3   On models with rear drum brakes, remove the brake drum and brake shoes, and disconnect the cable from the parking brake levers on the rear shoe (see Section 6). Remove the spring from the cable (see illustration), disconnect the cable ferrule from the

backing plate (see illustration) and pull the cable through the backing plate.
4   On models with rear disc brakes, remove the locking plate that attaches the cable to the bracket bolted to the caliper (see illustration) and disengage the cable from the toggle lever (see illustration).

14.4b . . . then disconnect the parking brake cable from the toggle lever

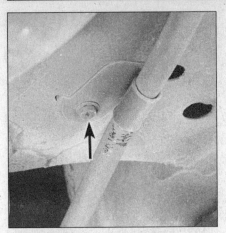

14.5a Rear bracket bolt (arrow) for left rear parking brake cable

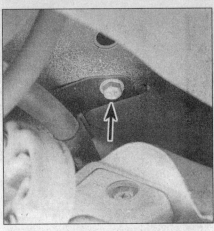

14.5b Center bracket bolt (arrow) for left rear parking brake cable

14.6a To gain access to the rear bracket bolt (arrow) for the right rear parking brake cable, remove this heat shield over the exhaust pipe

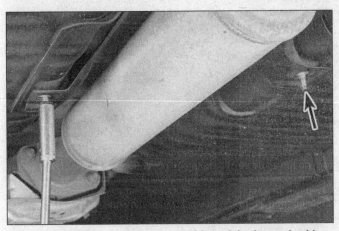

14.6b To get at the equalizer assembly and the forward cable, remove this heat shield

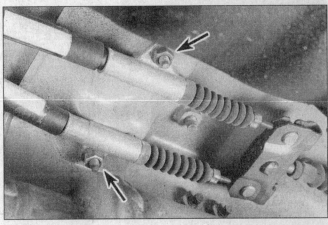

14.7 To disconnect the rear parking brake cables from the equalizer, remove these bracket nuts (arrows), slide the cables forward slightly and disengage the cable ends from the equalizer

5   Unbolt any cable brackets from the frame or the suspension pieces **(see illustrations)**.

6   Remove the exhaust pipe and catalytic converter heat shields **(see illustrations)**.

7   Remove the two cable housing nuts just behind the equalizer and disengage the cables from the equalizer **(see illustration)**.

8   Installation is the reverse of removal. Apply a light coat of grease to the portion of the cable end that engages with the equalizer.

9   Adjust the parking brake assembly when you're done (see Section 13).

Front cable

Refer to illustrations 14.11 and 14.13

10   Remove the center console (see Chapter 11).

11   Unplug the electrical connector **(see illustration)** for the parking brake warning light switch.

12   Remove the parking brake cable adjusting nut **(see illustration 13.4)**.

13   Remove the two bolts from the parking brake lever base **(see illustration)**.

14   Raise the rear of the vehicle and support it securely on jackstands. Block the front wheels. Remove the catalytic converter heat shield **(see illustration 14.6c)**.

15   Remove the two cable housing nuts just behind the equalizer and disengage the cables

from the equalizer **(see illustration 14.7)**.

16   Pull the front cable down through the floorpan.

17   Installation is the reverse of removal.

18   Adjust the parking brake assembly when you're done.

14.11 Unplug the electrical connector (arrow) for the parking brake warning light switch

14.13 To detach the parking brake lever (and the front cable) from the floorpan, remove these two bolts (arrows)

9

15.1 With the brake pedal fully released, measure the distance from the top of the pad to the floor

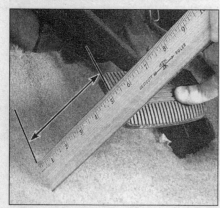

15.12 To measure brake pedal depressed height, start the engine, press the brake pedal all the way down and measure the distance between the pedal pad and the floor

15   Brake pedal - adjustment

Brake pedal released height

Refer to illustrations 15.1 and 15.3

1   With the brake pedal fully released, measure the distance from the top of the pad to the floor **(see illustration)**.
2   If the height is not as listed in the Specifications Section at the beginning of this Chapter it must be adjusted.
3   Loosen the locknut just in front of the power brake booster clevis **(see illustration)**.
4   Turn the booster input rod until the pedal height is correct.
5   Tighten the locknut.
6   After adjusting the pedal height, check the freeplay.

Brake pedal freeplay

Refer to illustration 15.7

7   Press down lightly on the brake pedal and, with a ruler, measure the distance that it moves freely before resistance is felt **(see illustration)**. The freeplay should be within the specified limits. If it isn't, it must be adjusted.

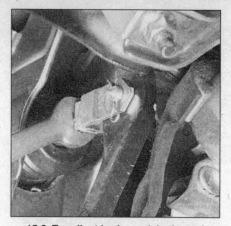

15.3 To adjust brake pedal released height, loosen the locknut in front of the brake booster clevis and turn the input rod until free height is correct (this procedure is also used to adjust brake pedal freeplay)

8   Loosen the locknut for the brake booster clevis **(see illustration 15.3)**.
9   Turn the booster input rod until the pedal freeplay is correct.
10   Tighten the locknut.

Brake pedal depressed height

Refer to illustration 15.12

11   After checking and, if necessary, adjusting the pedal released height and freeplay, the pedal depressed height must be checked.
12   With the engine running, press the brake pedal fully and measure the pedal pad-to-floor distance **(see illustration)**.
13   If the minimum depressed height is below that listed in the Specifications Section listed at the beginning of this Chapter, check the brake system for leaks or other damage.

16   Brake light switch - check and replacement

Check

Refer to illustration 16.1

1   The brake light switch **(see illustration)** is located on a bracket at the top of the brake pedal. The switch activates the brake lights at the rear of the vehicle when the pedal is depressed.
2   To check the brake light switch, simply note whether the brake lights come on when the pedal is depressed and go off when the pedal is released.
3   If the brake lights don't come on when the brake pedal is depressed, make sure the brake pedal is correctly adjusted (see Section 15). Then try adjusting the switch as follows.
4   Two locknuts - one in front of the switch mounting bracket and one behind it - secure the switch to the bracket. Loosen both locknuts, then move the switch forward or backward to provide a 0.012 to 0.039-inch clearance between the switch plunger and the

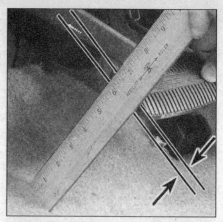

15.7 To measure brake pedal freeplay, press down lightly on the brake pedal and, with a ruler, measure the distance that it moves freely before resistance is felt

pedal stopper, with no pressure on the plunger, and tighten the locknuts. Recheck the clearance to verify that it didn't change when you tightened the locknuts. The switch should now function properly.
5   If the switch still doesn't work properly, either it isn't getting voltage, or the switch itself is defective. Use a voltmeter or test light to verify that there's voltage at the switch connector. With the pedal at rest, voltage should be present at one of the terminals of the switch. With the pedal depressed, voltage should be present at both terminals. If voltage isn't present at both terminals when the pedal is depressed, replace the switch.

Replacement

6   Unplug the electrical connector from the brake light switch.
7   Remove the front locknut, pull the switch to the rear and remove it.
8   Installation is the reverse of removal.
9   Adjust the brake pedal (see Section 15), then adjust the switch (see above).

16.1 The brake light switch (arrow) is located at the top of the brake pedal; it's attached to its mounting bracket with two locknuts, one in front of the bracket, one behind; to remove the switch, simply unplug the electrical connector, remove the front locknut and pull the switch to the rear

Chapter 10
Suspension and steering systems

Contents

Specifications

Torque specifications

Ft-lbs (unless otherwise indicated)

Front suspension

	Ft-lbs
Control arm	
Compression rod clamp	
Nut	97 to 117
Bolts	29 to 36
Control arm-to-steering knuckle balljoint nut	52 to 64
Pivot bolt nut	
1993 and 1994	65 to 87
1995 and later	87 to 108
Stabilizer bar	
Link-to-stabilizer nut	144 to 192 in-lbs
Link-to-control arm bolt	30 to 35
Stabilizer clamp bolts	144 to 192 in-lbs
Strut/coil spring assembly	
Strut-to-steering knuckle bolts/nuts	87 to 108
Strut-to-body upper mounting nuts	29 to 40
Piston rod nut	43 to 58

10

Torque specifications

Ft-lbs (unless otherwise indicated)

Rear suspension

Brake backing plate-to-rear knuckle bolts .... 29 to 38
Hub and bearing assembly retaining nut....................................... 138 to 188
Rear suspension arms
    Lateral links
        Lateral link-to-suspension member nuts .... 72 to 87
        Lateral link-to-rear knuckle nuts .... 62 to 72
    Radius rod bolts/nuts (both ends).. 62 to 72
Stabilizer bar
    Stabilizer bushing clamp nuts .... 30 to 35
    Stabilizer-to-connecting rod nuts .... 30 to 35
    Stabilizer connecting rod-to-strut nuts .... 30 to 35
Strut/coil spring assembly
    Strut-to-rear knuckle nuts .... 87 to 108
    Strut-to-body upper mounting nuts .... 31 to 40
    Piston rod nut.. 43 to 58

Steering

Airbag module Torx bolts .... 132 to 216 in-lbs
Steering gear mounting bracket bolts and nuts .... 54 to 72
Steering wheel nut .... 22 to 29
Tie-rod end-to-steering knuckle nut... 22 to 29

1.1 Typical front suspension components

1	Strut/coil spring assembly	4	Stabilizer bar	7	Outer CV joint
2	Control arm	5	Inner CV joint	8	Driveaxle assembly
3	Control arm bushing clamp	6	Vibration damper (right driveaxle only)	9	Steering gear

1 General information

Refer to illustrations 1.1 and 1.2

The front suspension system **(see illustration)** is a strut/coil spring design. The upper end of each strut is attached to the vehicle body. The lower end of the strut is connected to the upper end of the steering knuckle. The steering knuckle is attached to a balljoint mounted on the outer end of the control arm. The balljoint is an integral part of the control arm; if the balljoint is worn, the control arm must be replaced. A stabilizer bar is used on all models. The bar is attached to the frame with a pair of clamps and to the control arms with link rods.

The rear suspension system **(see illustration)** also uses strut/coil springs, a pair of lateral suspension arms and a trailing arm, or radius rod, at each corner. The upper ends of the struts are attached to the vehicle body and their lower ends are attached to the upper ends of the rear knuckles. The lower ends of the knuckles are attached to the outer ends of the lateral arms; the strut/knuckle/hub assemblies are positioned laterally by the trailing arms. A stabilizer bar is attached to the vehicle by a pair of brackets and to the struts by link rods.

The rack-and-pinion steering gear is located behind the engine/transaxle assembly at the bottom of the firewall (it's bolted to the lower rear crossmember). The steering gear actuates the tie-rods, which are attached to the steering knuckles. The inner ends of the tie-rods are protected by rubber boots which should be inspected periodically for secure attachment, tears and leaking lubricant.

The power assist system consists of a belt-driven pump and associated lines and hoses. The fluid level in the power steering pump reservoir should be checked periodically (see Chapter 1).

The steering wheel operates the steering shaft, which actuates the steering gear through universal joints. Looseness in the steering can be caused by wear in the steering shaft universal joints, the steering gear, the tie-rod ends and loose retaining bolts.

Frequently, when working on the suspension or steering system components, you may come across fasteners which seem impossible to loosen. These fasteners on the underside of the vehicle are continually subjected to water, road grime, mud, etc., and can become rusted or "frozen," making them extremely difficult to remove. In order to unscrew these stubborn fasteners without damaging them (or other components), be sure to use lots of penetrating oil and allow it to soak in for a while. Using a wire brush to clean exposed threads will also ease removal of the nut or bolt and prevent damage to the threads. Sometimes a sharp blow with a hammer and punch will break the bond between a nut and bolt threads, but care

must be taken to prevent the punch from slipping off the fastener and ruining the threads. Heating the stuck fastener and surrounding area with a torch sometimes helps too, but isn't recommended because of the obvious dangers associated with fire. Long breaker bars and extension, or "cheater," pipes will increase leverage, but never use an extension pipe on a ratchet - the ratcheting mechanism could be damaged. Sometimes tightening the nut or bolt first will help to break it loose. Fasteners that require drastic measures to remove should always be replaced with new ones.

Since most of the procedures dealt with in this Chapter involve jacking up the vehicle and working underneath it, a good pair of jackstands will be needed. A hydraulic floor jack is the preferred type of jack to lift the vehicle, and it can also be used to support certain components during various operations. **Warning:** *Never, under any circumstances, rely on a jack to support the vehicle while working on it. Whenever any of the suspension or steering fasteners are loosened or removed they must be inspected and, if necessary, replaced with new ones of the same part number or of original equipment quality and design. Torque specifications must be followed for proper reassembly and component retention. Never attempt to heat or straighten any suspension or steering components. Instead, replace any bent or damaged part with a new one.*

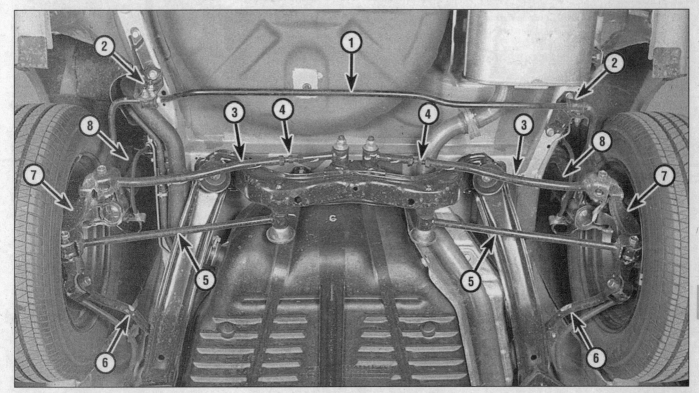

1.2 Typical rear suspension components

1 Stabilizer bar	4 Toe adjuster	7 Knuckle
2 Stabilizer bar bushing clamp	5 Front lateral arm	8 Strut/coil spring assembly
3 Rear lateral arm	6 Trailing arm (radius rod)	

10

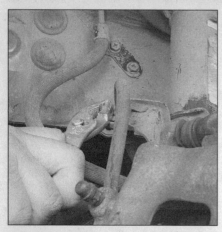

2.2 Remove the retaining clip with a pair of pliers and detach the brake hose from the strut

2.3 To detach the strut assembly from the steering knuckle, remove the ABS line bracket bolt (right arrow), remove the two nuts (left arrows), then drive out the strut-to-knuckle bolts with a hammer and punch

2.5 To detach the upper end of the strut assembly from the body, remove the upper mounting nuts (arrows)

2   Strut/coil spring assembly (front) - removal, inspection and installation

Removal

Refer to illustrations 2.2, 2.3 and 2.5

1   Loosen the front wheel lug nuts, raise the front of the vehicle and support it securely on jackstands. Remove the wheels.

2   Unclip the brake hose from the strut bracket **(see illustration)** and detach it from the bracket. If the vehicle is equipped with ABS, detach the speed sensor wiring harness from the strut by removing the clamp bracket bolt **(see illustration 2.3)**.

3   Remove the strut-to-knuckle nuts **(see illustration)** and knock the bolts out with a hammer and punch.

4   Separate the strut from the steering knuckle. Be careful not to overextend the inner CV joint and don't let the knuckle fall outward, as this could damage the brake hose.

5   Support the strut and spring assembly with one hand and remove the three strut-to-shock tower nuts **(see illustration)**. Remove the assembly out from the fenderwell.

Inspection

6   Check the strut body for leaking fluid, dents, cracks and other obvious damage which would warrant repair or replacement.

7   Check the coil spring for chips or cracks in the spring coating (this will cause premature spring failure due to corrosion). Inspect the spring seat for cuts, hardness and general deterioration.

8   If any undesirable conditions exist, proceed to the strut disassembly procedure (see Section 3).

Installation

9   Guide the strut assembly up into the fenderwell and insert the upper mounting studs through the holes in the shock tower. Once the studs protrude from the shock

tower, install the nuts so the strut won't fall back through. This is most easily accomplished with the help of an assistant, as the strut is quite heavy and awkward.

10   Slide the steering knuckle into the strut flange and insert two *new* bolts. Install new nuts and tighten them to the torque listed in this Chapter's Specifications.

11   Guide the brake hose through its bracket in the strut and install the retaining clip.

12   Install the wheel and lug nuts, then lower the vehicle and tighten the lug nuts to the torque listed in the Chapter 1 Specifications.

13   Tighten the upper mounting nuts to the torque listed in this Chapter's Specifications.

14   Drive the vehicle to an alignment shop to have the front end alignment checked, and if necessary, adjusted.

3   Strut/coil spring assembly - replacement

1   If the struts or coil springs exhibit the telltale signs of wear (leaking fluid, loss of damping capability, chipped, sagging or cracked coil springs) explore all options before beginning any work. The strut/shock absorber assemblies are not serviceable and must be replaced if a problem develops. However, strut assemblies complete with springs may be available on an exchange basis, which eliminates much time and work. Whichever route you choose to take, check on the cost and availability of parts before disassembling your vehicle. **Warning:** *Disassembling a strut is a potentially dangerous undertaking and utmost attention must be directed to the job, or serious injury may result. Use only a high-quality spring compressor and carefully follow the manufacturer's instructions furnished with the tool.*

3.3 Install the spring compressor according to the tool manufacturer's instructions and compress the spring until all pressure is relieved from the upper spring seat

After removing the coil spring from the strut assembly, set it aside in a safe, isolated area.

Disassembly

Refer to illustrations 3.3, 3.4, 3.5, 3.6 and 3.7

2   Remove the strut and spring assembly following the procedure described in the previous Section. Mount the strut assembly in a vise. Line the vise jaws with wood or rags to prevent damage to the unit and don't tighten the vise excessively.

3   Following the tool manufacturer's instructions, install the spring compressor (which can be obtained at most auto parts stores or equipment yards on a daily rental basis) on the spring and compress it sufficiently to relieve all pressure from the upper spring seat **(see illustration)**. This can be verified by wiggling the spring.

4   Remove the damper shaft nut **(see illustration)**.

5   Remove the upper suspension support **(see illustration)**. Inspect the bearing in the suspension support for smooth operation. If it

3.4 Remove the damper shaft nut

3.5 Lift the suspension support off the damper shaft

3.6 Remove the spring seat from the damper shaft

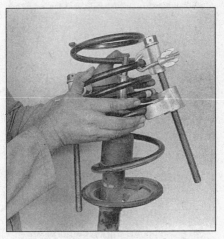

3.7 Remove the compressed spring assembly - keep the ends of the spring pointed away from your body

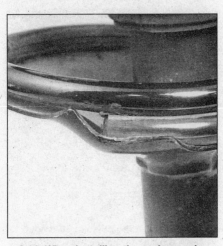

3.11 When installing the spring, make sure the end fits into the recessed portion of the lower seat (arrow)

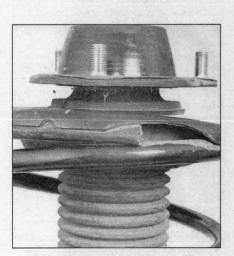

3.12a Make sure this cutout in the upper seat . . .

does not turn smoothly, replace the suspension support. Check the rubber portion of the suspension support for cracking and general deterioration. If there is any separation of the rubber, replace it.

6   Lift the spring seat and upper insulator from the damper shaft **(see illustration)**. Check the rubber spring seat for cracking and hardness, replacing it if necessary.

7   Carefully lift the compressed spring from the assembly **(see illustration)** and set it in a safe place. **Warning:** *Carry the spring carefully and never place any part of your body near the end of the spring!*

8   Slide the dust boot off the damper shaft.

9   Check the lower insulator (if equipped) for wear, cracking and hardness and replace it if necessary.

Reassembly

Refer to illustration 3.11, 3.12a and 3.12b

10   If the lower insulator is being replaced, set it into position with the dropped portion seated in the lowest part of the seat. Extend the damper rod to its full length and install

the dust boot.

11   Carefully place the coil spring onto the lower insulator, with the end of the spring resting in the lowest part of the insulator **(see illustration)**.

12   Install the upper insulator and the spring seat. Make sure the cutout on the spring seat is facing out (away from the vehicle), in line with the strut-to-knuckle attachment points **(see illustrations)**.

13   Install the dust seal and suspension support to the damper shaft.

14   Install the nut and tighten it to the torque listed in this Chapter's Specifications.

15   Install the strut/shock absorber and coil spring assembly (see Section 2).

4   Stabilizer bar (front) - removal and installation

Refer to illustrations 4.2 and 4.3

1   Loosen the wheel lug nuts, raise the front of the vehicle, support it securely on jackstands and remove the wheels.

10

3.12b . . . is facing out (toward the strut-to-knuckle flanges)

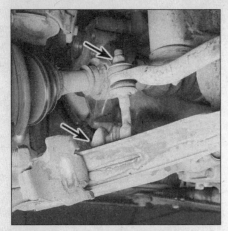

4.2 To disconnect the stabilizer link from the control arm, remove this nut (lower arrow) that attaches the link to the control arm; to disconnect the link from the stabilizer bar, remove this nut (upper arrow)

2   Disconnect the stabilizer links from both control arms **(see illustration)**.
3   Remove both stabilizer bar clamps **(see illustration)**.
4   Remove the stabilizer assembly.
5   Inspect the clamp bushings and the link bushings. If they're cracked or torn, replace them.
6   Installation is the reverse of removal. Be sure to tighten all fasteners to the torque listed in this Chapter's Specifications.

---

5   Control arm - removal, inspection and installation

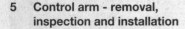

Removal

Refer to illustrations 5.3a, 5.3b, 5.3c, 5.3d, 5.4 and 5.5

1   Loosen the wheel lug nuts on the side to be dismantled, raise the front of the vehicle, support it securely on jackstands and remove

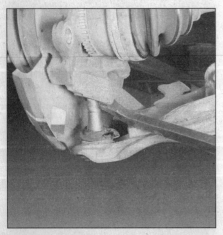

5.3d Separate the control arm from the steering knuckle by prying it down with a prybar or large screwdriver

4.3 To disconnect the stabilizer bar from the body, remove the bushing clamp nuts and bolts (arrows) from both sides

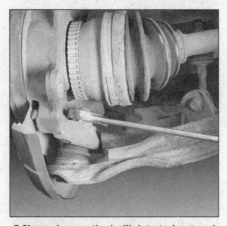

5.3b . . . loosen the balljoint stud nut and back it off as far as it will go (without actually removing it) . . .

the wheel.
2   Disconnect the stabilizer link from the control arm (see Section 4).
3   Remove the cotter pin and loosen the balljoint stud nut **(see illustrations)**. Separate the balljoint stud from the steering knuckle with a "picklefork"-type balljoint separator **(see illustration)**. Be sure to grease

5.4 Remove the front pivot stud nut (arrow)

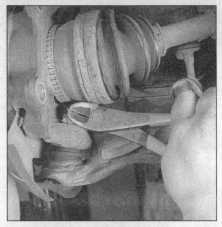

5.3a Remove the cotter pin . . .

5.3c . . . then pop the balljoint stud loose with a "picklefork" (be sure to grease the picklefork to protect the balljoint dust boot)

the picklefork to protect the balljoint dust boot. Separate the arm from the steering knuckle **(see illustration)**.
4   Remove the front pivot bolt nut and/or bolt **(see illustration)**.
5   Remove the rear bushing clamp bolts **(see illustration)**. Remove the control arm.

5.5 Remove the rear bushing clamp bolts (arrows)

Inspection

6   Inspect the front and rear bushings for cracks and tears. If either bushing is damaged or worn, replace the control arm; the bushings are not replaceable.

7   Inspect the control arm for straightness. If it's bent, replace it. Do not attempt to straighten a bent control arm.

Installation

8   Installation is the reverse of removal. Tighten all of the fasteners to the torque listed in this Chapter's Specifications. Be sure to install a new cotter pin through the balljoint stud.

9   Install the wheel and lug nuts, lower the vehicle and tighten the lug nuts to the torque listed in the Chapter 1 Specifications.

10   It's a good idea to have the front wheel alignment checked and, if necessary, adjusted after this job has been performed.

6   Balljoints - replacement

1   Loosen the wheel lug nuts, raise the vehicle and support it securely on jackstands. Remove the wheel.

2   The balljoint is an integral part of the control arm. If it's worn or damaged, replace the control arm (see Section 5).

3   Be sure to tighten all fasteners to the torque listed in this Chapter's Specifications. Also be sure to install a new cotter pin. If the cotter pin hole doesn't line up with the slots on the nut, tighten the nut until the next hole in the nut lines up with the hole in the balljoint stud.

4   Install the wheel and lug nuts. Lower the vehicle and tighten the lug nuts to the torque listed in the Chapter 1 Specifications.

5   It's a good idea to have the front wheel alignment checked, and if necessary, adjusted after this job has been performed.

7   Steering knuckle and hub - removal and installation

Warning: *Dust created by the brake system may contain asbestos, which is harmful to your health. Never blow it out with compressed air and don't inhale any of it. Do not, under any circumstances, use petroleum-based solvents to clean brake parts. Use brake system cleaner only.*

Removal

1   Loosen the wheel lug nuts, raise the vehicle and support it securely on jackstands. Remove the wheel.

2   Remove the brake caliper and the brake disc, and disconnect the brake hose from the strut (see Chapter 9).

3   If the vehicle is equipped with ABS, disconnect and remove the wheel speed sensor.

4   Remove the strut-to-steering knuckle

nuts, but don't remove the bolts yet (see Section 2).

5   Separate the tie-rod end from the steering knuckle arm (see Section 15).

6   Separate the balljoint from the steering knuckle (see Sections 6 and 7).

7   Push the driveaxle from the hub (see Chapter 8). Support the end of the driveaxle with a section of wire.

8   Remove the strut-to-knuckle bolts and separate the knuckle from the strut.

Installation

9   Guide the knuckle and hub assembly into position, inserting the driveaxle into the hub.

10   Push the knuckle into the strut flange and install the bolts and nuts, but don't tighten them yet.

11   Attach the control arm to the steering knuckle (see Section 5).

12   Attach the tie-rod end to the steering knuckle arm (see Section 15). Tighten the strut-to-knuckle nuts to the torque listed in this Chapter's Specifications.

13   Place the brake disc on the hub and install the caliper (see Chapter 9).

14   Install the driveaxle/hub nut and tighten it to the torque listed in the Chapter 8 Specifications.

15   Install the wheel and tighten the lug nuts but don't torque them yet.

16   Lower the vehicle and tighten the wheel lug nuts to the torque listed in the Chapter 1 Specifications.

8   Hub and bearing assembly (front) - removal and installation

Due to the special tools and expertise required to press the hub and bearing from the steering knuckle, this job should be left to a professional mechanic. However, the steering knuckle and hub may be removed and the assembly taken to a dealer service department or other repair shop. See Section 7 for the steering knuckle and hub removal procedure.

9.6a Pry off this rubber cover . . .

9   Strut/coil spring assembly (rear) - removal, inspection and installation

Removal

Refer to illustrations 9.3, 9.6a and 9.6b

1   Loosen the rear wheel lug nuts, raise the rear of the vehicle and support it securely on jackstands. Remove the wheels.

2   Unbolt the brake hose bracket from the strut. If the vehicle is equipped with ABS, detach the speed sensor wiring harness from the strut by removing the clamp bracket bolt.

3   Remove the strut-to-knuckle nuts **(see illustration)** and knock the bolts out with a hammer and punch.

4   Separate the strut from the knuckle. Don't allow the knuckle to fall outward, as this may damage the brake hose.

5   Remove the back seat (see Chapter 11) and the parcel shelf (see Section 12 in Chapter 12) to get at the upper mounting nuts.

6   Remove the strut rubber cover **(see illustration)**. Have an assistant support the strut and spring assembly while you remove the three strut-to-shock tower nuts **(see illustration)**. Remove the assembly through the fenderwell.

9.3 To disconnect the strut assembly from the rear knuckle, remove these two nuts and bolts (arrows)

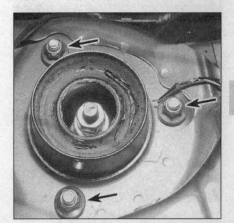

9.6b . . . and remove the strut upper mounting nuts (arrows)

10

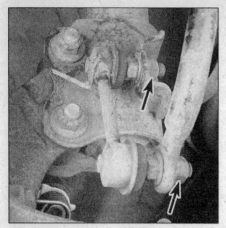

10.2 To disconnect the stabilizer link from the rear strut, remove the nut indicated by the upper arrow. To disconnect the stabilizer link from the stabilizer bar, remove the nut indicated by the lower arrow

Inspection

7   Inspect the strut as outlined in Steps 6 and 7 in Section 2. If any undesirable conditions exist, replace the strut (see Section 3).

Installation

8   Guide the strut assembly up into the fenderwell and insert the upper mounting studs through the holes in the shock tower. Make sure the strut upper spring seat is correctly aligned (the studs are not equidistant - they'll only fit through the holes in the shock tower one way). Once the studs protrude from the shock tower, install the nuts so the strut won't fall back through. This is most easily accomplished with the help of an assistant, as the strut is quite heavy and awkward.
9   Slide the steering knuckle into the strut flange and insert the two bolts. Install the nuts and tighten them to the torque listed in this Chapter's Specifications.
10   Connect the brake hose bracket to the strut and tighten the bolt securely. If the vehicle is equipped with ABS, install the speed sensor wiring harness bracket.

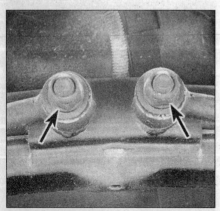

11.2b To disconnect the inner ends of the rear lateral arms, remove these nuts (arrows)

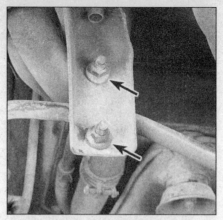

10.3 To disconnect the rear stabilizer bushing clamps, remove these nuts (arrows)

11   Install the wheel and lug nuts, then lower the vehicle and tighten the lug nuts to the torque listed in the Chapter 1 Specifications.
12   Tighten the upper mounting nuts to the torque listed in this Chapter's Specifications.
13   Drive the vehicle to an alignment shop to have the wheel alignment checked, and if necessary, adjusted.

10   Stabilizer bar (rear) - removal and installation

Refer to illustrations 10.2 and 10.3
1   Loosen the rear wheel lug nuts, raise the rear of the vehicle, support it securely on jackstands and remove the wheels.
2   Remove the nuts from the link rods that attach the stabilizer to the struts **(see illustration)**.
3   Remove the bushing clamp nuts **(see illustration)** and remove the stabilizer.
4   Inspect all clamp bushings. If they're cracked or torn, replace them.
5   Installation is the reverse of removal. Be sure to tighten all fasteners to the torque listed in this Chapter's Specifications.

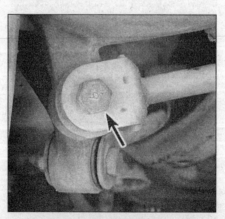

11.3 To disconnect the outer end of a lateral arm from the knuckle, remove the nut and bolt (arrow)

11.2a To disconnect the inner ends of a front lateral arm from the crossmember, remove this nut (arrow)

11   Rear suspension arms - removal and installation

1   Loosen the rear wheel lug nuts, raise the rear of the vehicle, support it securely on jackstands and remove the wheels.

Lateral arms

Refer to illustrations 11.2a, 11.2b and 11.3
2   Remove the inner pivot nuts **(see illustrations)**.
3   Remove the bolts and nuts that attach the lateral arms to the knuckle **(see illustration)**.
4   Remove the arms.
5   Installation is the reverse of removal. Be sure to tighten all fasteners to the torque listed in this Chapter's Specifications. **Note:** *Raise the rear suspension with a floor jack to simulate normal ride height before tightening the fasteners.*
6   When you're done, drive the vehicle to an alignment shop and have the rear-wheel toe adjusted.

Trailing arms

Refer to illustrations 11.7 and 11.8
7   Remove the pivot bolt and nut from the

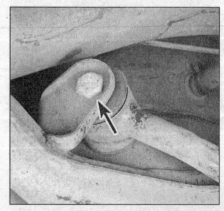

11.7 To disconnect the forward end of the trailing arm from the body, remove this nut and pivot bolt (arrow)

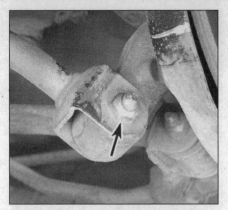

11.8 Remove the nut (arrow) and bolt that attach the rear end of the trailing arm to the knuckle

12.3 Remove the hub/bearing grease cap with a hammer and chisel

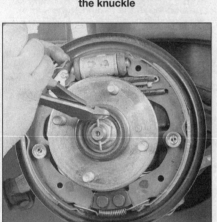

12.4 Remove the retaining nut cotter pin

12.5 Remove the bearing/hub retaining nut

12.6 Remove the hub/bearing assembly

forward end of the trailing arm **(see illustration)**.
8   Remove the bolt and nut that attach the rear end of the arm to the knuckle **(see illustration)**.
9   Remove the trailing arm.
10   Installation is the reverse of removal. Place a floor jack under the rear knuckle and raise the suspension to simulate normal ride height, then tighten the fasteners to the torque listed in this Chapter's Specifications.

All arms

11   Install the wheels and lug nuts, lower the vehicle and tighten the lug nuts to the torque listed in the Chapter 1 Specifications.
12   When you're done, drive the vehicle to an alignment shop and have the rear-wheel toe adjusted.

12 Hub and bearing assembly (rear) - removal and installation

Refer to illustrations 12.3, 12.4, 12.5 and 12.6
Note: *The illustrations accompanying this Section depict a vehicle with rear drum brakes. However, the procedure for removing the hub and bearing assembly on a model with rear disc brakes is identical.*

1   Loosen the rear wheel lug nuts, raise the rear of the vehicle, support it securely on jackstands and remove the wheels.
2   Remove the brake drum or disc (see Chapter 9).
3   Pry off the grease cap **(see illustration)**.
4   Remove the cotter pin **(see illustration)**.
5   Remove the wheel bearing locknut **(see illustration)**.
6   Remove the hub and bearing assembly **(see illustration)**.
7   Installation is the reverse of removal. Be sure to tighten all fasteners to the torque listed in this Chapter's Specifications.

13 Knuckle (rear) - removal and installation

Refer to illustrations 13.8a, 13.8b and 13.9
1   Loosen the rear wheel lug nuts, raise the rear of the vehicle, support it securely on jackstands and remove the wheels.
2   On models with rear drum brakes, remove the brake drum, brake shoes and wheel cylinder (see Chapter 9).
3   On models with rear disc brakes, remove the caliper and disc (see Chapter 9).
4   Remove the hub and bearing assembly

13.8a To remove the brake backing plate, remove these four bolts (arrows) (models with rear drum brakes)

(see Section 12).
5   On models with rear drum brakes, remove the brake assembly (see Chapter 9).
6   On models with rear drum brakes, disconnect the brake line fitting at the wheel cylinder (see Chapter 9).
7   On models with rear drum brakes, disconnect the parking brake cable housing bolts from the rear of the backing plate (see Chapter 9).
8   Remove the brake backing plate **(see**

13.8b To remove the backing plate, remove these three screws (arrows) (models with rear disc brakes)

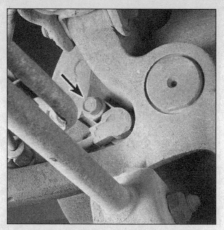

13.9 On models with ABS, be sure to remove this bolt (arrow), pull out the rear wheel speed sensor and carefully set it aside

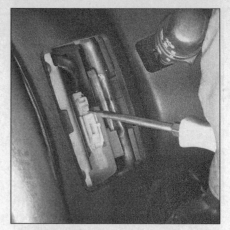

14.3 Pry off the access cover from the underside of the steering wheel and unplug the airbag module connector

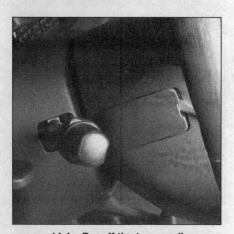

14.4a Pry off the two small side covers . . .

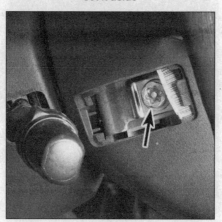

14.4b . . . and remove the left side Torx bolt (arrow); the Torx bolts are coated with a special bonding agent, so they must be discarded - be sure to replace them during reassembly

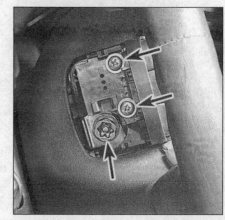

14.4c On the right side of the wheel, besides the Torx bolt (lower arrow), you'll also need to remove these two cruise control switch retaining screws (upper arrows)

illustrations).

9   On models with ABS, remove the rear wheel sensor **(see illustration)**.

10   Disconnect the stabilizer bar assembly from the knuckle (see Section 10).

11   Disconnect the suspension arms from the knuckle (see Section 11).

12   Disconnect the strut/coil spring assembly from the knuckle (see Section 9).

13   Remove the knuckle.

14   Installation is the reverse of removal. Be sure to tighten all fasteners to the torque listed in this Chapter's Specifications.

14   Steering wheel - removal and installation

Airbag models

Warning: *To prevent accidental deployment (and possible injury) when working near airbag components, disconnect the negative battery cable, then the positive battery cable and wait at least 10 minutes before beginning work (the system has a back-up capacitor that must fully discharge). For more information see Chapter 12.*

Removal

Refer to illustrations 14.3, 14.4a, 14.4b, 14.4c, 14.4d, 14.5 and 14.7

1   Disconnect the cable from the negative battery terminal, then the positive battery terminal and wait at least ten minutes before removing the steering wheel.

2   Turn the steering wheel so that the front wheels are pointing straight ahead.

3   Remove the small cover from the underside of the steering wheel and unplug the airbag module connector **(see illustration)**.

4   Remove the side covers and remove the Torx bolts behind them **(see illustrations)**, normally with a T50H Torx bit (it may be different on your model). The Torx bolts are coated with a special bonding agent, so they must be discarded; be sure to replace them with new ones during reassembly. Lift the airbag module off the steering wheel **(see illustration)**. **Warning:** *Handle the airbag module with care, carry the module with the trim cover side facing away from your body and store it in a safe location with the trim*

14.4d Carefully remove the airbag module

side facing up. See the precautions in Chapter 12.

5   Remove the steering wheel retaining nut, then mark the relationship of the steering wheel to the steering shaft **(see illustration)**.

14.5 Disconnect the electrical connector for the cruise control leads; after the steering wheel retaining nut has been removed, mark the relationship of the steering wheel to the steering shaft (arrows)

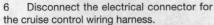

6   Disconnect the electrical connector for the cruise control wiring harness.
7   Use a steering wheel puller to separate the steering wheel from the steering shaft (see illustration). When removing the wheel, make sure the electrical leads for the airbag module and the cruise control system don't snag on the wheel. **Warning:** *Do not turn the steering shaft while the steering wheel is removed.*

Installation
Refer to illustration 14.8
8   Verify that the front wheels are pointing straight ahead, turn the spiral cable for the airbag clockwise by hand until it becomes hard to turn, then rotate the spiral cable counterclockwise about two and one-half turns until the two arrows are aligned (see illustration).
9   Pull the electrical leads for the airbag module and the cruise control system through the steering wheel and install the wheel. Make sure the spiral cable pin guides are properly engaged with their corresponding holes in the back of the steering wheel and pull the spiral cable through.
10   Install the steering wheel retaining nut and tighten it to the torque listed in this Chapter's Specifications.
11   Install the airbag module and secure it with new Torx bolts. Do not reuse the old bolts. Install the side covers.
12   Plug in the airbag module connector. Install the lower cover.
13   Verify that the airbag circuit is operational by turning the ignition key to the On or Start position. The "AIR BAG" warning light should illuminate for about seven seconds, then turn off.

Non-airbag models
Removal
14   Disconnect the cable from the negative terminal of the battery.
15   To remove the horn pad, insert a Phillips

14.7 Use a steering wheel puller to separate the steering wheel from the steering shaft

screwdriver into the hole on the lower side of the spoke and remove the clamps, then lift off the horn pad by hand.
16   The remainder of removal is similar to that for an airbag-equipped model (see Steps 6 and 7).

Installation
17   Before installing the wheel, lubricate the horn-contact slip ring and the sliding portion of the turn signal cancel pin with multi-purpose grease.
18   Install the steering wheel nut and tighten it to the torque listed in this Chapter's Specifications.
19   Connect the horn wire and install the horn pad.
20   Connect the negative battery cable.

---

15 Tie-rod ends - removal and installation

Removal
Refer to illustrations 15.2, 15.3a, 15.3b and 15.4
1   Loosen the wheel lug nuts. Raise the front of the vehicle, support it securely on jackstands, block the rear wheels and set the

15.3a Remove the cotter pin . . .

14.8 Verify that the front wheels are pointing straight ahead, turn the spiral cable clockwise by hand until it becomes hard to turn, then rotate the spiral cable counterclockwise about two and one-half turns, until the two arrows are aligned

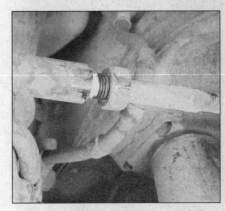

15.2 Loosen the jam nut, then mark the position of the tie-rod end in relation to the threads

parking brake. Remove the front wheel.
2   Loosen the jam nut enough to mark the position of the tie-rod end in relation to the threads (see illustration).
3   Remove the cotter pin and loosen, but don't remove, the nut on the tie-rod end stud (see illustrations).

15.3b . . . then loosen - but don't remove - the tie-rod end ballstud nut

10

15.4 Disconnect the tie-rod end from the steering knuckle arm with a puller

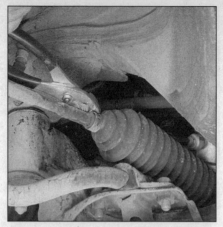

16.3a The outer end of each steering gear boot is secured by a spring-type clamp which can be slid off simply by pinching the ends together

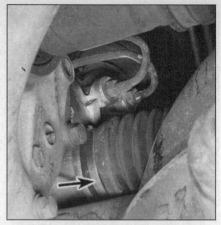

16.3b The inner end of each steering gear boot is retained by a clamp (arrow) which must be cut off and discarded

4   Disconnect the tie-rod end from the steering knuckle arm with a puller **(see illustration)**. Remove the nut and separate the tie-rod.
5   Unscrew the tie-rod end from the steering rod.

Installation

6   Thread the tie-rod end on to the marked position and insert the tie-rod stud into the steering knuckle arm. Tighten the jam nut securely.
7   Install the castle nut on the stud and tighten it to the torque listed in this Chapter's Specifications. Install a new cotter pin.
8   Install the wheel and lug nuts. Lower the vehicle and tighten the lug nuts to the torque listed in the Chapter 1 Specifications.
9   Have the alignment checked by a dealer service department or an alignment shop.

16   Steering gear boots - replacement

Refer to illustrations 16.3a and 16.3b
1   Loosen the lug nuts, raise the vehicle and support it securely on jackstands. Remove the wheel.
2   Remove the tie-rod end and jam nut (see Section 15).
3   Remove the outer steering gear boot clamp **(see illustration)** with a pair of pliers. Cut off the inner boot clamp **(see illustration)** with a pair of diagonal cutters. Slide the boot off.
4   Before installing the new boot, wrap the threads on the end of the steering rod with a layer of tape so the small end of the new boot isn't damaged.
5   Slide the new boot into position on the steering gear until it seats in the groove in the steering rod and install new clamps.
6   Remove the tape and install the tie-rod end (see Section 15).
7   Install the wheel and lug nuts. Lower the vehicle and tighten the lug nuts to the torque listed in the Chapter 1 Specifications.

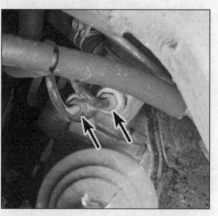

17.2 Disconnect the power steering pressure and return line fittings (arrows) from the steering gear

17   Steering gear - removal and installation

Warning: *Most models covered by this manual have airbags. Always disconnect the negative battery cable, then the positive cable and wait 10 minutes before working in the vicinity of the impact sensors, steering column or instrument panel to avoid the possibility of accidental deployment of the airbag, which could cause personal injury (see Chapter 12).*

Removal

Refer to illustrations 17.2, 17.3, 17.5a and 17.5b
1   Loosen the front wheel lug nuts, raise the front of the vehicle and support it securely on jackstands. Apply the parking brake and remove the wheels. Remove the engine splash shields.
2   Place a drain pan under the steering gear. Detach the power steering pressure and return lines **(see illustration)** and cap the ends to pre-

17.3 Remove the U-joint cover and mark the relationship of the universal joint to the steering gear input shaft, then remove the U-joint pinch bolt (arrow)

vent excessive fluid loss and contamination.
3   From inside the vehicle, remove the universal joint cover **(see illustration)**. Mark the relationship of the lower universal joint to the steering gear input shaft **(see illustration)**. Remove the lower intermediate shaft pinch bolt.
4   Separate the tie-rod ends from the steering knuckle arms (see Section 15).
5   Support the steering gear and remove the steering gear mounting bolts **(see illustrations)**. Separate the intermediate shaft from the steering gear input shaft and remove the steering gear assembly. **Warning:** *Do not turn the steering wheel while the steering gear is removed on a model equipped with an airbag. If the steering wheel is inadvertently turned, remove the steering wheel and center the spiral cable (see Section 14). To prevent the steering wheel from turning, loop the seat belt through the steering wheel and fasten it into its latch.*
6   Check the steering gear rubber mounts for excessive wear or deterioration, replacing them if necessary.

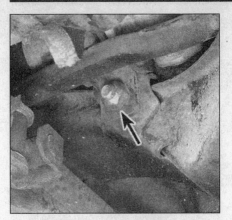

17.5a To remove the steering gear assembly, remove the left retaining nut (arrow) and bolt (directly above the nut, at upper end of bracket, not visible in this photo) . . .

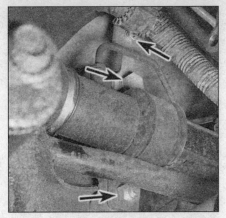

17.5b . . . and the right-side bolts (upper arrows) and nut (lower arrow)

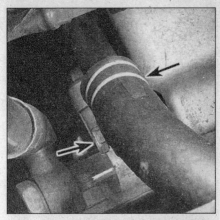

18.3 To detach the fluid return hose from the power steering pump, loosen the screw (lower arrow) for the hose clamp (upper arrow)

Installation

Note: *Make sure the steering gear is centered from side-to-side before installing it.*

7   Raise the steering gear into position and connect the U-joint, aligning the marks.
8   Install the mounting brackets and bolts and tighten them to the torque listed in this Chapter's Specifications.
9   Connect the tie-rod ends to the steering knuckle arms (see Section 15).
10   Install the U-joint pinch bolt and tighten it to the torque listed in this Chapter's Specifications.
11   If equipped with power steering, connect the power steering pressure and return hoses to the steering gear and fill the power steering pump reservoir with the recommended fluid (see Chapter 1).
12   Install the wheels and lug nuts, then lower the vehicle and tighten the lug nuts to the torque listed in the Chapter 1 Specifications. If equipped with power steering, bleed the steering system (see Section 19).

18   Power steering pump - removal and installation

Removal

Refer to illustrations 18.3, 18.4 and 18.6

1   Disconnect the cable from the negative battery terminal.
2   Using a large syringe or suction gun, suck as much fluid out of the power steering fluid reservoir as possible. Place a drain pan under the vehicle to catch any fluid that spills out when the hoses are disconnected.
3   Loosen the clamp and disconnect the fluid return hose from the pump **(see illustration)**.
4   Remove the pressure line-to-pump fitting or banjo bolt **(see illustration)**, then detach the line from the pump. Remove and discard the copper sealing washers. They must be replaced when installing the pump.

5   Loosen the alternator tensioner and remove the drivebelt (see Chapter 1).
6   Remove the pump mounting bolts **(see illustration)**, then remove the pump from the vehicle.

Installation

7   Installation is the reverse of removal. Be sure to tighten the pressure line fitting or banjo bolt to the torque listed in this Chapter's Specifications. Adjust the drivebelt tension following the procedure described in Chapter 1.
8   Top up the fluid level in the reservoir (see Chapter 1) and bleed the system (see Section 19).

19   Power steering system - bleeding

1   Following any operation in which the power steering fluid lines have been disconnected, the power steering system must be bled to remove all air and obtain proper

steering performance.
2   With the front wheels in the straight ahead position, check the power steering fluid level and, if low, add fluid until it reaches the Cold mark on the dipstick.
3   Start the engine and allow it to run at fast idle. Recheck the fluid level and add more if necessary to reach the Cold mark on the dipstick.
4   Bleed the system by turning the wheels from side to side, without hitting the stops. This will work the air out of the system. Keep the reservoir full of fluid as this is done.
5   When the air is worked out of the system, return the wheels to the straight ahead position and leave the vehicle running for several more minutes before shutting it off.
6   Road test the vehicle to be sure the steering system is functioning normally and noise free.
7   Recheck the fluid level to be sure it is up to the Hot mark on the dipstick while the engine is at normal operating temperature. Add fluid if necessary (see Chapter 1).

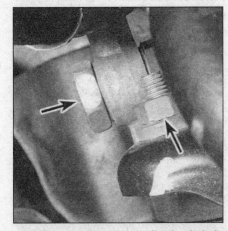

18.4 Remove the pressure line banjo bolt (left arrow) from the power steering pump; then remove the upper mounting nut and bolt (right arrow)

18.6 To gain access to the lower power steering pump mounting bolt (arrow), rotate the pump pulley until one of the holes in the pulley exposes the bolt

10

20   Wheels and tires - general information

Refer to illustration 20.1

1   All vehicles covered by this manual are equipped with metric-sized steel belted radial tires **(see illustration)**. Use of other size or type of tires may affect the ride and handling of the vehicle. Don't mix different types of tires, such as radials and bias belted, on the same vehicle as handling may be seriously affected. It's recommended that tires be replaced in pairs on the same axle, but if only one tire is being replaced, be sure it's the same size, structure and tread design as the other.

2   Because tire pressure has a substantial effect on handling and wear, the pressure on all tires should be checked at least once a month or before any extended trips (see Chapter 1).

3   Wheels must be replaced if they are bent, dented, leak air, have elongated bolt holes, are heavily rusted, out of vertical symmetry or if the lug nuts won't stay tight. Wheel repairs that use welding or peening are not recommended.

4   Tire and wheel balance is important in the overall handling, braking and performance of the vehicle. Unbalanced wheels can adversely affect handling and ride characteristics as well as tire life. Whenever a tire is installed on a wheel, the tire and wheel should be balanced by a shop with the proper equipment.

21   Wheel alignment - general information

Refer to illustration 21.1

A wheel alignment refers to the adjustments made to the wheels so they are in proper angular relationship to the suspension and the ground. Wheels that are out of proper alignment not only affect vehicle control, but also increase tire wear. The front end angles normally measured are camber, caster and toe-in **(see illustration)**. Camber and caster are preset at the factory on the vehicle covered by this manual; toe-in is the only adjustable angle on these vehicles. The rear toe-in can also be adjusted, but the camber and caster cannot (however, camber and caster are usually measured to check for bent or worn suspension parts).

Getting the proper wheel alignment is a very exacting process, one in which complicated and expensive machines are necessary to perform the job properly. Because of this, you should have a technician with the proper equipment perform these tasks. We will, however, use this space to give you a basic idea of what is involved with a wheel alignment so you can better understand the process and deal intelligently with the shop that does the work.

Toe-in is the turning in of the wheels.

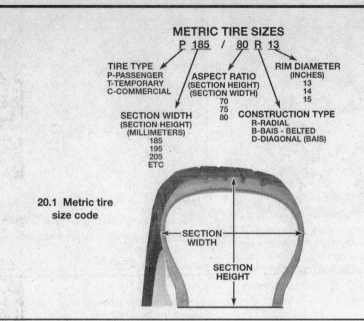

20.1   Metric tire size code

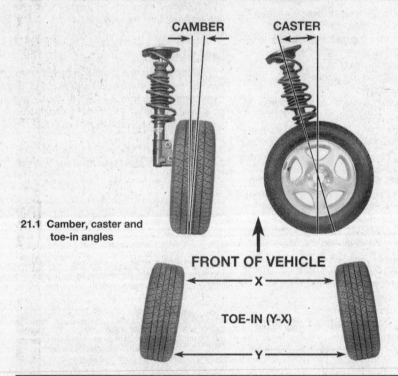

21.1   Camber, caster and toe-in angles

The purpose of a toe specification is to ensure parallel rolling of the wheels. In a vehicle with zero toe-in, the distance between the front edges of the wheels will be the same as the distance between the rear edges of the wheels. The actual amount of toe-in is normally only a fraction of an inch. On the front end, toe-in is controlled by the tie-rod end position on the tie-rod. On the rear end, it's controlled by a threaded adjuster on the inner end of the rear (number two) suspension arm. Incorrect toe-in will cause the tires to wear improperly by making them scrub against the road surface.

Camber is the tilting of the wheels from vertical when viewed from one end of the vehicle. When the wheels tilt out at the top, the camber is said to be positive (+). When the wheels tilt in at the top the camber is negative (-). The amount of tilt is measured in degrees from vertical and this measurement is called the camber angle. This angle affects the amount of tire tread which contacts the road and compensates for changes in the suspension geometry when the vehicle is cornering or traveling over an undulating surface.

Caster is the tilting of the front steering axis from the vertical. A tilt toward the rear is positive caster and a tilt toward the front is negative caster.

Chapter 11   Body

Contents

1   General information

These models feature a "unibody" construction, using a floor pan with front and rear frame side rails which support the body components, front and rear suspension systems and other mechanical components. Certain components are particularly vulnerable to accident damage and can be unbolted and repaired or replaced. Among these parts are the body moldings, bumpers, hood and trunk lids and all glass.

Only general body maintenance practices and body panel repair procedures within the scope of the do-it-yourselfer are included in this Chapter.

2   Body - maintenance

1   The condition of your vehicle's body is very important, because the resale value depends a great deal on it. It's much more difficult to repair a neglected or damaged body than it is to repair mechanical components. The hidden areas of the body, such as the wheel wells, the frame and the engine compartment, are equally important, although they don't require as frequent attention as the rest of the body.

2   Once a year, or every 12,000 miles, it's a good idea to have the underside of the body steam cleaned. All traces of dirt and oil will be removed and the area can then be inspected carefully for rust, damaged brake lines, frayed electrical wires, damaged cables and other problems. The front suspension components should be greased after completion of this job.

3   At the same time, clean the engine and the engine compartment with a steam cleaner or water soluble degreaser.

4   The wheel wells should be given close attention, since undercoating can peel away and stones and dirt thrown up by the tires can cause the paint to chip and flake, allowing rust to set in. If rust is found, clean down to the bare metal and apply an anti-rust paint.

5   The body should be washed about once a week. Wet the vehicle thoroughly to soften the dirt, then wash it down with a soft sponge and plenty of clean soapy water. If the surplus dirt is not washed off very carefully, it can wear down the paint.

6   Spots of tar or asphalt thrown up from the road should be removed with a cloth soaked in solvent.

7   Once every six months, wax the body and chrome trim. If a chrome cleaner is used to remove rust from any of the vehicle's plated parts, remember that the cleaner also removes part of the chrome, so use it sparingly.

3   Vinyl trim - maintenance

Don't clean vinyl trim with detergents, caustic soap or petroleum-based cleaners. Plain soap and water works just fine, with a soft brush to clean dirt that may be ingrained. Wash the vinyl as frequently as the rest of the vehicle.

After cleaning, application of a high quality rubber and vinyl protectant will help prevent oxidation and cracks. The protectant can also be applied to weatherstripping, vacuum lines and rubber hoses (which often fail as a result of chemical degradation) and to the tires.

11

These photos illustrate a method of repairing simple dents. They are intended to supplement *Body repair - minor damage* in this Chapter and should not be used as the sole instructions for body repair on these vehicles.

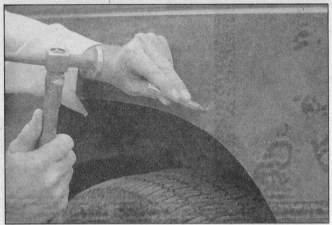

1   If you can't access the backside of the body panel to hammer out the dent, pull it out with a slide-hammer-type dent puller. In the deepest portion of the dent or along the crease line, drill or punch hole(s) at least one inch apart . . .

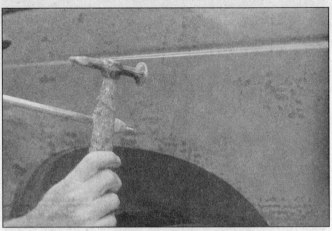

2   . . . then screw the slide-hammer into the hole and operate it. Tap with a hammer near the edge of the dent to help 'pop' the metal back to its original shape. When you're finished, the dent area should be close to its original contour and about 1/8-inch below the surface of the surrounding metal

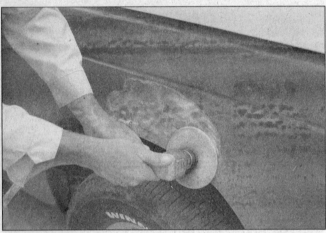

3   Using coarse-grit sandpaper, remove the paint down to the bare metal. Hand sanding works fine, but the disc sander shown here makes the job faster. Use finer (about 320-grit) sandpaper to feather-edge the paint at least one inch around the dent area

4   When the paint is removed, touch will probably be more helpful than sight for telling if the metal is straight. Hammer down the high spots or raise the low spots as necessary. Clean the repair area with wax/silicone remover

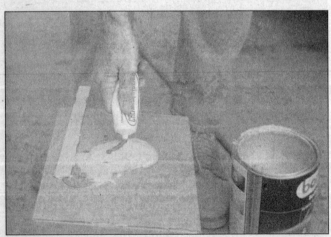

5   Following label instructions, mix up a batch of plastic filler and hardener. The ratio of filler to hardener is critical, and, if you mix it incorrectly, it will either not cure properly or cure too quickly (you won't have time to file and sand it into shape)

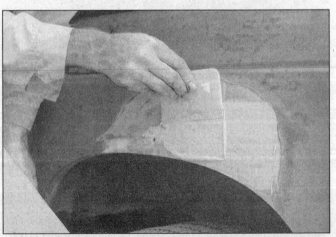

6   Working quickly so the filler doesn't harden, use a plastic applicator to press the body filler firmly into the metal, assuring it bonds completely. Work the filler until it matches the original contour and is slightly above the surrounding metal

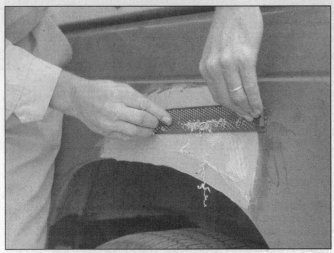

7   Let the filler harden until you can just dent it with your fingernail. Use a body file or Surform tool (shown here) to rough-shape the filler

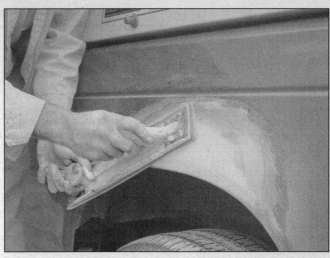

8   Use coarse-grit sandpaper and a sanding board or block to work the filler down until it's smooth and even. Work down to finer grits of sandpaper - always using a board or block - ending up with 360 or 400 grit

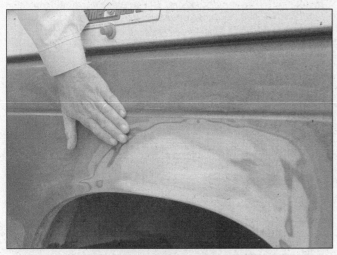

9   You shouldn't be able to feel any ridge at the transition from the filler to the bare metal or from the bare metal to the old paint. As soon as the repair is flat and uniform, remove the dust and mask off the adjacent panels or trim pieces

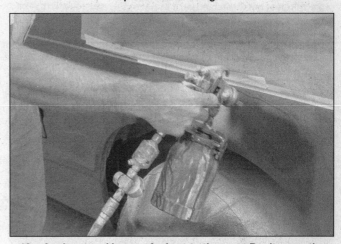

10   Apply several layers of primer to the area. Don't spray the primer on too heavy, so it sags or runs, and make sure each coat is dry before you spray on the next one. A professional-type spray gun is being used here, but aerosol spray primer is available inexpensively from auto parts stores

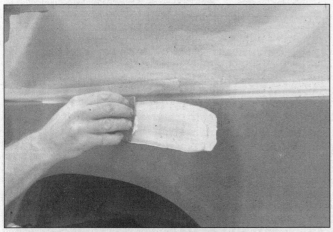

11   The primer will help reveal imperfections or scratches. Fill these with glazing compound. Follow the label instructions and sand it with 360 or 400-grit sandpaper until it's smooth. Repeat the glazing, sanding and respraying until the primer reveals a perfectly smooth surface

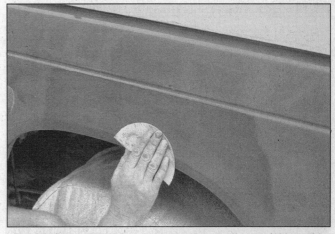

12   Finish sand the primer with very fine sandpaper (400 or 600-grit) to remove the primer overspray. Clean the area with water and allow it to dry. Use a tack rag to remove any dust, then apply the finish coat. Don't attempt to rub out or wax the repair area until the paint has dried completely (at least two weeks)

4   Upholstery and carpets - maintenance

1   Every three months remove the carpets or mats and clean the interior of the vehicle (more frequently if necessary). Vacuum the upholstery and carpets to remove loose dirt and dust.
2   Leather upholstery requires special care. Stains should be removed with warm water and a very mild soap solution. Use a clean, damp cloth to remove the soap, then wipe again with a dry cloth. Never use alcohol, gasoline, nail polish remover or thinner to clean leather upholstery.
3   After cleaning, regularly treat leather upholstery with a leather wax. Never use car wax on leather upholstery.
4   In areas where the interior of the vehicle is subject to bright sunlight, cover leather seats with a sheet if the vehicle is to be left out for any length of time.

5   Body repair - minor damage

See photo sequence

Repair of minor scratches

1   If the scratch is superficial and does not penetrate to the metal of the body, repair is very simple. Lightly rub the scratched area with a fine rubbing compound to remove loose paint and built-up wax. Rinse the area with clean water.
2   Apply touch-up paint to the scratch, using a small brush. Continue to apply thin layers of paint until the surface of the paint in the scratch is level with the surrounding paint. Allow the new paint at least two weeks to harden, then blend it into the surrounding paint by rubbing with a very fine rubbing compound. Finally, apply a coat of wax to the scratch area.
3   If the scratch has penetrated the paint and exposed the metal of the body, causing the metal to rust, a different repair technique is required. Remove all loose rust from the bottom of the scratch with a pocket knife, then apply rust inhibiting paint to prevent the formation of rust in the future. Using a rubber or nylon applicator, coat the scratched area with glaze-type filler. If required, the filler can be mixed with thinner to provide a very thin paste, which is ideal for filling narrow scratches. Before the glaze filler in the scratch hardens, wrap a piece of smooth cotton cloth around the tip of a finger. Dip the cloth in thinner and then quickly wipe it along the surface of the scratch. This will ensure that the surface of the filler is slightly hollow. The scratch can now be painted over as described earlier in this Section.

Repair of dents

4   When repairing dents, the first job is to pull the dent out until the affected area is as close as possible to its original shape. There is no point in trying to restore the original shape completely as the metal in the damaged area will have stretched on impact and cannot be restored to its original contours. It is better to bring the level of the dent up to a point which is about 1/8-inch below the level of the surrounding metal. In cases where the dent is very shallow, it is not worth trying to pull it out at all.
5   If the back side of the dent is accessible, it can be hammered out gently from behind using a soft-face hammer. While doing this, hold a block of wood firmly against the opposite side of the metal to absorb the hammer blows and prevent the metal from being stretched.
6   If the dent is in a section of the body which has double layers, or some other factor makes it inaccessible from behind, a different technique is required. Drill several small holes through the metal inside the damaged area, particularly in the deeper sections. Screw long, self-tapping screws into the holes just enough for them to get a good grip in the metal. Now the dent can be pulled out by pulling on the protruding heads of the screws with locking pliers.
7   The next stage of repair is the removal of paint from the damaged area and from an inch or so of the surrounding metal. This is done with a wire brush or sanding disk in a drill motor, although it can be done just as effectively by hand with sandpaper. To complete the preparation for filling, score the surface of the bare metal with a screwdriver or the tang of a file, or drill small holes in the affected area. This will provide a good grip for the filler material. To complete the repair, see the subsection on filling and painting later in this Section.

Repair of rust holes or gashes

8   Remove all paint from the affected area and from an inch or so of the surrounding metal using a sanding disk or wire brush mounted in a drill motor. If these are not available, a few sheets of sandpaper will do the job just as effectively.
9   With the paint removed, you will be able to determine the severity of the corrosion and decide whether to replace the whole panel, if possible, or repair the affected area. New body panels are not as expensive as most people think and it is often quicker to install a new panel than to repair large areas of rust.
10   Remove all trim pieces from the affected area except those which will act as a guide to the original shape of the damaged body, such as headlight shells, etc. Using metal snips or a hacksaw blade, remove all loose metal and any other metal that is badly affected by rust. Hammer the edges of the hole in to create a slight depression for the filler material.
11   Wire brush the affected area to remove the powdery rust from the surface of the metal. If the back of the rusted area is accessible, treat it with rust inhibiting paint.
12   Before filling is done, block the hole in some way. This can be done with sheet metal riveted or screwed into place, or by stuffing the hole with wire mesh.
13   Once the hole is blocked off, the affected area can be filled and painted. See the following subsection on filling and painting.

Filling and painting

14   Many types of body fillers are available, but generally speaking, body repair kits which contain filler paste and a tube of resin hardener are best for this type of repair work. A wide, flexible plastic or nylon applicator will be necessary for imparting a smooth and contoured finish to the surface of the filler material. Mix up a small amount of filler on a clean piece of wood or cardboard (use the hardener sparingly). Follow the manufacturer's instructions on the package, otherwise the filler will set incorrectly.
15   Using the applicator, apply the filler paste to the prepared area. Draw the applicator across the surface of the filler to achieve the desired contour and to level the filler surface. As soon as a contour that approximates the original one is achieved, stop working the paste. If you continue, the paste will begin to stick to the applicator. Continue to add thin layers of paste at 20-minute intervals until the level of the filler is just above the surrounding metal.
16   Once the filler has hardened, the excess can be removed with a body file. From then on, progressively finer grades of sandpaper should be used, starting with a 180-grit paper and finishing with 600-grit wet-or-dry paper. Always wrap the sandpaper around a flat rubber or wooden block, otherwise the surface of the filler will not be completely flat. During the sanding of the filler surface, the wet-or-dry paper should be periodically rinsed in water. This will ensure that a very smooth finish is produced in the final stage.
17   At this point, the repair area should be surrounded by a ring of bare metal, which in turn should be encircled by the finely feathered edge of good paint. Rinse the repair area with clean water until all of the dust produced by the sanding operation is gone.
18   Spray the entire area with a light coat of primer. This will reveal any imperfections in the surface of the filler. Repair the imperfections with fresh filler paste or glaze filler and once more smooth the surface with sandpaper. Repeat this spray-and-repair procedure until you are satisfied that the surface of the filler and the feathered edge of the paint are perfect. Rinse the area with clean water and allow it to dry completely.
19   The repair area is now ready for painting. Spray painting must be carried out in a warm, dry, windless and dust free atmosphere. These conditions can be created if you have access to a large indoor work area, but if you are forced to work in the open, you will have to pick the day very carefully. If you are working indoors, dousing the floor in the work area with water will help settle the dust

which would otherwise be in the air. If the repair area is confined to one body panel, mask off the surrounding panels. This will help minimize the effects of a slight mismatch in paint color. Trim pieces such as chrome strips, door handles, etc., will also need to be masked off or removed. Use masking tape and several thickness of newspaper for the masking operations.

20   Before spraying, shake the paint can thoroughly, then spray a test area until the spray painting technique is mastered. Cover the repair area with a thick coat of primer. The thickness should be built up using several thin layers of primer rather than one thick one. Using 600-grit wet-or-dry sandpaper, rub down the surface of the primer until it is very smooth. While doing this, the work area should be thoroughly rinsed with water and the wet-or-dry sandpaper periodically rinsed as well. Allow the primer to dry before spraying additional coats.

21   Spray on the top coat, again building up the thickness by using several thin layers of paint. Begin spraying in the center of the repair area and then, using a circular motion, work out until the whole repair area and about two inches of the surrounding original paint is covered. Remove all masking material 10 to 15 minutes after spraying on the final coat of paint. Allow the new paint at least two weeks to harden, then use a very fine rubbing compound to blend the edges of the new paint into the existing paint. Finally, apply a coat of wax.

6   Body repair - major damage

1   Major damage must be repaired by an auto body shop specifically equipped to perform unibody repairs. These shops have the specialized equipment required to do the job properly.

2   If the damage is extensive, the body must be checked for proper alignment or the vehicle's handling characteristics may be adversely affected and other components may wear at an accelerated rate.

3   Due to the fact that all of the major body components (hood, fenders, etc.) are separate and replaceable units, any seriously damaged components should be replaced rather than repaired. Sometimes the components can be found in a wrecking yard that specializes in used vehicle components, often at considerable savings over the cost of new parts.

7   Hinges and locks - maintenance

Once every 3000 miles, or every three months, the hinges and latch assemblies on the doors, hood and trunk should be given a few drops of light oil or lock lubricant. The door latch strikers should also be lubricated with a thin coat of grease to reduce wear and ensure free movement. Lubricate the door

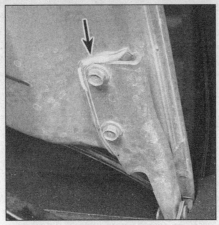

9.2 Before removing the hood, draw a mark (arrows) around the hinge plate

and trunk locks with spray-on graphite lubricant.

8   Windshield and fixed glass - replacement

Replacement of the windshield and fixed glass requires the use of special fast-setting adhesive/caulk materials and some specialized tools. It is recommended that these operations be left to a dealer or a shop specializing in glass work.

9   Hood - removal, installation and adjustment

Note: *The hood is heavy and somewhat awkward to remove and install - at least two people should perform this procedure.*

Removal and installation
Refer to illustration 9.2 and 9.4

1   Use blankets or pads to cover the cowl area of the body and fenders. This will protect the body and paint as the hood is lifted off.

2   Make marks or scribe a line around the hood hinge to ensure proper alignment during installation **(see illustration)**.

3   Disconnect any cables or wires that will

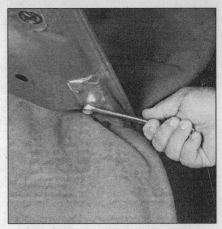

9.4 Remove the four retaining bolts and lift off the hood with the help of an assistant

interfere with removal.

4   Have an assistant support the hood. Remove the hinge-to-hood screws or bolts **(see illustration)**.

5   Lift off the hood.

6   Installation is the reverse of removal.

Adjustment
Refer to illustrations 9.10 and 9.11

7   Fore-and-aft and side-to-side adjustment of the hood is done by moving the hinge plate slot after loosening the bolts or nuts.

8   Scribe a line around the entire hinge plate so you can determine the amount of movement **(see illustration 9.2)**.

9   Loosen the bolts or nuts and move the hood into correct alignment. Move it only a little at a time. Tighten the hinge bolts and carefully lower the hood to check the position.

10   If necessary after installation, the entire hood latch assembly can be adjusted up-and-down as well as from side-to-side on the radiator support so the hood closes securely and flush with the fenders. To make the adjustment, scribe a line or mark around the hood latch mounting bolts to provide a reference point, then loosen them and reposition the latch assembly, as necessary **(see illustration)**. Following adjustment, retighten the mounting bolts.

9.10 To adjust the hood latch, loosen the retaining bolts (arrows), move the latch and retighten bolts, then close the hood to check the fit

11

9.11 Adjust the hood closing height by turning the hood bumpers in or out

10.2 Pry out the cable retainer from the backside of the hood latch assembly, then disengage the cable

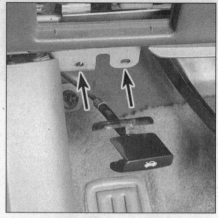

10.6 Remove the hood release lever retaining screws (arrows) and pull the cable rearward into the passenger compartment

11   Finally, adjust the hood bumpers on the radiator support so the hood, when closed, is flush with the fenders **(see illustration)**.

12   The hood latch assembly, as well as the hinges, should be periodically lubricated with white, lithium-base grease to prevent binding and wear.

10   Hood release latch and cable - removal and installation

Latch

Refer to illustration 10.2

1   Scribe a line around the latch to aid alignment when installing, then detach the latch retaining bolts to the radiator support **(see illustration 9.10)** and remove the latch.

2   Disconnect the hood release cable by disengaging the cable from the latch assembly **(see illustration)**.

3   Installation is the reverse of the removal procedure. **Note:** *Adjust the latch so the hood engages securely when closed and the hood bumpers are slightly compressed.*

Cable

Refer to illustration 10.6

4   Disconnect the hood release cable from the latch assembly as described above.

5   Attach a piece of stiff wire to the end of the cable, then follow the cable back to the firewall and detach all the cable retaining clips.

6   Working in the passenger compartment, remove the lower steering column trim cover (see Section 24). Then remove the two release lever mounting bolts and detach the hood release lever **(see illustration)**.

7   Pull the cable and grommet rearward into the passenger compartment until you can see the wire. Ensure that the new cable has a grommet attached, then remove the old cable from the wire and replace it with the new cable.

8   Working from engine compartment pull the wire back through the firewall.

9   Installation is the reverse of the removal **Note:** *Push on the grommet with your fingers*

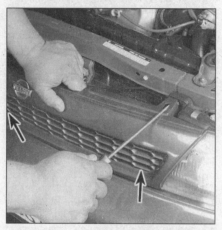

11.1 Pull out on the grille while using a screwdriver to release the grille retaining clips (arrows)

11.3 Remove the clips from the radiator support and insert them back into the grille

from the passenger compartment to seat the grommet in the firewall correctly.

11   Radiator grille - removal and installation

Refer to illustrations 11.1 and 11.3

Warning: *The models covered by this manual are equipped with airbags. Always disconnect the negative battery cable, then the positive cable and wait 10 minutes before working in the vicinity of the impact sensors, steering column or instrument panel to avoid the possibility of accidental deployment of the airbag, which could cause personal injury (see Chapter 12). The airbag circuits are easily identified by yellow insulation covering the entire wiring harness or just prior to the wire harness connectors. Do not use electrical test equipment on any of these wires or tamper with them in any way.*

Note: *On 1998 and later models, the grille is an integral part of the front bumper and cannot be removed separately.*

1   Using a screwdriver, rotate the two

upper grille retaining clips a quarter turn counterclockwise while pulling outward on the radiator grille **(see illustration)**.

2   Looking through the lower section of the grille, detach the two lower retaining clips and remove the grille.

3   Remove the grille retaining clips from the radiator support **(see illustration)** and reinstall the clips back into the grille.

4   To install, position the grille in place and press on the grille until the clips snap into place on the radiator support.

12   Trunk lid - removal, installation and adjustment

Note: *The trunk lid is heavy and somewhat awkward to remove and install - at least two people should perform this procedure.*

Removal and installation

Refer to illustrations 12.3 and 12.4

1   Open the trunk lid and cover the edges of the trunk compartment with pads or cloths to protect the painted surfaces when the lid is removed.

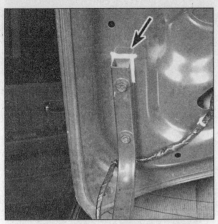

12.3 Scribe a mark around the hinge plate (arrow) for realignment of the trunk lid on installation

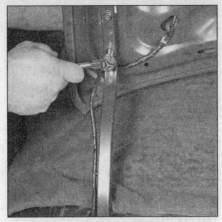

12.4 With an assistant holding the trunk lid - remove the four retaining bolts and lift off the trunk lid

12.9a Scribe a mark around the striker assembly (arrow) as a reference point to aid in the adjustment procedure

12.9b Loosen the bolts and move the striker assembly as necessary (arrows) to adjust the trunk lid flush with the quarter panels in the closed position

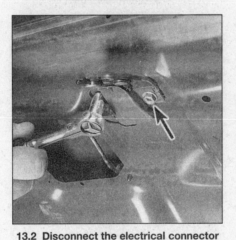

13.2 Disconnect the electrical connector and detach the two retaining bolts (arrow) to remove the trunk lid latch

13.7 Remove the cable retaining bolt then twist the cable end to remove it from the striker assembly

2   Disconnect any cables or wire harness connectors attached to the trunk lid that would interfere with removal.

3   Make alignment marks around the hinge mounting bolts with a marking pen **(see illustration)**.

4   While an assistant supports the trunk lid, remove the lid-to-hinge bolts **(see illustration)**.
on both sides and lift it off.

5   Installation is the reverse of removal.
Note: *When reinstalling the trunk lid, align the lid-to-hinge bolts with the marks made during removal.*

Adjustment

Refer to illustrations 12.9a and 12.9b

6   Fore-and-aft and side-to-side adjustment of the trunk lid is accomplished by moving the lid in relation to the hinge after loosening the bolts or nuts.

7   Scribe a line around the entire hinge plate as described earlier in this Section so you can determine the amount of movement.

8   Loosen the bolts or nuts and move the trunk lid into correct alignment. Move it only a little at a time. Tighten the hinge bolts or nuts

and carefully lower the trunk lid to check the alignment.

9   If necessary after installation, the entire trunk lid striker assembly can be adjusted up and down as well as from side to side on the trunk lid so the lid closes securely and is flush with the rear quarter panels. To do this, scribe a line around the trunk lid striker assembly to provide a reference point **(see illustration)**. Then loosen the bolts and reposition the striker as necessary **(see illustration)**. Following adjustment, retighten the mounting bolts.

10   The trunk lid latch assembly, as well as the hinges, should be periodically lubricated with white lithium-base grease to prevent sticking and wear.

13   Trunk lid latch, release cable, and lock cylinder - removal and installation

Trunk lid latch

Refer to illustration 13.2

1   Open the trunk and scribe a line around

the trunk lid latch assembly for a reference point to aid the installation procedure.

2   The trunk lid latch is retained by two bolts **(see illustration)**. See Section 12 for adjustment procedures.

3   Disconnect the electrical connections.

4   Detach the two retaining bolts and remove the latch.

5   Installation is the reverse of removal.

Trunk and fuel door release cable

Refer to illustrations 13.7, 13.8a, 13.8b and 13.11

6   Working in the trunk, pry out the plastic clips securing the drivers side and rear inside finishing panels to allow access to the striker assembly.

7   Scribe a line around the trunk release cable bracket to aid in adjustment during the installation procedure. Then remove the retaining bolt and twist the cable free from the striker assembly **(see illustration)**.

8   Open the fuel door and remove the

11

13.8a Open the fuel door and remove the opening mechanism retaining nut (arrow)

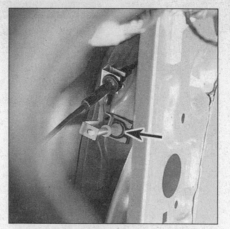

13.8b To manually open the fuel door remove the driver's side trunk finishing panel and pull upward on the mechanism (arrow)

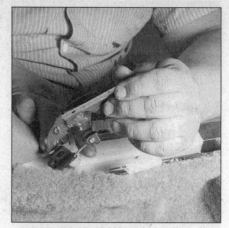

13.11 Pull back the carpeting to access the release cable and lever retaining screws (arrows)

13.14 Working through the trunk lid access hole, remove the lock cylinder retaining clip (arrow)

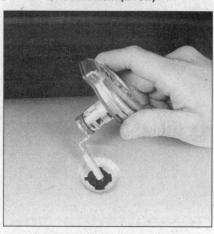

13.15 Grasp the emblem and twist outward to remove the lock cylinder

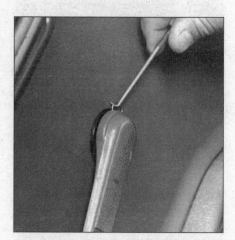

14.2 Use a hooked tool like this to remove the window crank retaining clip

opening mechanism retaining nut **(see illustration)**. **Note:** *If the cable is broken the fuel door can be opened manually by reaching between the inner fender support and rear quarter panel and pulling upward on the fuel door opening mechanism* **(see illustration)**.

9   Working in the drivers side passenger compartment, pry off the front and rear kick panels at the bottom of the door openings.

10 Remove the rear seat as described in Section 27. Then detach the front seat belt trim cover and the plastic handle from the trunk release lever.

11 Peel back the carpeting to access the release cable and lever retaining bolts. Then remove the lever retaining bolts **(see illustration)** and all the cable retaining clips. Attach a piece of thin wire to the end of the cable.

12 Working in the trunk compartment, pull the cable assembly towards the rear of the vehicle until you can see the wire.

13 Installation is the reverse of removal.

Trunk lock cylinder

Refer to illustrations 13.14 and 13.15

14 Open the trunk. Look upward through the trunk lid access hole and remove the lock cylinder retaining clip **(see illustration)**.

15 Working from the outside of the trunk lid, grasp the lock cylinder emblem and twist it outward to remove **(see illustration)**.

16 Installation is the reverse of removal.

14 Door trim panel - removal and installation

Refer to illustrations 14.2, 14.3, 14.4a, 14.4b, 14.5a, 14.5b, 14.6 and 14.7

1   Disconnect the negative cable from the battery.

2   On manual window equipped models, remove the window crank, using a hooked tool to remove the retainer clip **(see illustration)**. A special tool is available for this purpose, but it's not essential. With the clip removed, pull off the handle.

3   On power window equipped models, pry out the armrest switch control plate **(see illustration)** and disconnect the electrical connections.

4   Detach the armrest pull handle retaining screw **(see illustration)** then remove the inside door handle trim bezel **(see illustration)**.

5   Insert a wide putty knife, a thin screw-

14.3 Using a small screwdriver, pry out the armrest switch control plate

14.4a Detach the armrest/pull handle retaining screw (typical)

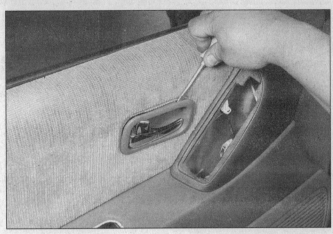

14.4b Pry out the inside door handle trim bezel (typical)

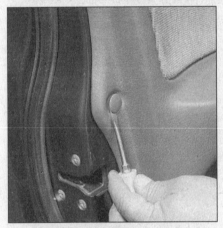

14.5a Detach the plastic clips securing the door trim panel ends (typical)

14.5b Remove the retaining screws along the lower edge of the trim panel (typical)

driver or a special trim panel removal tool between the trim panel and the head of the retaining clip to disengage the door panel retaining clips (see illustration). then remove the remaining door panel retaining screws at the lower edge of the door panel (see illustration).

6   Once all of the clips and screws are dis-
engaged, detach the trim panel, disconnect any electrical connectors and remove the trim panel from the vehicle by gently pulling it up and out (see illustration).

7   For access to the inner door remove the door panel support bracket (see illustration). Then peel back the watershield, taking care not to tear it. To install the trim panel, first

press the watershield back into place. If necessary, add more sealant to hold it in place.

8   Installation is the reverse of removal.

15   Door - removal, installation and adjustment

Note: *The door is heavy and somewhat awkward to remove and install - at least two people should perform this procedure.*

Removal and installation

Refer to illustrations 15.6, 15.8a and 15.8b

1   Lower the window completely in the door and then disconnect the negative cable from the battery.

2   Open the door all the way and support it on jacks or blocks covered with rags to prevent damaging the paint.

3   Remove the door trim panel and water deflector as described in Section 14.

4   Disconnect all electrical connections, ground wires and harness retaining clips from the door. **Note:** *It is a good idea to label all connections to aid the reassembly process.*

5   From the door side, detach the rubber conduit between the body and the door.

14.6   Once all the retaining clips and screws have been removed, detach the trim panel by gently pulling up and out

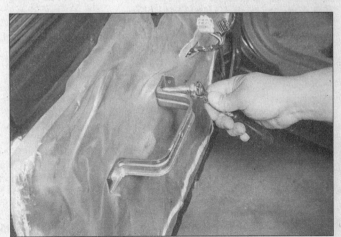

14.7   Remove the door panel support bracket and carefully peel back the watershield to access the inner door

11

15.6 Gently tap the pin for the door stop strut in the direction shown

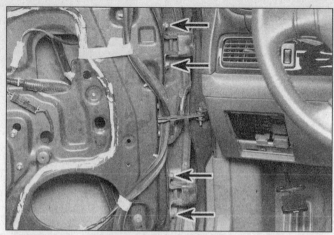

15.8a Before loosening the door retaining bolts (arrows), draw a line around the hinge plate for a reinstallation reference

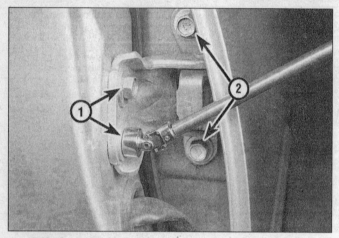

15.8b Open the front door to access the rear door hinge to body (1) and hinge to door (2) bolts (arrows)

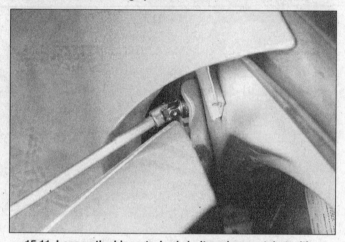

15.11 Loosen the hinge-to-body bolts using a ratchet with a swivel socket to adjust the doors

Then pull the wiring harness through conduit hole and remove from the door.

6   Remove the door stop strut center pin **(see illustration)**.

7   Mark around the door hinges with a pen or a scribe to facilitate realignment during reassembly.

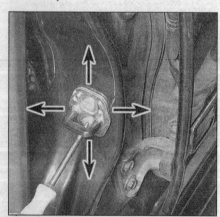

15.13 Adjust the door lock striker by loosening the mounting screws and gently tapping the striker in the desired direction (arrows)

8   With an assistant holding the door, remove the hinge to door bolts **(see illustrations)** and lift the door off.

9   Installation is the reverse of removal.

Adjustment

Refer to illustrations 15.11, and 15.13

10   Having proper door to body alignment is a critical part of a well functioning door assembly. First check the door hinge pins for excessive play. Fully open the door and lift up and down on the door without lifting the body. If a door has 1/16-inch or more excessive play, the hinges should be replaced.

11   Door-to-body alignment adjustments are made by loosening the hinge-to-body bolts **(see illustration)** or hinge-to-door bolts and moving the door. Proper body alignment is achieved when the top of the doors are parallel with the roof section, the front door is flush with the fender, the rear door is flush with the rear quarter panel and the bottom of the doors are aligned with the lower rocker panel. If these goals can't be reached by adjusting the hinge-to-body or hinge-to-door bolts, body alignment shims may have to be purchased and inserted behind the hinges to achieve correct alignment.

12   To adjust the door closed position, scribe a line or mark around the striker plate to provide a reference point, then check that the door latch is contacting the center of the latch striker. If not adjust the up and down position first.

13   Finally adjust the latch striker sideways position, so that the door panel is flush with the center pillar or rear quarter panel and provides positive engagement with the latch mechanism **(see illustration)**.

16   Door latch, lock cylinder and handles - removal and installation

Door latch

Refer to illustrations 16.2 and 16.4

1   To remove the latch assembly, raise the window then remove the door trim panel and watershield as described in Section 14.

2   Working through the large access hole, disengage the outside door handle-to-latch rod, outside door lock-to-latch rod, bell crank-to-lock rod and the inside handle-to-latch rod **(see illustration)**.

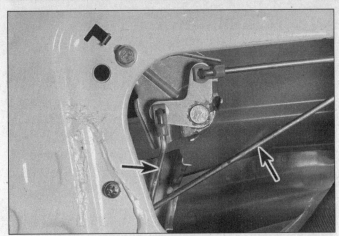

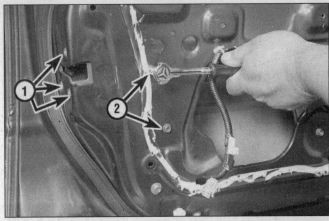

16.2 Detach the plastic clips on the two rods (arrows) leading to the latch

16.4 Remove the latch screws (1) from the end of the door, then detach the lock solenoid retaining bolts (2) and pull the latch assembly through the access hole

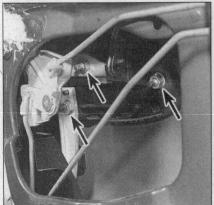

16.8a To remove the lock cylinder, detach the plastic clip securing the lock rod then pry off the lock cylinder retaining clip

16.8b Working from the outside of the door, pull the lock cylinder free

16.10 The outside handle retaining nuts (arrows) can be reached through the access hole in the door frame

3   All door locking rods are attached by plastic clips. The plastic clips can be removed by unsnapping the portion engaging the connecting rod and then by pulling the rod out of its locating hole.

4   Remove the screws securing the latch to the door. On models with power locks detach the lock solenoid retaining bolts **(see illustration)**, then remove the latch assembly from the door.

5   Installation is the reverse of removal.

Door lock cylinder and outside handle

Refer to illustrations 16.8a, 16.8b and 16.10

6   To remove the lock cylinder, raise the window and remove the door trim panel and watershield as described in Section 14.

7   Working through the large access hole, disengage the plastic clip that secures the lock cylinder to latch rod.

8   Using a screwdriver, slide the lock cylinder retaining clip out of engagement and remove the lock cylinder from the door **(see illustrations)**.

9   To remove the outside handle, work through the access hole and disengage the

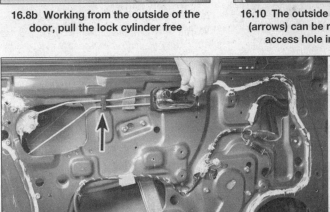

plastic clip that secures the outside handle-to-latch rod.

10   Remove the outside handle retaining nuts **(see illustration)** and pull the handle from the door.

11   Installation is the reverse of removal.

Inside handle

Refer to illustration 16.13

12   Remove the door trim panel as de-

scribed in Section 14 and peel back the upper half of the watershield.

13   Unclip the door actuating rod guide, then remove the door handle retaining screw **(see illustration)**.

14   Pull the handle free, disconnect the actuating rods from the back side of the handle control and remove the handle from the door.

15   Installation is the reverse of removal.

16.13 Remove the handle retaining screw, then rotate the handle out and detach the actuating rods from the backside

11

17.4 Raise the window just enough to access the glass retaining bolts (arrows) through the hole in the door frame

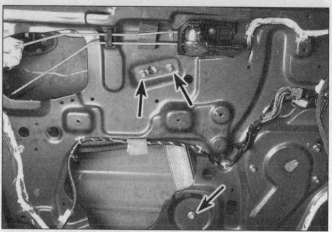

18.4a Detach the window equalizer mounting bolts (arrows) . . .

17 Door window glass - removal and installation

Refer to illustration 17.4

1   Remove the door trim panel and the plastic watershield (see Section 14).
2   Lower the window glass all the way down into the door.
3   Carefully pry the inner weatherstrip out of the door window opening.
4   Raise the window just enough to access the window retaining bolts through the hole in the door frame **(see illustration)**.
5   Place a rag over the glass to help prevent scratching the glass and remove the two glass mounting bolts.
6   Remove the glass by pulling it up and out.
7   Installation is the reverse of removal.

18 Door window glass regulator - removal and installation

Refer to illustrations 18.4a and 18.4b

1   Remove the door trim panel and the plastic watershield (see Section 14).
2   Remove the window glass assembly (see Section 17).
3   On power operated windows, disconnect the electrical connector from the window regulator motor.
4   Remove the equalizer arm bracket and the regulator mounting bolts **(see illustrations)**.
5   Pull the equalizer arm and regulator assemblies through the service hole in the door frame to remove it.
6   Installation is the reverse of removal.

19 Bumpers - removal and installation

Warning: *The models covered by this manual are equipped with airbags. Always disconnect the negative battery cable, then the positive cable and wait 10 minutes before working in the vicinity of the impact sensors, steering column or instrument panel to avoid the possibility of accidental deployment of the airbag, which could cause personal injury (see Chapter 12). The airbag circuits are easily identified by yellow insulation covering the* entire wiring harness or just prior to the wire harness connectors. Do not use electrical test equipment on any of these wires or tamper with them in any way.

Front bumper (1993 through 1997 models)

Refer to illustrations 19.3, 19.6a, 19.6b and 19.6c

1   Apply the parking brake, raise the vehicle and support it securely on jackstands.
2   Disconnect the negative battery cable. If equipped with airbag(s), disconnect the positive battery cable and wait 10 minutes before proceeding any further.
3   Detach the left and right fog lamps **(see illustration)**.
4   Remove the radiator grille (see Section 11).
5   Working under the vehicle, detach the plastic trim securing the bumper cover to access the bumper cover retaining bolts.
6   Detach the six retaining nuts securing the bumper cover to the right and left fender, then remove three retaining screws in the radiator grille area and pull off the bumper cover **(see illustrations)**.

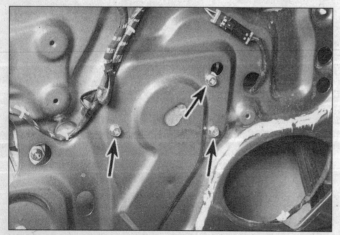

18.4b . . . then remove the window regulator bolts (arrows)

19.3 Remove the fog lamp retaining bolts (arrows) and disconnect the electrical connector from the back side

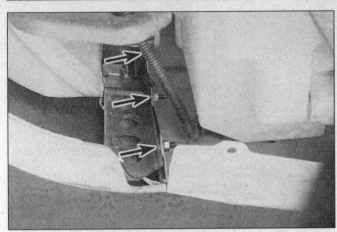

19.6a The side marker light must be removed to access the front bumper cover-to-fender retaining nuts (arrows)

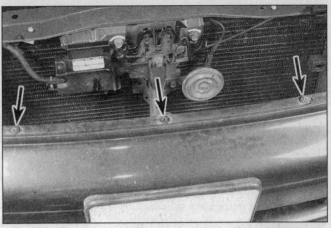

19.6b Detach the retaining screws (arrows) in the radiator grille area . . .

19.6c . . . then detach the retaining nut and clips (arrows) from beneath the license plate and remove the front bumper cover

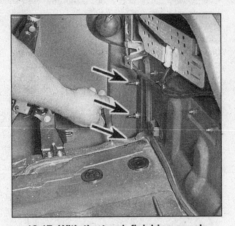

19.17 With the trunk finishing panels removed - detach the rear bumper cover to fender retaining nuts (arrows) located in each corner of the trunk

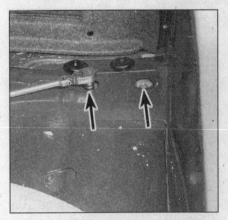

19.19 Pry out the four plastic floor pan plugs and remove the bumper retaining bolts

7 Remove two bumper retaining bolts and pull the bumper assembly out and away from the vehicle.

8 Installation is the reverse of removal.

Front bumper (1998 and later models)

9 Apply the parking brake, raise the vehicle and support it securely on jackstands.

10 Disconnect the negative battery cable. If equipped with airbag(s), disconnect the positive battery cable and wait 10 minutes before proceeding any further.

11 Working under the vehicle, detach the screws and clips securing the front bumper to the right and left front fender protectors.

12 Remove the bolts securing the bumper to the bumper reinforcement and the support and pull the bumper assembly out and away from the vehicle.

13 Installation is the reverse of removal.

Rear bumper (1993 through 1997 models)

Refer to illustrations 19.12 and 19.14

14 Apply the parking brake, raise the vehicle and support it securely on jackstands.

15 Disconnect the negative battery cable. If equipped with airbag(s), disconnect the positive battery cable and wait 10 minutes before proceeding any further.

16 Working in the trunk, pry out the plastic clips securing the drivers side, passenger side, and rear inside trunk finishing panels to allow access to the bumper retaining bolts.

17 Detach the six retaining nuts securing the bumper cover to the right and left quarter panels **(see illustration)**, then remove the two bolts securing the license plate section of the bumper cover.

18 Working under the vehicle, remove the three bolts securing the bumper cover to the bumper and pull off the bumper cover.

19 Working in the trunk, pry out the plastic floor pan plugs located behind the taillight assemblies and remove the bumper retaining bolts **(see illustration)**.

20 Pull the bumper assembly out and away from the vehicle.

21 Installation is the reverse of removal.

Rear bumper (1998 and later models)

22 Apply the parking brake, raise the vehi-

cle and support it securely on jackstands.

23 Disconnect the negative battery cable. If equipped with airbag(s), disconnect the positive battery cable and wait 10 minutes before proceeding any further.

24 Working in the rear wheel fenderwells, remove the screws securing the rear bumper to the right and left rear fender panels.

25 Working under the vehicle, detach the clips securing the rear bumper to the right and left rear fender panels.

26 Detach the clips securing the rear bumper to the bumper reinforcement.

27 Working in the trunk, detach the clips securing the rear bumper to the body panel below the trunk opening and pull the bumper assembly out and away from the vehicle.

28 Installation is the reverse of removal.

20 Outside mirrors - removal and installation

Refer to illustrations 20.2a, 20.2b and 20.4

1 Remove the door trim panel and the plastic watershield (see Section 14).

11

20.2a Pry out the mirror trim cover . . .

2   Pry off the mirror trim cover, then detach the trim retaining screws **(see illustrations)**.
3   Disconnect the electrical connector from the mirror.
4   Remove the three mirror retaining bolts and detach the mirror from the vehicle **(see illustration)**.
5   Installation is the reverse of removal.

---

21   Center console - removal and installation

Warning: *The models covered by this manual are equipped with airbags. Always disconnect the negative battery cable, then the positive cable and wait 10 minutes before working in the vicinity of the impact sensors, steering column or instrument panel to avoid the possibility of accidental deployment of the airbag, which could cause personal injury (see Chapter 12). The airbag circuits are easily identified by yellow insulation covering the entire wiring harness or just prior to the wire harness connectors. Do not use electrical test equipment on any of these wires or tamper with them in any way.*

Floor console (1993 through 1997 models)

Refer to illustrations 21.2, 21.4a, 21.4b, 21.6 and 21.9

1   Disconnect the negative battery cable. If

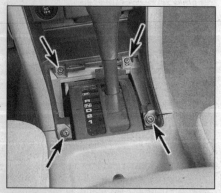

21.4a Remove the lower radio trim bezel screws (arrows) located under the gear selector trim plate

20.2b . . . then remove the trim retaining screws

equipped with airbag(s), disconnect the positive battery cable and wait 10 minutes before proceeding any further.
2   Pry out the gear selector trim bezel **(see illustration)**.
3   Remove the drivers side lower steering column trim cover (see Section 24).
4   Detach the screws securing the radio trim bezel **(see illustrations)**.
5   Apply the parking brake, then move the gear selector to the towards the rear of the vehicle and remove the radio trim bezel.
6   Slide the front seats all the way forward to access the floor console glove box retaining screws **(see illustration)**. Then detach the two remaining screws in gear selector area and remove the rear half of the floor console.
7   Remove the dashboard center trim panel (see Section 23).
8   Remove the radio (see Chapter 12) and the ash tray.
9   Remove the front console screws and lift the console up and over the shift lever **(see illustration)**.Disconnect any electrical connections and remove the console from the vehicle.
10   Installation is the reverse of removal.

Floor console (1998 and later models)

11   Disconnect the negative battery cable. If equipped with airbag(s), disconnect the positive battery cable and wait 10 minutes before proceeding any further.

21.4b Detach the upper radio trim bezel screws (arrows) and remove the bezel

20.4 Disconnect the electrical connector (1), then detach the three mirror retaining bolts (2) and remove the mirror from the vehicle

21.2 Use a screwdriver to pry out the gear selector trim bezel

12   Pry out the gear selector trim bezel.
13   Remove the drivers side lower steering column trim cover (see Section 24).
14   Detach the screws securing the radio trim bezel.
15   Apply the parking brake, then move the gear selector toward the rear of the vehicle and remove the radio trim bezel.
16   Detach the front two screws and single center screw securing the floor console.
17   Slide the seats all the way forward to access the screws securing the sides of the floor console and remove the floor console.
18   Installation is the reverse of removal.

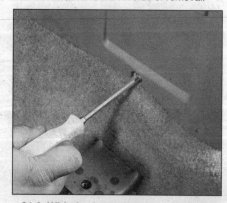

21.6 With the front seats moved all the way forward, remove the screws securing the rear of the floor console

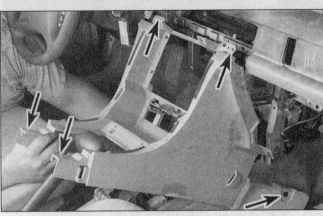

21.9 Detach the floor console front half retaining screws (arrows), then remove the console from the vehicle

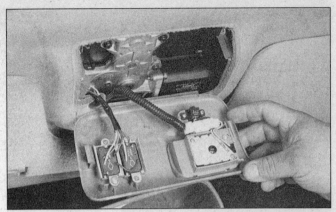

21.19 Grasp the overhead console on the passengers side and pull downward

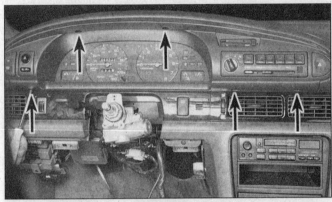

22.2 Remove the bezel retaining screws (arrows) (steering wheel removed for clarity) (1995 through 1997 models)

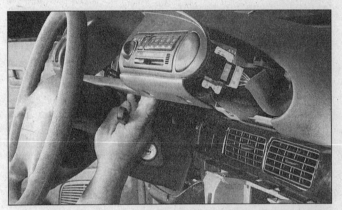

22.3 Pull outward on the instrument cluster bezel and remove the electrical connectors from the backside (1995 through 1997 models)

Overhead console

Refer to illustration 21.11

19   To remove the overhead console grasp the passengers side of the console and pull downward to unclip it **(see illustration)**.
20   Disconnect the electrical connections and remove the console from the vehicle.
21   Installation is the reverse of removal.

22 Instrument cluster bezel - removal and installation

Refer to illustrations 22.2 and 22.3
Warning: *The models covered by this manual are equipped with airbags. Always disconnect the negative battery cable, then the positive cable and wait 10 minutes before working in the vicinity of the impact sensors, steering column or instrument panel to avoid the possibility of accidental deployment of the airbag, which could cause personal injury (see Chapter 12). The airbag circuits are easily identified by yellow insulation covering the entire wiring harness or just prior to the wire harness connectors. Do not use electrical test equipment on any of these wires or tamper with them in any way.*
1   Disconnect the negative battery cable. If equipped with airbag(s), disconnect the positive battery cable and wait 10 minutes before

proceeding any further.
2   On 1993 through 1997 models, remove the five bezel retaining screws **(see illustration)**. On 1998 and later models, remove the two bezel retaining screws.
3   Tilt the steering wheel down and pull the instrument cluster bezel outward to access the electrical connections on the backside **(see illustration)**.
4   Disconnect all electrical connections from the backside of the cluster bezel and remove the bezel from the vehicle.
5   Installation is the reverse of removal.

23 Dashboard trim panels - removal and installation

Warning: *The models covered by this manual are equipped with airbags. Always disconnect the negative battery cable, then the positive cable and wait 10 minutes before working in the vicinity of the impact sensors, steering column or instrument panel to avoid the possibility of accidental deployment of the airbag, which could cause personal injury (see Chapter 12). The airbag circuits are easily identified by yellow insulation covering the entire wiring harness or just prior to the wire harness connectors. Do not use electrical test equipment on any of these wires or tamper with them in any way.*
1   Disconnect the negative battery cable. If equipped with airbag(s), disconnect the positive battery cable and wait 10 minutes before

proceeding any further.

Lower steering column trim cover

Refer to illustration 23.2
2   Working in the drivers side passenger compartment, detach the retaining screws along the lower edge of the steering column trim cover **(see illustration)**.

23.2 Remove the retaining screws along the bottom edge of the lower steering column trim cover, then remove it from the vehicle

11

23.7a Grasp the left side center trim panel with two hands and unsnap it from the dashboard assembly

23.7b Pull the left side center trim panel forward enough to disconnect the electrical connectors from the backside

23.8 Detach the two screws securing the right side center trim panel and remove it from the vehicle

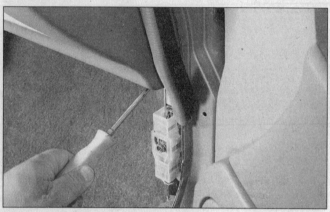

23.10 After detaching the right side kick panel, remove the two screws securing the lower edge of the glove box compartment

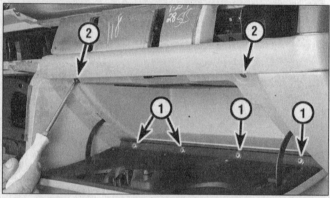

23.11 Detach the glove box door retaining screws (1), detach the screws at the top edge (2) and remove the glove box assembly from the vehicle

3   Pull outward on the lower edge of the steering column trim cover and detach it from the vehicle.

4   Installation is the reverse of removal.

Center trim panel (1993 through 1997 models)

Refer to illustrations 23.7a, 23.7b and 23.8

Note: *The 1998 and later models are not equipped with the center trim panel.*

5   Remove the instrument cluster bezel (see Section 22)

6   Detach the radio trim bezel (see Section 21)

7   Unsnap the left side center trim panel and disconnect the electrical connectors from the backside **(see illustrations)**.

8   Detach the two retaining screws and unsnap the right side center trim panel **(see illustration)**.

9   Installation is the reverse of removal.

Glove box

Refer to illustrations 23.10 and 23.11

10   Detach the right side kick panel and remove the screws along the lower edge of the glove box assembly **(see illustration)**.

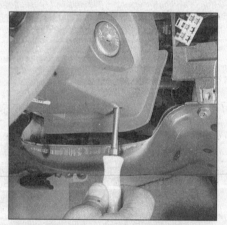

24.2a Remove the steering column cover screws

11   Open the glove box door, then detach the door retaining screws and remove the door from the glove box assembly **(see illustration)**.

12   Detach the remaining two screws securing the upper edge of the glove box. Disconnect the lamp from the backside of the glove box and remove the assembly from the vehicle.

13   Installation is the reverse of removal.

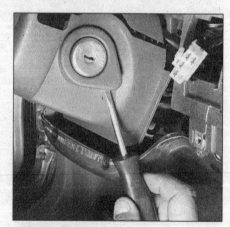

24.2b Pry off the lock cylinder cover

24   Steering column cover - removal and installation

Refer to illustrations 24.2a, 24.2b and 24.3

1   Remove the lower steering column trim cover and the left side center trim panel (see Section 24).

2   Remove the steering column cover screws. Then pry off the lock cylinder cover **(see illustrations)**.

24.3 Separate the steering column cover halves to remove them from the steering column

25.6a Detach the lower dash reinforcement panel retaining bolts (arrows) (1993 through 1997 models)

25.6b Remove the steering column retaining bolts and lower it away from the instrument panel

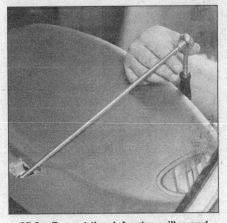

25.8a Pry out the defroster grilles and remove the instrument panel retaining screws

Section 10).

6   On 1993 through 1997 models, remove the lower dash reinforcement panel. On all models, detach the bolts securing the steering column and lower it away from the instrument panel **(see illustrations)**.

7   On 1993 through 1997 models, remove the two nuts securing the lower corners of the instrument panel. On 1998 and later models, remove the three lower nuts and one screw securing the instrument panel.

8   Pry out the defroster grilles and remove the screws securing the upper edge of the instrument panel, then remove the screws retaining the instrument panel to the cowl support tube **(see illustrations)**.

9   Detach any electrical connectors interfering with removal, then pull the instrument panel towards the rear of the vehicle to remove it.

10   Installation is the reverse of removal.

3   Separate the cover halves and detach them from the steering column **(see illustration)**.

4   Installation is the reverse of removal.

25 Instrument panel - removal and installation

Refer to illustrations 25.6a, 25.6b, 25.8a and 25.8b

Warning: *The models covered by this manual are equipped with airbags. Always disconnect the negative battery cable, then the positive cable and wait 10 minutes before working in the vicinity of the impact sensors, steering column or instrument panel to avoid the possibility of accidental deployment of the airbag, which could cause personal injury (see Chapter 12). The airbag circuits are easily identified by yellow insulation covering the entire wiring harness or just prior to the wire harness connectors. Do not use electrical test equipment on any of these wires or tamper with them in any way.*

1   Disconnect the negative battery cable. If equipped with airbag(s), disconnect the positive battery cable and wait 10 minutes before

proceeding any further.

2   Remove the dashboard trim panels (see Section 23) and the center floor console (see Section 21).

3   Remove the instrument cluster (see Chapter 12).

4   Disconnect the passenger side air bag (if equipped).

5   Detach the nuts and bolts securing the fuse box and the hood release handle (see

26 Cowl cover - removal and installation

Refer to illustrations 26.1 and 26.2

1   Pry off the plastic trim cap on the windshield wiper arms, then detach the wiper arm

25.8b Remove the screws retaining the instrument panel to the cowl support tube (arrows) (typical)

11

26.1 Pry off the wiper arm trim caps, loosen the retaining nut and remove the wiper arm

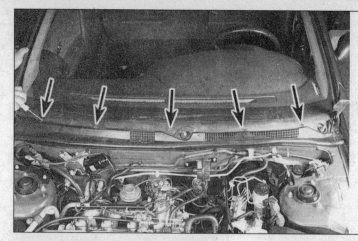

26.2 Remove the screws from the rear of the cowl cover

retaining nuts and remove the wiper arms **(see illustration)**.

2   Remove the retaining screws securing the rear of the cowl cover **(see illustration)**.

3   Carefully pry out the clips securing the hood seal and remove the cowl cover from the vehicle. The hood seal clips are approximately six to seven inches apart - pry directly underneath the clip being careful not to tear the hood seal.

4   Installation is the reverse of removal.

27   Seats - removal and installation

Front seat

Refer to illustration 27.2

1   Position the seat all the way forward or all the way to the rear to access the front seat retaining bolts.

2   Detach any bolt trim covers and remove the retaining bolts **(see illustration)**.

3   Tilt the seat upward to access the underneath, then disconnect any electrical connectors and lift the seat from the vehicle.

4   Installation is the reverse of removal.

Rear seat

Refer to illustrations 27.5, 27.6a and 27.6b

5   Remove the seat cushion retaining bolts **(see illustration)**. Then lift up on the front edge and remove the cushion from the vehicle.

6   Detach the retaining bolts at the lower edge of the seat back and at the middle of the trunk access door **(see illustrations)**.

7   Lift up on the lower edge of the seat back and remove it from the vehicle.

8   Installation is the reverse of removal.

28   Seat belts - check

1   Check the seat belts, buckles, latch plates and guide loops for any obvious damage or signs of wear.

27.2 Detach the trim covers to access the front seat retaining bolts

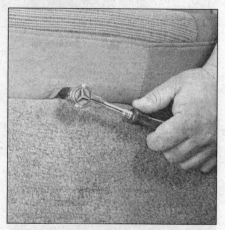

27.5 Using a socket and ratchet, remove the rear seat cushion retaining bolts

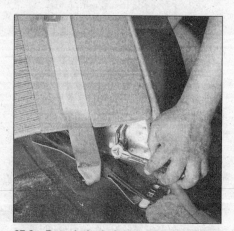

27.6a Detach the bolts securing the lower corners of the seat back . . .

27.6b . . . then detach the bolts in the trunk access door opening and remove the seat from the vehicle

2   Make sure the seat belt reminder light comes on when the key is turned on.

3   The seat belts are designed to lock up during a sudden stop or impact, yet allow free movement during normal driving. The retractors should hold the belt against your chest while driving and rewind the belt when the buckle is unlatched.

4   If any of the above checks reveal problems with the seat-belt system, replace parts as necessary. **Note:** *Before replacing any seat belt components, check with your local dealership service department, the seat belt system could be covered under the vehicle warranty.*

Chapter 12
Chassis electrical system

Contents

1 General information

The electrical system is a 12-volt, negative ground type. Power for the lights and all electrical accessories is supplied by a lead/acid-type battery which is charged by the alternator.

This Chapter covers repair and service procedures for the various electrical components not associated with the engine. Information on the battery, alternator, distributor and starter motor can be found in Chapter 5.

It should be noted that when portions of the electrical system are serviced, the cable should be disconnected from the negative battery terminal to prevent electrical shorts and/or fires.

2 Electrical troubleshooting - general information

A typical electrical circuit consists of an electrical component, any switches, relays, motors, fuses, fusible links or circuit breakers related to that component and the wiring and electrical connectors that link the component to both the battery and the chassis. To help you pinpoint an electrical circuit problem, wiring diagrams are included at the end of this Chapter.

Before tackling any troublesome electrical circuit, first study the appropriate wiring diagrams to get a complete understanding of what makes up that individual circuit. Trouble spots, for instance, can often be narrowed down by noting if other components related to the circuit are operating properly. If several components or circuits fail at one time, chances are the problem is in a fuse or ground connection, because several circuits are often routed through the same fuse and ground connections.

Electrical problems usually stem from simple causes, such as loose or corroded connections, a blown fuse, a melted fusible link or a bad relay. Visually inspect the condition of all fuses, wires and connections in a problem circuit before troubleshooting it.

If testing instruments are going to be utilized, use the diagrams to plan ahead of time where you will make the necessary connections in order to accurately pinpoint the trouble spot.

The basic tools needed for electrical troubleshooting include a circuit tester or voltmeter (a 12-volt bulb with a set of test leads can also be used), a continuity tester, which includes a bulb, battery and set of test leads, and a jumper wire, preferably with a circuit breaker incorporated, which can be used to bypass electrical components. Before attempting to locate a problem with test instruments, use the wiring diagram(s) to decide where to make the connections.

Voltage checks

Voltage checks should be performed if a circuit is not functioning properly. Connect one lead of a circuit tester to either the negative battery terminal or a known good ground. Connect the other lead to a electrical connector in the circuit being tested, preferably nearest to the battery or fuse. If the bulb of the tester lights, voltage is present, which means that the part of the circuit between the electrical connector and the battery is problem free. Continue checking the rest of the circuit in the same fashion. When you reach a point at which no voltage is present, the problem lies between that point and the last test point with voltage. Most of the time the problem can be traced to a loose connection. **Note:** Keep in mind that some circuits receive voltage only when the ignition key is in the Accessory or Run position.

Finding a short

One method of finding shorts in a circuit is to remove the fuse and connect a test light or voltmeter in its place. There should be no voltage present in the circuit. Move the wiring harness from side to side while watching the test light. If the bulb goes on, there is a short to ground somewhere in that area, probably where the insulation has rubbed through. The same test can be performed on each component in the circuit, even a switch.

Ground check

Perform a ground test to check whether a component is properly grounded. Disconnect the battery and connect one lead of a self-powered test light, known as a continuity

12

3.1a The interior fuse box is located under the left side (driver's side) of the instrument panel, behind the fuse panel cover

tester, to a known good ground. Connect the other lead to the wire or ground connection being tested. If the bulb goes on, the ground is good. If the bulb does not go on, the ground is not good.

Continuity check

A continuity check is done to determine if there are any breaks in a circuit - if it is passing electricity properly. With the circuit off (no power in the circuit), a self-powered continuity tester can be used to check the circuit. Connect the test leads to both ends of the circuit (or to the "power" end and a good ground), and if the test light comes on the circuit is passing current properly. If the light doesn't come on, there is a break somewhere in the circuit. The same procedure can be used to test a switch, by connecting the continuity tester to the power in and power out

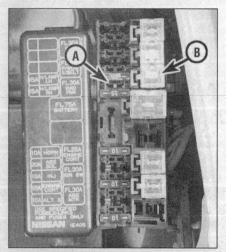

3.1c The engine compartment fuse box is located in the left front corner of the engine compartment - it contains fuses (A) as well as fusible links (B)

sides of the switch. With the switch turned On, the test light should come on.

Finding an open circuit

When diagnosing for possible open circuits, it is often difficult to locate them by sight because oxidation or terminal misalignment are hidden by the electrical connectors. Merely wiggling an electrical connector on a sensor or in the wiring harness may correct the open circuit condition. Remember this when an open circuit is indicated when troubleshooting a circuit. Intermittent problems may also be caused by oxidized or loose connections.

Electrical troubleshooting is simple if you keep in mind that all electrical circuits are basically electricity running from the battery, through the wires, switches, relays, fuses and fusible links to each electrical component (light bulb, motor, etc.) and to ground, from which it is passed back to the battery. Any electrical problem is an interruption in the flow of electricity to and from the battery.

3 Fuses - general information

Refer to illustrations 3.1a, 3.1b, 3.1c and 3.3

The electrical circuits of the vehicle are protected by a combination of fuses, circuit breakers and fusible links. The fuse blocks are located under the instrument panel on the left side of the dashboard, and in the engine compartment next to the battery **(see illustrations)**.

Each of the fuses is designed to protect a specific circuit, and the various circuits are identified on the fuse panel cover.

Miniaturized fuses are employed in the fuse blocks. These compact fuses, with blade terminal design, allow fingertip removal and replacement. If an electrical component fails, always check the fuse first. A blown fuse is easily identified through the clear plastic body. Visually inspect the element for evidence of damage **(see illustration)**.

Be sure to replace blown fuses with the correct type. Fuses of different ratings are physically interchangeable, but only fuses of the proper rating should be used. Replacing a fuse with one of a higher or lower value than

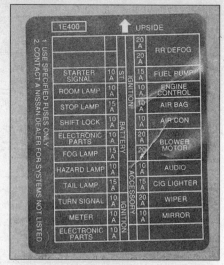

3.1b Interior fuse panel identification

specified is not recommended. Each electrical circuit needs a specific amount of protection. The amperage value of each fuse is molded into the fuse body.

If the replacement fuse immediately fails, don't replace it again until the cause of the problem is isolated and corrected. In most cases, this will be a short circuit in the wiring caused by a broken or deteriorated wire.

4 Fusible links - general information

Some circuits are protected by fusible links. The links are used in circuits which are not ordinarily fused, such as the ignition circuit.

The fusible links are located in the engine compartment fuse block **(see illustration 3.1c)** and are similar to fuses.

To replace a fusible link, first disconnect the negative cable from the battery. Unplug the burned-out link from the fuse block and replace it with a new one (available from your dealer or auto parts store). Always determine the cause for the overload which melted the fusible link before installing a new one.

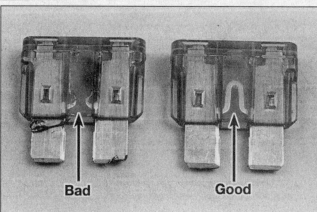

3.3 When a fuse blows, the element between the terminals melts - the fuse on the left is blown, the fuse on the right is good

Bad   Good

5.2 The circuit breakers are located behind the interior fuse panel in the drivers side passenger compartment

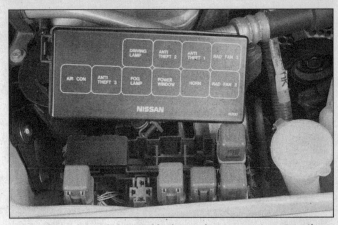

6.1a Relay box 1 is located in the engine compartment on the right side inner fenderwell

6.1b Relay box 2 is located in the engine compartment between the engine and the battery

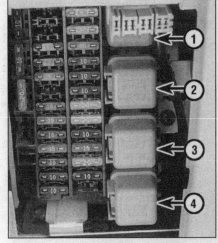

6.1c Interior fuse box relay details

1 *Rear window defogger relay*
2 *Ignition relay 2*
3 *Accessory relay*
4 *Ignition relay 1*

5 Circuit breakers - general information

Refer to illustration 5.2

Circuit breakers protect components such as sunroof, power windows, power door locks and headlights.

On some models the circuit breaker resets itself automatically, so an electrical overload in a circuit breaker protected system will cause the circuit to fail momentarily, then come back on. If the circuit doesn't come back on, check it immediately **(see illustration)**. Once the condition is corrected, the circuit breaker will resume its normal function. Some circuit breakers must be reset manually.

6 Relays - general information and testing

General information

Refer to illustrations 6.1a, 6.1b, 6.1c, 6.1d and 6.1e

1 Several electrical accessories in the vehicle, such as the fuel injection system, power windows, power door locks, power seats, cruise control and air conditioning use relays to transmit the electrical signal to the

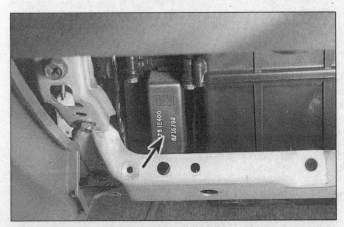

6.1d The blower motor High relay is located behind the glove box

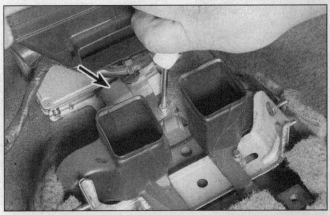

6.1e The ECM relay (arrow) is mounted on the ECM, which is located in front of the floor console by the accelerator pedal

12

component. Relays use a low-current circuit (the control circuit) to open and close a high-current circuit (the power circuit). If the relay is defective, that component will not operate properly. The various relays are mounted in engine compartment and several locations throughout the vehicle **(see illustrations)**. If a faulty relay is suspected, it can be removed and tested using the procedure below or by a dealer service department or a repair shop. Defective relays must be replaced as a unit.

Testing

Refer to illustration 6.4

2   It's best to refer to the wiring diagram for the circuit to determine the proper hook-ups for the relay you're testing. However, if you're not able to determine the correct hook-up from the wiring diagrams, you may be able to determine the test hook-ups from the information that follows.

Relays with four or six terminals

3   On most relays with four or six terminals, two of the terminals are the relay's control circuit (they connect to the relay coil which, when energized, closes the large contacts to complete the circuit). The other terminals are the power circuit (they are connected together within the relay when the control-circuit coil is energized).

4   Relays are sometimes marked as an aid to help you determine which terminals are the control circuit and which are the power circuit **(see illustration)**. As a general rule, the two thicker wires connected to the relay are the power circuit; the thinner wires are the control circuit.

5   Remove the relay from the vehicle and check for continuity between the relay power circuit terminals. There should be no continuity.

6   Connect a fused jumper wire between one of the two control circuit terminals and the positive battery terminal. Connect another jumper wire between the other control circuit terminal and ground. When the connections are made, the relay should click. On some relays, polarity may be critical, so, if the relay doesn't click, try swapping the jumper wires on the control circuit terminals.

7   With the jumper wires connected, check for continuity between the power circuit terminals. Now, there should be continuity.

8   If the relay fails any of the above tests, replace it.

7   Turn signal/hazard flashers - check and replacement

Refer to illustration 7.5

Warning: *The models covered by this manual are equipped with airbags. Always disconnect the negative battery cable, then the positive cable and wait 10 minutes before working in the vicinity of the impact sensors, steering column or instrument panel to avoid the possibility of accidental deployment of the*

6.4 Most relays are marked on the outside to easily identify the control circuit and power circuits

7.5 The combination flasher unit for the turn signals and hazard flashers is located to the left of the floor console behind the lower steering column trim cover

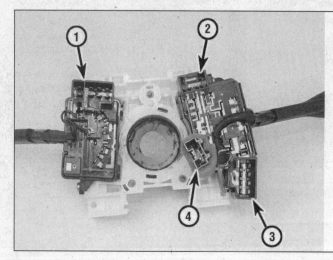

8.5a Combination switch connector identification guide

1   *Windshield wiper switch*
2   *Turn signal and cornering light switch*
3   *Headlight switch*
4   *Fog light switch*

airbag, which could cause personal injury (see Section 27). The airbag circuits are easily identified by yellow insulation covering the entire wiring harness or just prior to the wire harness connectors. Do not use electrical test equipment on any of these wires or tamper with them in any way.*

Check

1   The models covered by this manual use a single combination turn signal and hazard flasher unit.

2   When the flasher unit is functioning properly, an audible click can be heard during its operation. If the turn signals fail on one side or the other and the flasher unit does not make its characteristic clicking sound, a faulty turn signal bulb is indicated.

3   If both turn signals fail to blink, the problem may be due to a blown fuse, a faulty flasher unit, a broken switch or a loose or open connection. If a quick check of the fuse box indicates that the turn signal fuse has blown, check the wiring for a short before installing a new fuse.

Replacement

4   Remove the lower steering column trim cover (see Chapter 11).

5   Disconnect the electrical connector and remove the flasher unit from its mounting bracket **(see illustration)**.

6   Make sure that the replacement unit is identical to the original. Compare the old one to the new one before installing it.

7   Installation is the reverse of removal.

8   Steering column switches - check and replacement

Warning: *The models covered by this manual are equipped with airbags. Always disconnect the negative battery cable, then the positive cable and wait 10 minutes before working in the vicinity of the impact sensors, steering column or instrument panel to avoid the possibility of accidental deployment of the airbag, which could cause personal injury (see Section 27). The airbag circuits are easily identified by yellow insulation covering the entire wiring harness or just prior to the wire harness connectors. Do not use electrical test equipment on any of these wires or tamper with them in any way.*

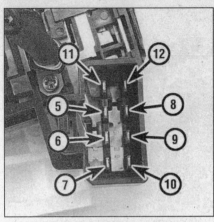
8.5b Headlight switch terminal guide

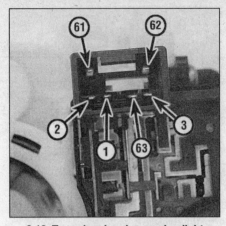
8.10 Turn signal and cornering light switch terminal guide

8.16 Fog light switch terminal guide

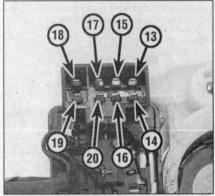
8.19 Windshield wiper switch terminal guide

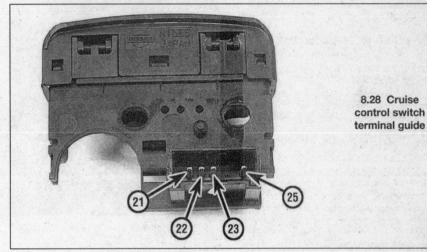
8.28 Cruise control switch terminal guide

Check

1   Disconnect the negative battery cable, then the positive cable and wait 10 minutes before proceeding any further.
2   Remove the lower steering column trim cover, left hand center trim panel, and steering column cover (see Chapter 11). Then remove the switch to be tested from the switch body (see illustration 8.38).
3   Using an ohmmeter, check for continuity between the indicated terminals with the various switches in each of the indicated positions.

Lighting switch

Refer to illustration 8.5a, 8.5b, 8.10 and 8.16
4   Turn the headlight switch ON and to the low beam position.
5   Check for continuity between terminals 5 and 7, and between 8 and 10 for proper operation of the low beam switch (see illustrations).
6   Check for continuity between terminals 11 and 12 for proper operation of the lights ON warning chime switch.
7   Push the headlight lever to the high beam position.
8   Check for continuity between terminals 5 and 6 and between 8 and 9 for proper operation of the high beam switch.
9   Move the left hand turn signal to the ON position.

10   Check for continuity between terminals 1 and 3 for proper operation of the left hand turn signal (see illustration).
11   Check for continuity between terminals 61 and 63 for proper operation of the left hand cornering light (if equipped).
12   Move the right hand turn signal to the ON position.
13   Check for continuity between terminals 1 and 2 for proper operation of the right hand turn signal.
14   Check for continuity between terminals 61 and 62 for proper operation of the right hand cornering light (if equipped).
15   Turn the fog light switch to the ON position.
16   Check for continuity between terminals 1 and 2 for proper operation of the fog light switch (see illustration).
17   If continuity does not exist at the specified terminals, replace the defective switch.
Note: *The headlight, turn signal, cornering lights and fog lamp switches are referred to as the lighting switch. If any part of the switch malfunctions it must be replaced as a unit.*

Windshield wiper switch

Refer to illustration 8.19
18   Move the wiper switch lever to the intermittent position.
19   Check for continuity between termi-

nals 13 and 14, then between 15 and 17 for proper operation in the intermittent position (see illustration).
20   Check for continuity between terminals 19 and 20 while moving the wiper volume switch back and forth. Notice the resistance change on the ohmmeter, indicating proper operation of the wiper volume switch.
21   Move the wiper lever to the low position.
22   Check for continuity between terminals 14 and 17 for proper operation in the low position.
23   Move the wiper lever to the high position.
24   Check for continuity between terminals 16 and 17 for proper operation in the high position.
25   If continuity does not exist at the specified terminals, replace the wiper switch.

Cruise control and horn switches

Refer to illustration 8.28
26   Pry out the cruise control switch trim cover located on the right backside of the steering wheel. Then detach the switch retaining screws and remove the switch.
27   Press the Resume/Accel switch to the ON position.
28   Check for continuity between terminals 21 and 23 for proper operation of the resume switch (see illustration).

12

8.38 Remove the retaining screws (arrows) from the defective switch and detach it from the switch body

8.42 Pry out the horn button and remove the spring and retaining screw

29   Press the Set/Coast switch to the ON position.
30   Check for continuity between terminals 21 and 22 for proper operation of the coast switch.
31   Press the cancel switch to the ON position.
32   Check for continuity between terminals 21, 22 and 23 for proper operation of the cancel switch.
33   Check for continuity between terminals 21 and 25 for proper operation of the horn circuit.
34   To test the horn switches, connect an ohmmeter to ground and terminal 25 of the cruise control switch connector in the steering wheel. Then check for continuity while holding each horn switch in the ON position.
35   .If continuity does not exist when depressing a switch, replace the defective switch.

Replacement

Refer to illustrations 8.38 and 8.42

36   Disconnect the negative battery cable, then the positive cable and wait 10 minutes before proceeding any further.
37   Remove the lower steering column trim cover, left hand center trim panel, and steering column cover (see Chapter 11).
38   Remove the wiper switch retaining screws **(see illustration)**.
39   Disconnect the electrical connectors from the backside and remove the switch or switches on the vehicle.
40   To remove the cruise control switch refer to Step 26.
41   To remove the horn switches, detach the drivers side airbag (see Chapter 10).
42   Pry out the horn switch buttons, then detach the retaining screws and electrical connectors and remove the switch **(see illustration)**.
43   Installation is the reverse of removal.

---

9   Ignition switch and key lock cylinder - check and replacement

Warning: *The models covered by this manual*

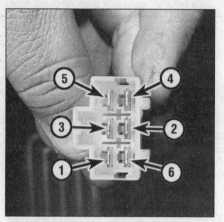

9.6 Ignition/key switch terminal guide

are equipped with airbags. Always disconnect the negative battery cable, then the positive cable and wait 10 minutes before working in the vicinity of the impact sensors, steering column or instrument panel to avoid the possibility of accidental deployment of the airbag, which could cause personal injury (see Section 27). The airbag circuits are easily identified by yellow insulation covering the entire wiring harness or just prior to the wire harness connectors. Do not use electrical test equipment on any of these wires or tamper with them in any way.

1   Disconnect the negative battery cable, then the positive cable and wait 10 minutes before proceeding any further.
2   Remove the lower steering column trim cover, left hand center trim panel, and steering column cover (see Chapter 11).

Check

Refer to illustration 9.6

3   Trace the wire from the ignition switch to the main wiring harness connector and disconnect the connector.
4   Use an ohmmeter to check for continuity at the indicated terminals with the switch in each indicated positions.
5   Turn the ignition key to the accessory (ACC) position.

9.13 To remove the ignition/lock cylinder assembly, drill out the two retaining bolts (arrows)

6   Check for continuity between terminals 1 and 2 for proper operation in the accessory position **(see illustration)**.
7   Turn the ignition key to the ON position.
8   Check for continuity between terminals 1, 2, 3 and 4 for proper operation in the ON position.
9   Turn the ignition key to the start (ST) position and hold it.
10   Check for continuity between terminals 1, 3, 5 and 6 for proper operation in the start position.
11   If continuity does not exist at the specified terminals, replace the ignition switch.

Replacement

Refer to illustration 9.13

12   Remove the steering wheel (see Chapter 10).
13   Remove the shear-head bolts retaining the ignition switch/lock cylinder assembly and separate the bracket halves from the steering column. This can be accomplished by drilling out the screws or using a small chisel and hammer to make slots in the them so they can be unscrewed with a screwdriver **(see illustration)**.
14   Place the new switch assembly in position, install the new shear-head bolts and

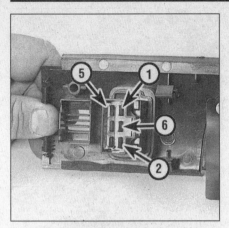

10.4 Rear window defogger switch terminal guide

tighten them until the heads snap off. If equipped with an airbag, be sure to center the spiral cable before installing the steering wheel (see Chapter 10).

15   The remainder of the installation is the reverse of removal.

10   Rear window defogger switch - check and replacement

Refer to illustration 10.4

Warning: *The models covered by this manual are equipped airbags. Always disconnect the negative battery cable, then the positive cable and wait 10 minutes before working in the vicinity of the impact sensors, steering column or instrument panel to avoid the possibility of accidental deployment of the airbag, which could cause personal injury (see Section 27). The airbag circuits are easily identified by yellow insulation covering the entire wiring harness or just prior to the wire harness connectors. Do not use electrical test equipment on any of these wires or tamper with them in any way.*

11   Rear window defogger - check and repair

1   The rear window defogger consists of a number of horizontal elements baked onto the glass surface.

11.4 When measuring the voltage at the rear window defogger grid, wrap a piece of aluminum foil around the negative probe of the voltmeter and press the foil against the wire with your finger

1   Disconnect the negative battery cable, then the positive cable and wait 10 minutes before proceeding any further.
2   Remove the left hand center trim panel (see Chapter 11).
3   Press the defogger switch to the ON position.
4   Check for continuity between terminals 1 and 2 and between 5 and 6 for proper operation of the rear window defogger switch **(see illustration)**.
5   If continuity does not exist at the specified terminals, unsnap the switch from the center trim panel and replace the defogger switch.

2   Small breaks in the element can be repaired without removing the rear window.

Check

Refer to illustrations 11.4, 11.5 and 11.7

3   Turn the ignition switch and defogger system switches to the ON position.
4   When measuring voltage during the next two tests, wrap a piece of aluminum foil around the tip of the voltmeter negative probe and press the foil against the heating element with your finger **(see illustration)**.
5   Check the voltage at the center of each heating element **(see illustration)**. If the voltage is 6-volts, the element is okay (there is no break). If the voltage is 12-volts, the element is broken between the center of the element and the positive end. If the voltage is 0-volts the element is broken between the center of the element and ground.
6   Connect the negative lead to a good body ground. The reading should stay the same.
7   To find the break, place the voltmeter positive lead against the defogger positive terminal. Place the voltmeter negative lead with the foil strip against the heating element at the positive terminal end and slide it toward the negative terminal end. The point at which the voltmeter deflects from zero to several volts is the point at which the heating element is broken **(see illustration)**.

Repair

Refer to illustration 11.13

8   Repair the break in the element using a repair kit specifically recommended for this purpose, available at most auto parts stores. Plastic conductive epoxy.should be included in any such kit
9   Prior to repairing a break, turn off the system and allow it to cool off for a few minutes.
10   Lightly buff the element area with fine steel wool, then clean it thoroughly with rubbing alcohol.

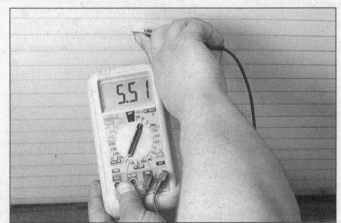

11.5 To determine if a heating element has broken, check the voltage at the center of each element - if the voltage is 6-volts, the element is unbroken - if the voltage is 12-volts, the element is broken between the center and the positive end - if there is no voltage, the element is broken between the center and ground

11.7 To find the break, place the voltmeter positive lead against the defogger positive terminal, place the voltmeter negative lead with the foil strip against the heating element at the positive terminal end and slide it toward the negative terminal end - the point at which the voltmeter reading changes abruptly is the point at which the element is broken

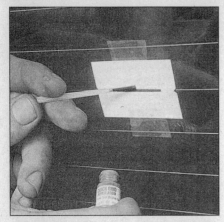

11.13 To use a defogger repair kit, apply masking tape to the inside of the window at the damaged area, then brush on the special conductive coating

12.3a Remove the retaining screws (arrows) and pull the radio out

12.3b Disconnect the antenna lead and electrical connectors, then remove the unit

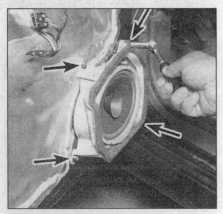

12.6 Remove the retaining bolts (arrows), pull the speaker out and disconnect it

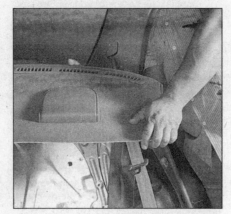

12.9 Using a screwdriver or trim removal tool, pry up the clips securing the rear parcel shelf and remove it from the vehicle

13.1 The antenna mast retaining nut can be removed with a pair of snap-ring pliers or a similar tool

11   Use masking tape to mask off the area being repaired.

12   Thoroughly mix the epoxy, following the instructions provided with the repair kit.

13   Apply the epoxy material to the slit in the masking tape, overlapping the undamaged area about 3/4-inch on either end **(see illustration)**.

14   Allow the repair to cure for 24 hours before removing the tape and using the system.

12   Radio and speakers - removal and installation

Warning: *The models covered by this manual are equipped airbags. Always disconnect the negative battery cable, then the positive cable and wait 10 minutes before working in the vicinity of the impact sensors, steering column or instrument panel to avoid the possibility of accidental deployment of the airbag, which could cause personal injury (see Section 27). The airbag circuits are easily identified by yellow insulation covering the entire wiring harness or just prior to the wire harness connectors. Do not use electrical test*

equipment on any of these wires or tamper with them in any way.

1   Disconnect the negative battery cable, then the positive cable and wait 10 minutes before proceeding any further.

Radio

Refer to illustrations 12.3a and 12.3b

2   Remove the radio trim bezel (see Chapter 11).

3   Remove the retaining screws and pull the radio outward to access the backside, then disconnect the electrical connectors and the antenna lead and lift the radio out of the vehicle **(see illustrations)**.

4   Installation is the reverse of removal.

Front speakers

Refer to illustration 12.6

5   Remove the front door trim panel (see Chapter 11).

6   Remove the speaker retaining bolts. Disconnect the electrical connector and remove the speaker from the vehicle **(see illustration)**.

7   Installation is the reverse of removal.

Rear speakers

Refer to illustration 12.9

8   Remove the rear seat from the vehicle (see Chapter 11).

9   Pry up the plastic clips securing the rear parcel shelf. Then lift up and out to remove it from the vehicle **(see illustration)**.

10   Remove the speaker retaining screws. Disconnect the electrical connector and remove the speaker from the vehicle.

11   Installation is the reverse of removal.

13   Power antenna - removal and installation

Antenna motor

Refer to illustrations 13.1 and 13.3

1   Remove the antenna mast retaining nut **(see illustration)**.

2   Working in the trunk, pry out the plastic clips securing the passenger side trunk finishing panels to allow access to the antenna motor.

3   Detach the motor retaining bolts **(see**

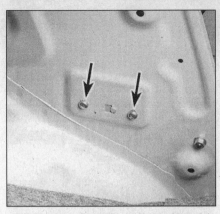

13.3 Remove the retaining bolts (arrows), pull the antenna motor out and disconnect it

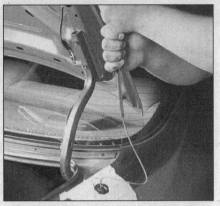

13.6 With the ignition key and the radio in the ON position, guide the antenna mast out of the motor assembly

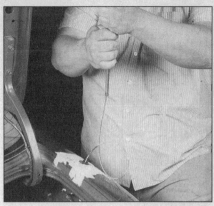

13.7 Insert the new antenna so that the antenna cable teeth face the rear of the vehicle

illustration). Disconnect the electrical connector and remove the antenna motor from the vehicle.

4   Installation is the reverse of removal.

Antenna mast

Refer to illustrations 13.6 and 13.7

Note: *At least two people should perform this task.*

5   Remove the antenna mast retaining nut **(see illustration 13.1).**

6   With one person controlling the ignition switch and the second person holding the antenna mast, turn the ignition key and the radio to the ON position. This will enable the antenna mast to unwind itself from the motor assembly **(see illustration).**

7   When installing the antenna mast insert the antenna cable with the teeth facing the rear of the vehicle. Then have your assistant turn the ignition key and the radio to the ON position. This will enable the antenna mast to wind itself back into the motor assembly **(see illustration).**

8   The remainder of the installation is the reverse of removal.

14 Headlights - replacement

Refer to illustration 14.4

Warning: *The halogen gas-filled headlight bulbs are under pressure and may shatter if the surface is damaged or the bulb is dropped. Wear eye protection and handle the bulbs carefully, grasping only the base whenever possible. Do not touch the surface of the bulb with your fingers because the oil from your skin could cause it to overheat and fail prematurely. If you do touch the bulb surface, clean it with rubbing alcohol.*

1993 through 1999 models

1   Detach the mounting bolts and remove the battery.

2   Remove the coolant recovery reservoir located at the far right hand corner of the engine compartment.

3   Locate the bulb retaining rings on the

14.4 Rotate the bulb retaining ring counterclockwise - remove the rubber boot (arrow), then pull the bulb assembly from the housing

back of the headlight housings.

4   Rotate the headlight bulb retaining ring counterclockwise (viewed from the rear). Then remove the rubber boot **(see illustration).**

5   Withdraw the bulb assembly from the headlight housing.

6   Disconnect the bulb holder from the electrical connector.

7   Without touching the glass with your bare fingers, insert the new bulb assembly into the headlight housing, install the rubber boot and tighten the retaining ring.

8   Plug in the electrical connector. Test headlight operation, then close the hood.

2000 and later models

9   Remove the headlight housing (see Section 16).

10   Rotate the bulb retaining ring counterclockwise (viewed from the rear) and remove it.

11   Withdraw the bulb assembly from the headlight housing.

12   Without touching the glass with your fingers, insert the new bulb assembly into the headlight housing.

13   Install the retaining ring and tighten securely.

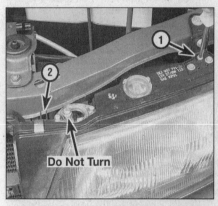

15.2 The adjuster screw closest to the outer fender (1) controls the vertical movement and the one closest to the radiator (2) controls the horizontal movement - turn the adjusting screws to center the bubble in each level gauge

14   Install the headlight housing (see Section 16). Test the bulb operation.

15 Headlights - adjustment

Refer to illustrations 15.2 and 15.8

Note: *It is important that the headlights are aimed correctly. If adjusted incorrectly they could blind the driver of an oncoming vehicle and cause a serious accident or seriously reduce your ability to see the road. The headlights should be checked for proper aim every 12 months and any time a new headlight is installed or front end body work is performed.*

1   Adjustment should be made with the vehicle sitting level, the gas tank half-full and no unusually heavy load in the vehicle.

2   The headlights have two adjustment screws located on the top of each headlight housing. The vertical (up and down) adjustment screws are located next to the front fenders and the horizontal (left to right) adjusting screws are located next to the radiator grille **(see illustration).**

3   Adjustments are made by turning the

screws to center the level gauges located on top of the headlight housings.

4 If head light housing has been replaced or the vehicle has suffered front end damage refer to following procedure.

5 This method requires a blank wall, masking tape and a level floor.

6 Position masking tape vertically on the wall in reference to the vehicle centerline and the centerlines of both headlights.

7 Position a horizontal tape line in reference to the centerline of all the headlights. **Note:** *It may be easier to position the tape on the wall with the vehicle parked only a few inches away.*

8 Adjustment should be made with the vehicle parked 25 feet from the wall, sitting level, the gas tank half-full and no unusually heavy load in the vehicle **(see illustration)**.

9 Starting with the low beam adjustment, position the high intensity zone so it is two inches below the horizontal line and two inches to the right of the headlight vertical line. Adjustments are made by turning the screws located on top of the headlight housings **(see illustration 15.2)**.

10 With the high beams on, the high intensity zone should be vertically centered with the exact center just below the horizontal line. **Note:** *It may not be possible to position the headlight aim exactly for both high and low beams. If a compromise must be made, keep in mind that the low beams are the most used and have the greatest effect on safety.*

11 Have the headlights adjusted by a dealer service department or service station at the earliest opportunity.

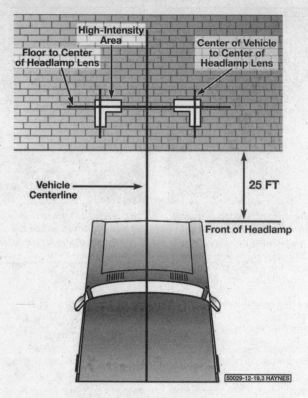

15.8 Headlight adjustment details

---

16 Headlight housing - removal and installation

Refer to illustrations 16.4a, 16.4b and 16.4c

1993 through 1999 models

1 Disconnect the negative battery cable.

2 Remove the headlight bulb (see Section 14) and cornering light bulb.

3 Remove the radiator grille (Chapter 11).

4 Remove the retaining nuts and bolts, detach the housing and withdraw it from the vehicle **(see illustrations)**.

5 Installation is the reverse of removal.

2000 and later models

6 Disconnect the battery negative cable.

7 Remove the four screws securing the front bumper protector. Pull the bumper protector aside and remove the single nut securing the headlight housing.

8 Working under the hood, remove the three remaining nuts and single bolt securing the headlight housing.

9 Pull the headlight housing part way out and disconnect the electrical harness from the bulb assembly. Withdraw the headlight housing from the vehicle.

10 Installation is the reverse of removal.

---

17 Bulb replacement

Front parking/turn signal, side marker and cornering lights

Refer to illustrations 17.1, 17.2 and 17.3

1 The front parking/turn signal lights are mounted in the front bumper. Remove the

16.4a Remove the bolt securing the headlight housing to the radiator support

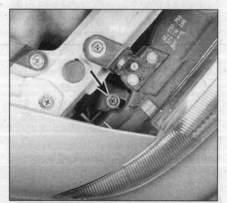

16.4b Using a ratchet, socket and long extension remove the lower bolt (arrow) from the outer end of the headlight housing

16.4c The retaining nuts securing the inner end of the headlight housing are accessible from behind the radiator support

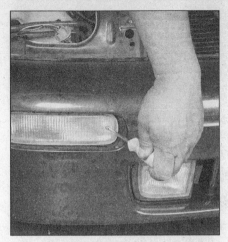

17.1 The front parking/turn signal light bulb is accessible after removing the lens retaining screw

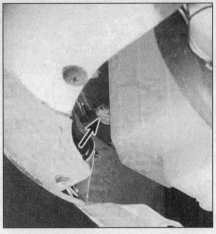

17.2 Working from inside of the fender, remove the side marker light bulb and socket assembly

17.3 The cornering light bulb (arrow), is accessed from the inside of the fenderwell. The bulb is removed by turning the holder counterclockwise, then pushing in and turning the bulb

two screws retaining the lens and replace the bulb **(see illustration)**.

2   Working under the vehicle on the inside of the fender, remove the bulb holder from the side marker light housing **(see illustration)**. Push inward and rotate the bulb to remove it from the holder.

3   To access the cornering light bulbs, first remove the battery and the coolant recovery reservoir. Turn the bulb holder counterclockwise and remove it from the back of the housing **(see illustration)**. Push in and rotate the bulb counterclockwise to remove it from the holder.

Rear turn signal, brake, tail and back-up lights

Refer to illustration 17.5

4   Remove the plastic clips securing the rear trunk compartment finishing panel.

5   Detach the tail light bulb cluster from the rear tail light housing **(see illustration)**. The defective bulb can then be pulled out of the socket and replaced.

License plate light

Refer to illustration 17.6

6   Pry out the plastic trim ring surrounding the lens **(see illustration)**.

7   Remove the lens and replace the defective bulb.

High-mounted brake light

Refer to illustration 17.8

8   The brake light cover is retained by screws. Remove the cover and replace the bulb **(see illustration)**.

Interior lights

Refer to illustration 17.9

9   Using a small screwdriver, remove the lens and replace the bulb **(see illustration)**.

Instrument cluster illumination

Refer to illustration 17.10

10   To gain access to the instrument cluster illumination lights, the instrument cluster will

17.5 Press the bulb cluster retaining clip inward then detach it from the tail light housing

17.6 Using a screwdriver, pry out the trim ring surrounding the license plate light lens

17.8 Open the trunk lid for access to the high-mounted brake light retaining screws (arrows)

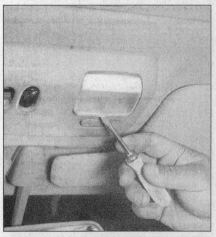

17.9 Use a small screwdriver to pry out the interior light lens

12

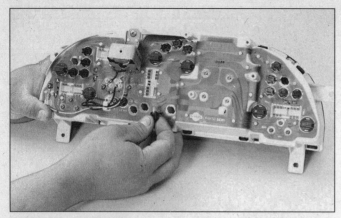

17.10 To remove an instrument cluster bulb, depress the bulb and rotate it counterclockwise

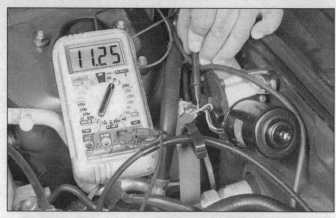

19.2 Use a voltmeter or test light to check for battery power at the wiper motor

have to be removed (see Section 20). The bulbs can then be removed and replaced from the rear of the cluster **(see illustration)**.

18 Daytime Running Lights (DRL) - general information

The Daytime Running Lights (DRL) system used on Canadian models illuminates the headlights whenever the engine is running. The only exception is with the engine running and the parking brake engaged. Once the parking brake is released, the lights will remain on as long as the ignition switch is on, even if the parking brake is later applied.

The DRL system supplies reduced power to the headlights so they won't be too bright for daytime use, while prolonging headlight life.

19 Wiper motor - check and replacement

Wiper motor circuit check

Refer to illustration 19.2
Note: *Refer to the wiring diagrams for wire colors and locations in the following checks. Keep in mind that power wires are generally larger in diameter and brighter colors, where ground wires are usually smaller in diameter and darker colors. When checking for voltage, probe a grounded 12-volt test light to each terminal at a connector until it lights; this verifies voltage (power) at the terminal.*

1   If the wipers work slowly, make sure the battery is in good condition and has a strong charge (see Chapter 1). If the battery is in good condition, remove the wiper motor (see below) and operate the wiper arms by hand. Check for binding linkage and pivots. Lubricate or repair the linkage or pivots as necessary. Reinstall the wiper motor. If the wipers still operate slowly, check for loose or corroded connections, especially the ground connection. If all connections look OK,

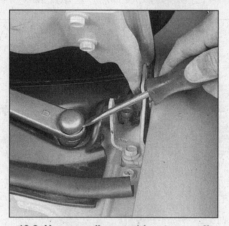

19.8 Use a small screwdriver to pry off the wiper arm nut cover, then remove the nut and pull the arm straight off its splined shaft

replace the motor.
2   If the wipers fail to operate when activated, check the fuse. If the fuse is OK, connect a jumper wire between the wiper motor and ground, then retest. If the motor works now, repair the ground connection. If the motor still doesn't work, turn on the wipers and check for voltage at the motor **(see illustration)**. If there's no voltage at the motor, remove the motor and check it off the vehicle with fused jumper wires from the battery. If the motor now works, check for binding linkage (see Step 1 above). If the motor still doesn't work, replace it. If there's no voltage at the motor, check for voltage at the switch. If there's no voltage at the switch, check the wiring between the switch and fuse panel for continuity. If the wiring is OK, the switch is probably bad.
3   If the wipers only work on one speed, check the continuity of the wires between the switch and motor. If the wires are OK, replace the switch.
4   If the interval (delay) function is inoperative, check the continuity of all the wiring between the switch and motor. If the wiring is OK, replace the interval module.

19.10 Detach the wiper motor shield and unscrew the spindle nut (arrow)

5 .   If the wipers stop at the position they're in when the switch is turned off (fail to park), check for voltage at the wiper motor when the wiper switch is OFF but the ignition is ON. If voltage is present, the limit switch in the motor is malfunctioning. Replace the wiper motor. If no voltage is present, trace and repair the limit switch wiring between the fuse panel and wiper motor.
6   If the wipers won't shut off unless the ignition is OFF, disconnect the wiring from the wiper control switch. If the wipers stop, replace the switch. If the wipers keep running, there's a defective limit switch in the motor; replace the motor.
7   If the wipers won't retract below the hoodline, check for mechanical obstructions in the wiper linkage or on the vehicle's body which would prevent the wipers from parking. If there are no obstructions, check the wiring between the switch and motor for continuity. If the wiring is OK, replace the wiper motor.

Wiper motor replacement

Refer to illustrations 19.8, 19.10 and 19.11
8   Pry off the cover for the wiper arm nuts, unscrew the nuts and remove both wiper arms **(see illustration)**.

19.11 Remove the wiper motor retaining bolts (arrows)

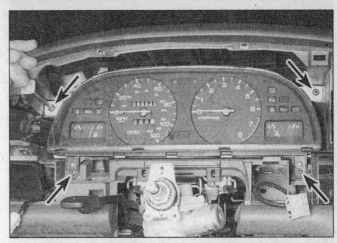

20.3 Remove the instrument cluster screws (arrows)

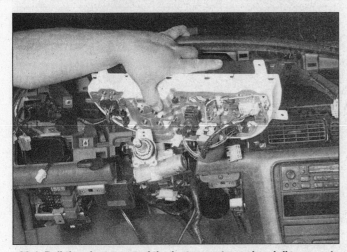

20.4 Pull the cluster out of the instrument panel and disconnect the electrical connectors from the backside

21.3 Check for power at the horn terminal with the horn button depressed

9   Remove the screws and detach the cowl cover (see Chapter 11).
10   Remove the wiper motor shield and motor spindle nut **(see illustration)**.
11   Remove the wiper motor retaining bolts **(see illustration)**, detach the wiper arm linkage then pull the motor out from the firewall.
12   Disconnect the electrical connector and remove the motor from the vehicle.
13   Installation is the reverse of removal.

20   Instrument cluster - removal and installation

Refer to illustrations 20.3 and 20.4
Warning: *The models covered by this manual are equipped with airbags. Always disconnect the negative battery cable, then the positive cable and wait 10 minutes before working in the vicinity of the impact sensors, steering column or instrument panel to avoid the possibility of accidental deployment of the airbag, which could cause personal injury (see Section 27). The airbag circuits are easily identified by yellow insulation covering the entire wiring harness or just prior to the wire*

harness connectors. Do not use electrical test equipment on any of these wires or tamper with them in any way.
1   Disconnect the negative battery cable, then the positive cable and wait 10 minutes before proceeding any further.
2   Tilt the steering column to it lowest position and remove the instrument cluster bezel (see Chapter 11).
3   Remove the retaining screws and pull the cluster forward **(see illustration)**.
4   Disconnect the electrical connectors and remove the cluster from the vehicle **(see illustration)**.
5   Installation is the reverse of removal.

21   Horn - check and replacement

Check

Refer to illustration 21.3
Note: *Check the fuses before beginning electrical diagnosis.*
1   Disconnect the electrical connector from the horn.
2   To test the horn, connect battery voltage to the two terminals with a pair of jumper

wires. If the horn doesn't sound, replace it.
3   If the horn does sound, check for voltage at the terminal when the horn button is depressed **(see illustration)**. If there's voltage at the terminal, check for a bad ground at the horn.
4   If there's no voltage at the horn, check the relay (see Section 6). Note that most horn relays are either the four-terminal or externally grounded three-terminal type.
5   If the relay is OK, check for voltage to the relay power and control circuits. If either of the circuits is not receiving voltage, inspect the wiring between the relay and the fuse panel.
6   If both relay circuits are receiving voltage, depress the horn button and check the circuit from the relay to the horn button for continuity to ground. If there's no continuity, check the circuit for an open. If there's no open circuit, replace the horn button.
7   If there's continuity to ground through the horn button, check for an open or short in the circuit from the relay to the horn.

Replacement

Refer to illustration 21.8
8   To replace the horn, disconnect the

12

21.8 Disconnect the electrical connector, remove the bolt (arrow) and detach the horn

22.5a The cruise control servo is located on the firewall - make sure the vacuum hose (arrow) is connected securely

22.5b Make sure the cruise control cable (1), accelerator cable (2) and associated linkage are not damaged and operate smoothly when the throttle is opened

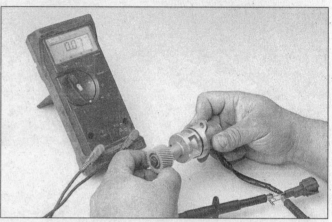

22.6 Using a voltmeter set on the AC scale, check the vehicle speed sensor to make sure the voltage pulses as the gear is turned

electrical connector and remove the bracket bolt **(see illustration)**.

9   Installation is the reverse of removal.

22   Cruise control system - description and check

Refer to illustrations 22.5a, 22.5b and 22.6

1   The cruise control system maintains vehicle speed with a vacuum actuated servo motor located on the firewall in the engine compartment, which is connected to the throttle linkage by a cable. The system consists of the servo motor, brake switch, control switches, a relay and associated vacuum hoses. Some features of the system require special testers and diagnostic procedures which are beyond the scope of this manual. Listed below are some general procedures that may be used to locate common problems.

2   Locate and check the fuse (see Section 3).

3   Have an assistant operate the brake lights while you check their operation (voltage from the brake light switch deactivates the cruise control).

4   If the brake lights don't come on or

don't shut off, correct the problem and retest the cruise control.

5   Visually inspect the vacuum hose connected to the servo and check the control linkage between the cruise control servo and the throttle linkage and replace as necessary **(see illustrations)**.

6   The cruise control system uses a speed sensing device. The speed sensor is located in the transaxle. To test the speed sensor, remove the speed sensor from the transaxle, connect a digital voltmeter to the speed sensor electrical terminals and rotate the gear **(see illustration)**. The voltage should pulse from 0-volts to approximately 0.5-volts as the sensor rotates. If not, the sensor is defective.

7   Test drive the vehicle to determine if the cruise control is now working. If it isn't, take it to a dealer service department or an automotive electrical specialist for further diagnosis and repair.

23   Power window system - description and check

Refer to illustration 23.12

1   The power window system operates

electric motors, mounted in the doors, which lower and raise the windows. The system consists of the control switches, the motors, regulators, glass mechanisms and associated wiring.

2   The power windows can be lowered and raised from the master control switch by the driver or by remote switches located at the individual windows. Each window has a separate motor which is reversible. The position of the control switch determines the polarity and therefore the direction of operation.

3   The circuit is protected by a fuse and a circuit breaker. Each motor is also equipped with an internal circuit breaker, this prevents one stuck window from disabling the whole system.

4   The power window system will only operate when the ignition switch is ON. In addition, many models have a window lockout switch at the master control switch which, when activated, disables the switches at the rear windows and, sometimes, the switch at the passenger's window also. Always check these items before troubleshooting a window problem.

5   These procedures are general in nature, so if you can't find the problem using them, take the vehicle to a dealer service depart-

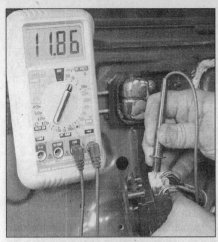

23.12 If no voltage is present at the motor with the switch depressed, check for voltage at the switch

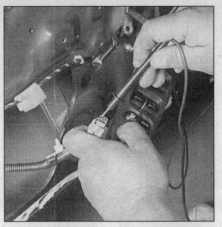

24.6 Check for voltage at the lock solenoid while the lock switch is operated

25.6 Carefully pry out the side view mirror switch

ment or other properly equipped repair facility.

6   If the power windows won't operate, always check the fuse and circuit breaker first.

7   If only the rear windows are inoperative, or if the windows only operate from the master control switch, check the rear window lockout switch for continuity in the unlocked position. Replace it if it doesn't have continuity.

8   Check the wiring between the switches and fuse panel for continuity. Repair the wiring, if necessary.

9   If only one window is inoperative from the master control switch, try the other control switch at the window. **Note:** *This doesn't apply to the drivers door window.*

10   If the same window works from one switch, but not the other, check the switch for continuity.

11   If the switch tests OK, check for a short or open in the circuit between the affected switch and the window motor.

12   If one window is inoperative from both switches, remove the trim panel from the affected door and check for voltage at the switch **(see illustration)**. and at the motor while the switch is operated.

13   If voltage is reaching the motor, disconnect the glass from the regulator (see Chapter 11). Move the window up and down by hand while checking for binding and damage. Also check for binding and damage to the regulator. If the regulator is not damaged and the window moves up and down smoothly, replace the motor. If there's binding or damage, lubricate, repair or replace parts, as necessary.

14   If voltage isn't reaching the motor, check the wiring in the circuit for continuity between the switches and motors. You'll need to consult the wiring diagram for the vehicle. If the circuit is equipped with a relay, check that the relay is grounded properly and receiving voltage.

15   Test the windows after you are done to confirm proper repairs.

24 Power door lock system - description and check

Refer to illustration 24.6

The power door lock system operates the door lock actuators mounted in each door. The system consists of the switches, actuators, a control unit and associated wiring. Diagnosis can usually be limited to simple checks of the wiring connections and actuators for minor faults which can be easily repaired. Since this system uses an electronic control unit in-depth diagnosis should be left to a dealership service department. The door lock control unit is located behind the instrument panel, to the right of the fuse box.

Power door lock systems are operated by bi-directional solenoids located in the doors. The lock switches have two operating positions: Lock and Unlock. When activated, the switch sends a ground signal to the door lock control unit to lock or unlock the doors. Depending on which way the switch is activated, the control unit reverses polarity to the solenoids, allowing the two sides of the circuit to be used alternately as the feed (positive) and ground side.

Some vehicles may have an anti-theft systems incorporated into the power locks. If you are unable to locate the trouble using the following general Steps, consult your a dealer service department.

1   Always check the circuit protection first. Some vehicles use a combination of circuit breakers and fuses.

2   Operate the door lock switches in both directions (Lock and Unlock) with the engine off. Listen for the click of the solenoids operating.

3   Test the switches for continuity. Replace the switch if there's not continuity in both switch positions.

4   Check the wiring between the switches, control unit and solenoids for continuity. Repair the wiring if there's no continuity.

5   Check for a bad ground at the switches or the control unit.

6   If all but one lock solenoids operate,

remove the trim panel from the affected door (see Chapter 11). and check for voltage at the solenoid while the lock switch is operated **(see illustration)**. One of the wires should have voltage in the Lock position; the other should have voltage in the Unlock position.

7   If the inoperative solenoid is receiving voltage, replace the solenoid.

8   If the inoperative solenoid isn't receiving voltage, check for an open or short in the wire between the lock solenoid and the control unit. **Note:** *It's common for wires to break in the portion of the harness between the body and door (opening and closing the door fatigues and eventually breaks the wires).*

25 Electric side view mirrors - description and check

Refer to illustration 25.6

1   Most electric rear view mirrors use two motors to move the glass; one for up and down adjustments and one for left-right adjustments.

2   The control switch has a selector portion which sends voltage to the left or right side mirror. With the ignition ON but the engine OFF, roll down the windows and operate the mirror control switch through all functions (left-right and up-down) for both the left and right side mirrors.

3   Listen carefully for the sound of the electric motors running in the mirrors.

4   If the motors can be heard but the mirror glass doesn't move, there's probably a problem with the drive mechanism inside the mirror. Remove and disassemble the mirror to locate the problem.

5   If the mirrors don't operate and no sound comes from the mirrors, check the fuse (see Chapter 1).

6   If the fuse is OK, remove the mirror control switch **(see illustration)** from its mounting without disconnecting the wires attached to it. Turn the ignition ON and check for voltage at the switch. There should be voltage at one terminal. If there's no voltage at the switch, check for an open or short in the cir-

12

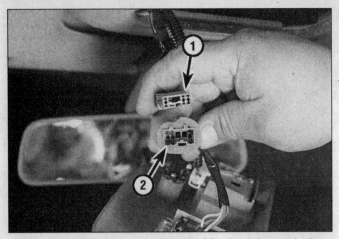

26.6 Check for voltage at connector (1) - check for continuity at connector (2)

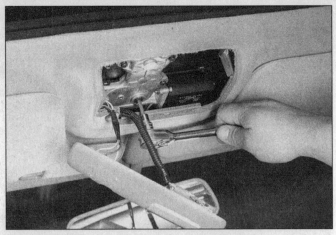

26.9 To close the sunroof manually, insert an Allen wrench in the motor shaft and rotate it clockwise

cuit between the fuse panel and the switch.

7   If there's voltage at the switch, disconnect it. Check the switch for continuity in all its operating positions. If the switch does not have continuity, replace it.

8   Re-connect the switch. Locate the wire going from the switch to ground. Leaving the switch connected, connect a jumper wire between this wire and ground. If the mirror works normally with this wire in place, repair the faulty ground connection.

9   If the mirror still doesn't work, remove the mirror and check the wires at the mirror for voltage. Check with ignition ON and the mirror selector switch on the appropriate side. Operate the mirror switch in all its positions. There should be voltage at one of the switch-to-mirror wires in each switch position (except the neutral "off" position).

10   If there's not voltage in each switch position, check the circuit between the mirror and control switch for opens and shorts.

11   If there's voltage, remove the mirror and test it off the vehicle with jumper wires. Replace the mirror if it fails this test.

26 Electric sunroof - description and check

Refer to illustrations 26.6 and 26.9

1   The electric sunroof is powered by a single motor located in the roof behind the overhead console. The power circuit is protected by a circuit breaker.

2   The control switches (tilt and slide) send a ground signal to the sunroof motor when the switches are pressed. Power is supplied to the motor from the ignition relay 2. With the ignition ON but the engine OFF, operate the sunroof control switch through the tilt and slide functions.

3   Listen carefully for the sound of the sunroof motor running in the roof.

4   If the motors can be heard but the sunroof glass doesn't move, there's probably a problem with the drive mechanism or drive cables.

5   If the sunroof does not operate and no sound comes from the motor, check the fuse (see Chapter 1).

6   If the fuse is OK, remove the control switches (see Chapter 11). Disconnect the wires attached to it. Turn the ignition ON and check for voltage at the switch **(see illustration)**. There should be voltage at four terminals. If there's no voltage at the switch, check for power and ground at the motor. If power and ground exist at the motor and there's still no voltage at the switch replace the motor. If there's no voltage at the motor, check the ignition relay or an open or short in the wiring between the ignition relay and the motor.

7   If there's voltage at the switch, disconnect it. Check the switch for continuity in all its operating positions. If the switch does not have continuity, replace it.

8   If the switch has continuity re-connect the switch. Locate the wire going from the switch to ground. Leaving the switch connected, connect a jumper wire between this wire and ground. If the motor works normally with this wire in place, repair the faulty ground connection.

9   The sunroof can be closed manually by inserting an Allen wrench into the motor shaft and rotating it clockwise **(see illustration)**.

27.1a Early model front crash sensor is located behind the radiator grille

27 Airbag system - general information

Refer to illustrations 27.1a, 27.1b and 27.1c

Most models are equipped with a Supplemental Restraint System (SRS), more commonly known as an airbag. This system is designed to protect the driver, and on 1994 and later models, the passenger from serious injury in the event of a head-on or frontal collision. It consists of an airbag module in the center of the steering wheel, a passenger airbag module on the right side of the dash above the glove box (1994 and later models), a crash sensor mounted at the front of the vehicle (1993 and 1994 models), and a diagnostic/control unit and safing sensors located inside the passenger compartment **(see illustrations)**.

Airbag module

The airbag module contains a housing incorporating the cushion (airbag) and inflator unit. The inflator assembly is mounted on the back of the housing over a hole through which gas is expelled, inflating the bag almost instantaneously when an electrical signal is sent from the system. The specially

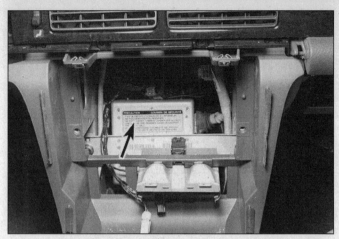

27.1b Early model airbag control unit is located behind the radio

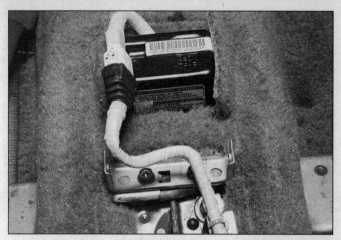

27.1c Later model control/sensor unit is located underneath the floor console

wound wire that carries this signal to the drivers module is called a spiral cable. The spiral cable is a flat, ribbon-like electrically conductive tape which is wound many times so that it can transmit an electrical signal regardless of steering wheel position.

Sensors

Early models (1993 and 1994) have three sensors: a crash sensor at the front of the vehicle behind the bumper and a safing sensor in the center of the vehicle. Later models (1995 through 1997) have one sensor unit which consists of a drive circuit, CPU, G-sensor and safing sensor located in the center of the vehicle under the floor console.

The front crash sensor on early models is basically a pressure sensitive switch that completes an electrical circuit during an impact of sufficient G-force. The electrical signal from the crash sensors is sent to the control unit and safing sensors, which then completes the circuit and inflates the airbag(s).

Diagnostic/control unit and safing sensors

The diagnostic/control unit contains an on-board microprocessor which monitors the operation of the system and sensors. It checks this system every time the vehicle is started, causing the AIRBAG light to go on then off, if the system is operating properly. If

there is a fault in the system, the light will go on and stay on and the unit will store fault codes indicating the nature of the fault. If the AIRBAG light goes on and stays on, the vehicle should be taken to your dealer immediately for service.

Precautions

Warning: *Failure to follow these precautions could result in accidental deployment of the airbag and personal injury.*

Whenever working in the vicinity of the steering wheel, steering column or any of the other SRS system components, the system must be disarmed. To disarm the system:

a) *Point the wheels straight ahead and turn the key to the Lock position.*
b) *Disconnect the cable from the negative battery terminal and then the positive terminal.*
c) *Wait at least 10 minutes for the back-up power supply to be depleted.*

Whenever handling an airbag module, always keep the airbag opening pointed away from your body. Never place the airbag module on a bench of other surface with the airbag opening facing the surface. Always place the airbag module in a safe location with the airbag opening facing up.

Never measure the resistance of any SRS component. An ohmmeter has a built-in battery supply that could accidentally deploy the airbag.

Never use electrical welding equipment on a vehicle equipped with an airbag without first disconnecting the yellow airbag connector, located under the steering column near the combination switch connector (drivers airbag) and behind the glove box (passengers airbag).

Never dispose of a live airbag module. Return it to your dealer for safe deployment, using special equipment, and disposal.

28 Wiring diagrams - general information

Since it isn't possible to include a complete wiring diagram for every year covered by this manual, the following diagrams are those that are typical and most commonly needed.

Prior to troubleshooting any circuits, check the fuse and circuit breakers (if equipped) to make sure they are in good condition. Make sure the battery is properly charged and has clean, tight cable connections (see Chapter 1).

When checking the wiring system, make sure that all electrical connectors are clean, with no broken or loose pins. When unplugging an electrical connector, do not pull on the wires, only on the connector housings themselves.

12

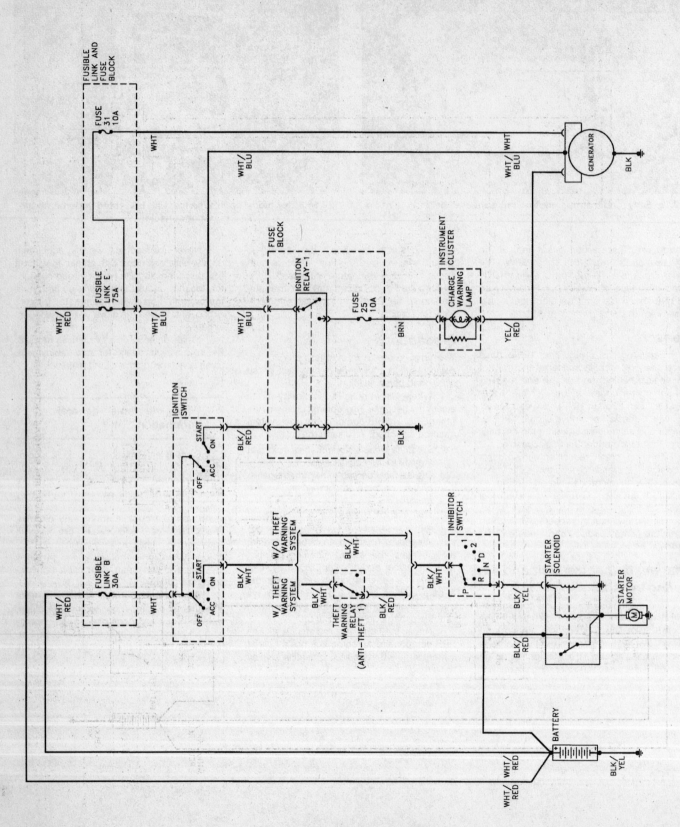

Typical starting and charging system (automatic transmission)

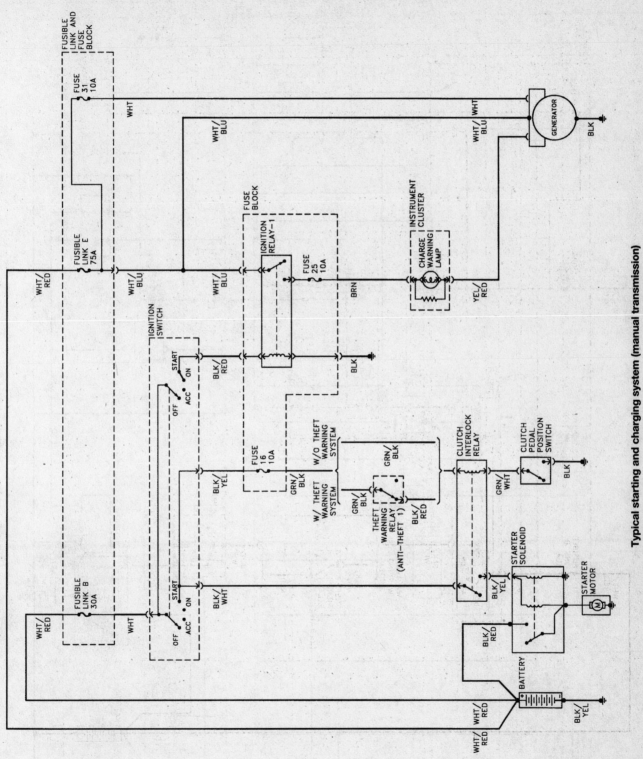

Typical starting and charging system (manual transmission)

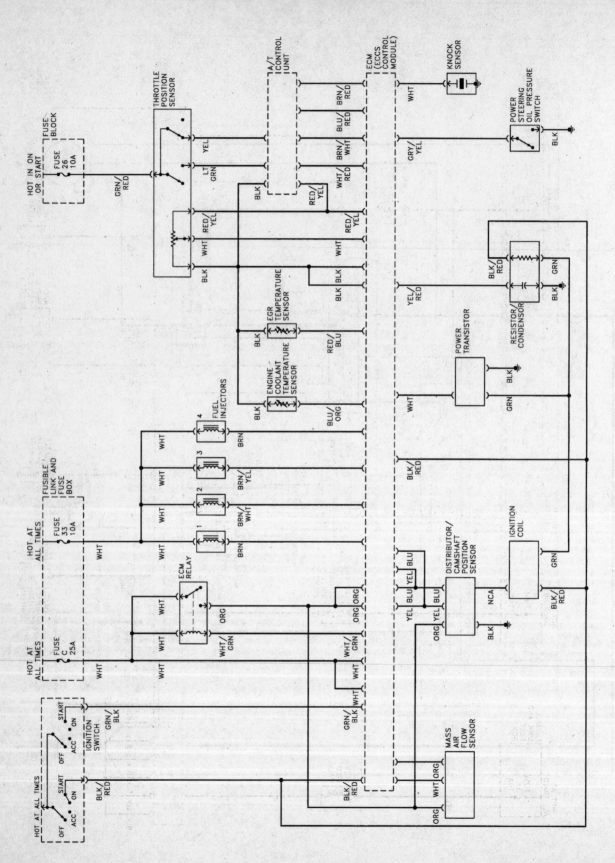

Typical 1993 and 1994 engine control system (part 1 of 2)

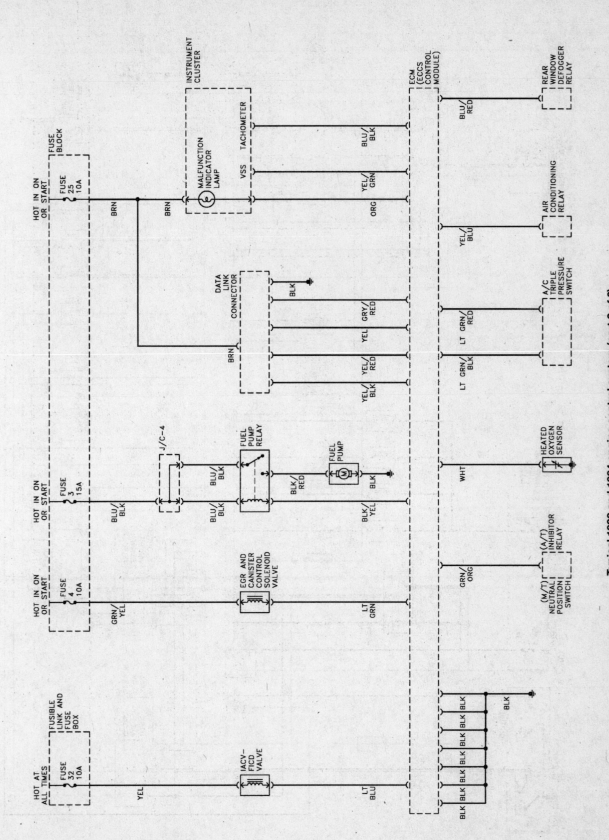

Typical 1993 and 1994 engine control system (part 2 of 2)

12

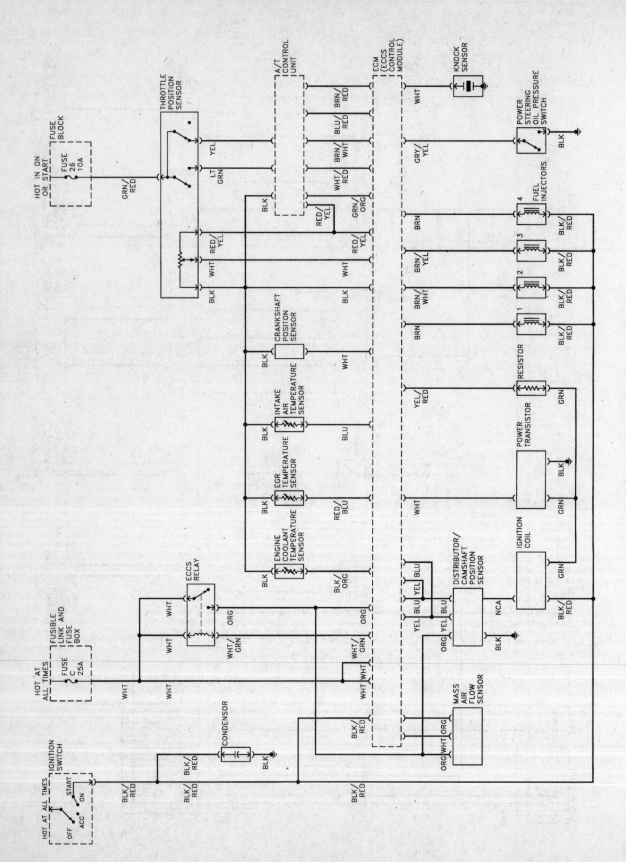

Typical 1995 and later engine control system (part 1 of 2)

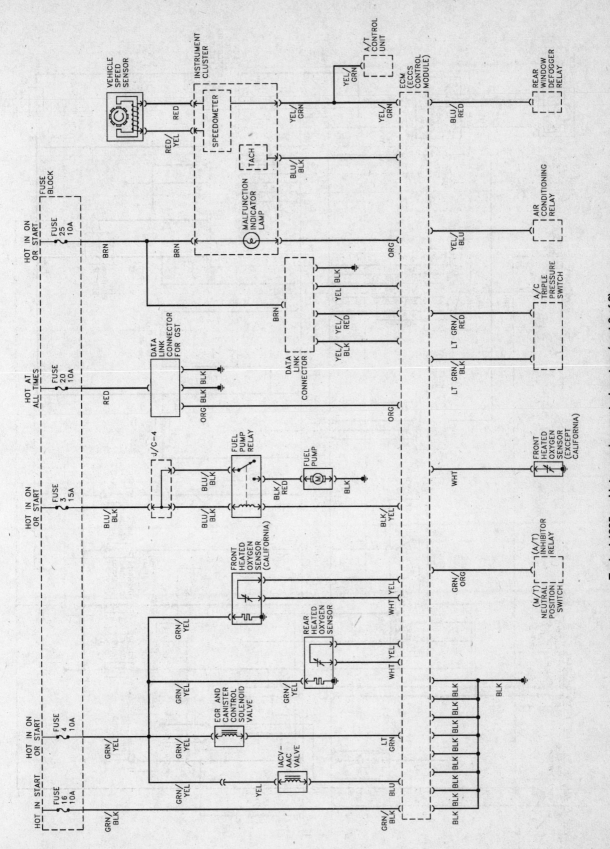

Typical 1995 and later engine control system (part 2 of 2)

12

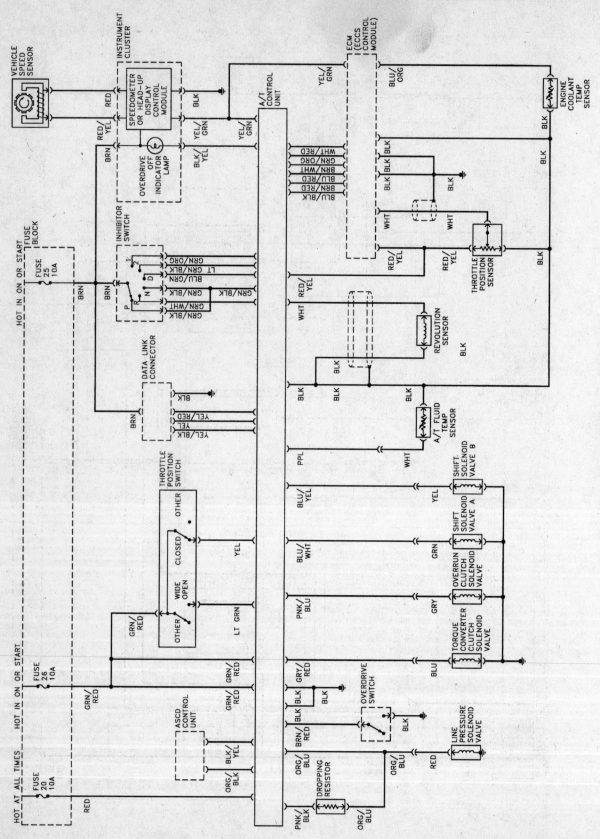

Typical automatic transmission control system

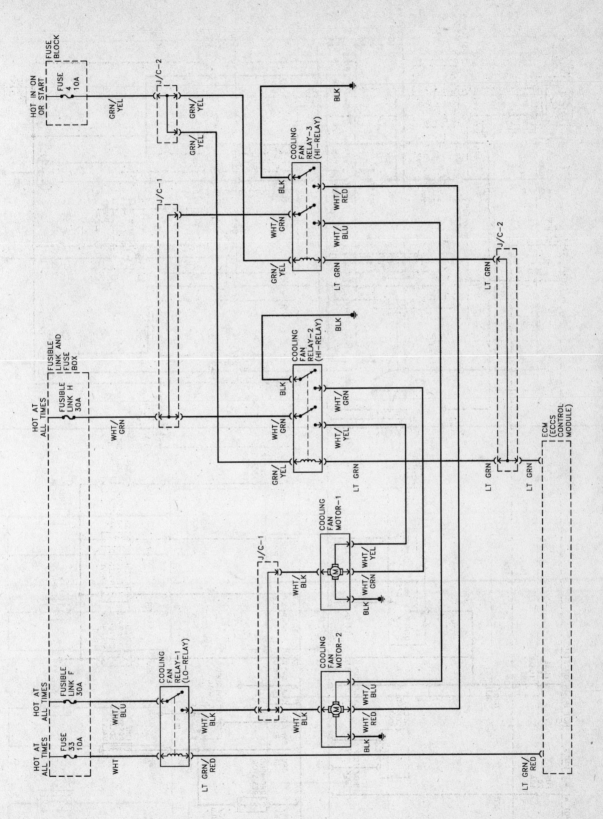

Typical cooling fan control system

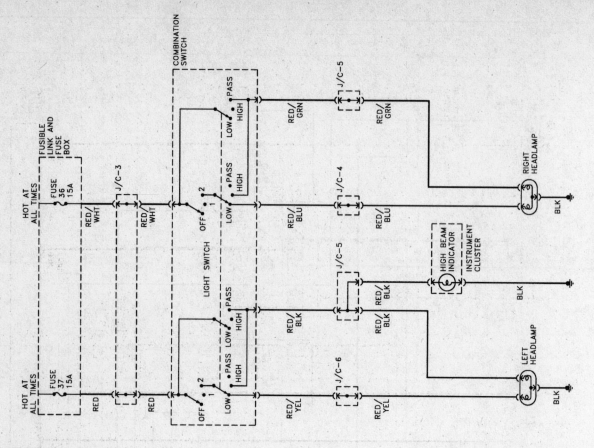

Typical headlight system

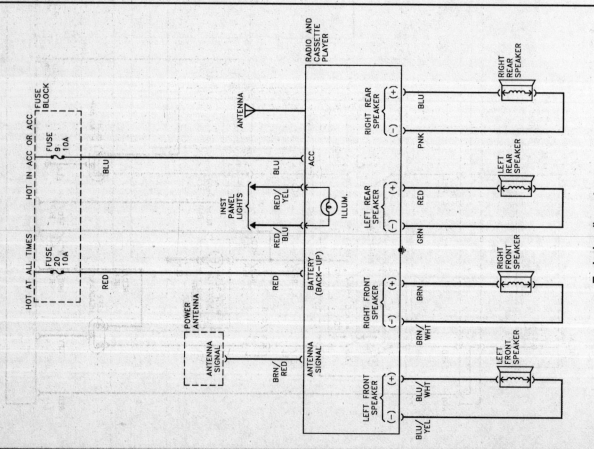

Typical audio system

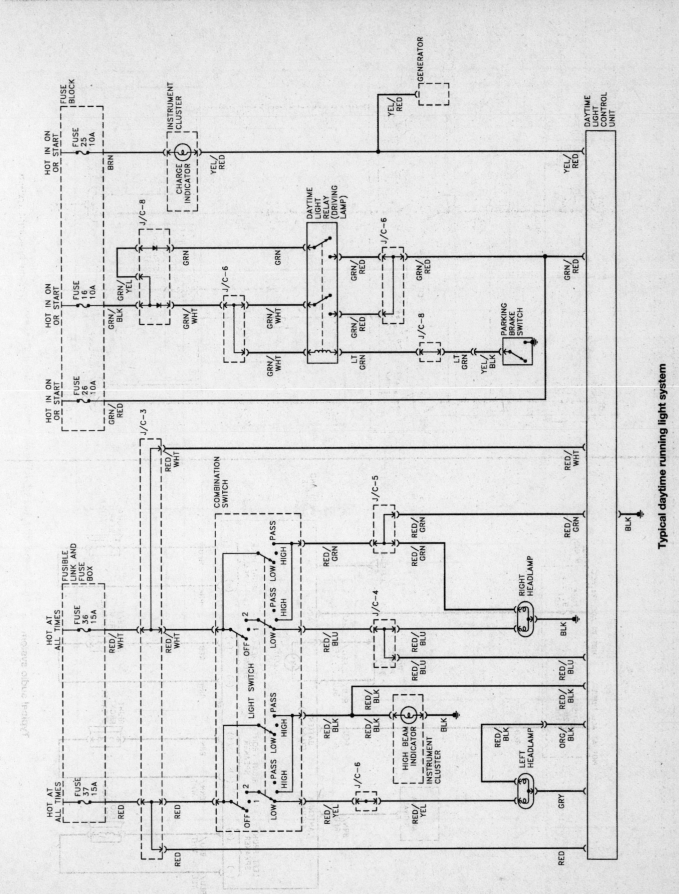

Typical daytime running light system

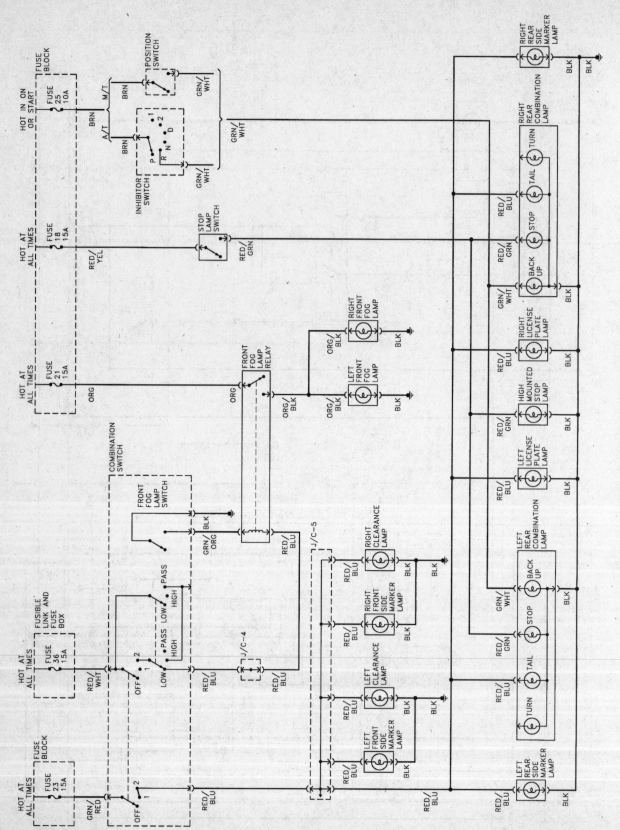

Typical brake light, tail light, fog light and running light system

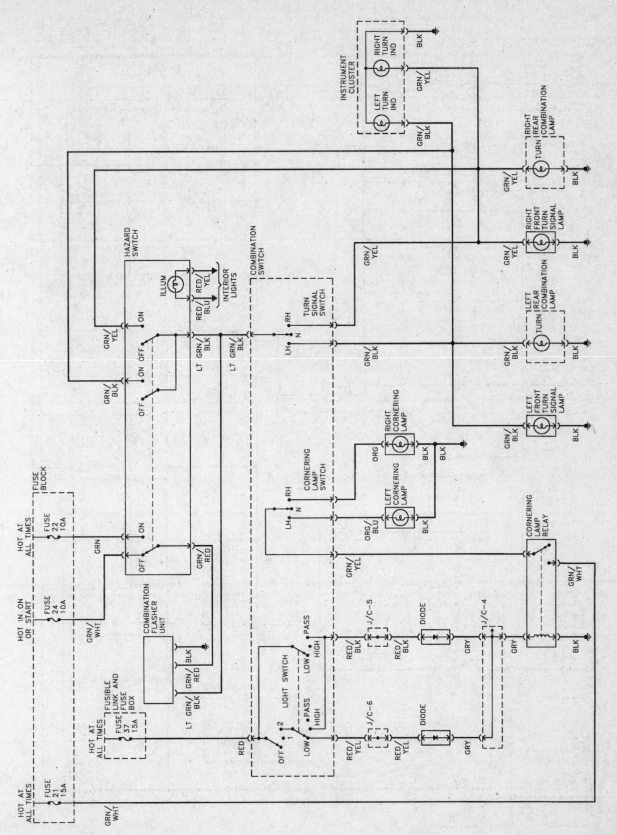

Typical turn signal system

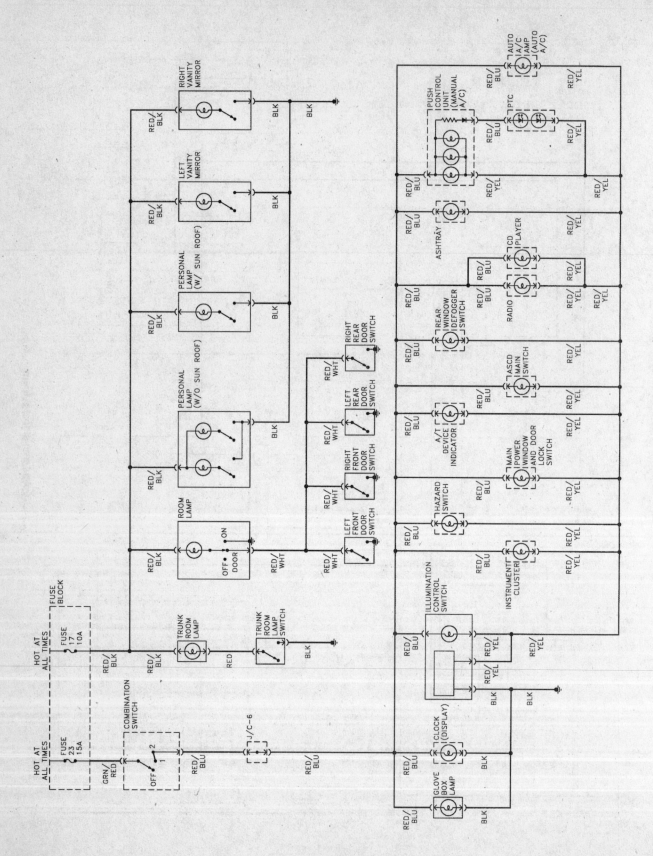

Typical interior lighting system

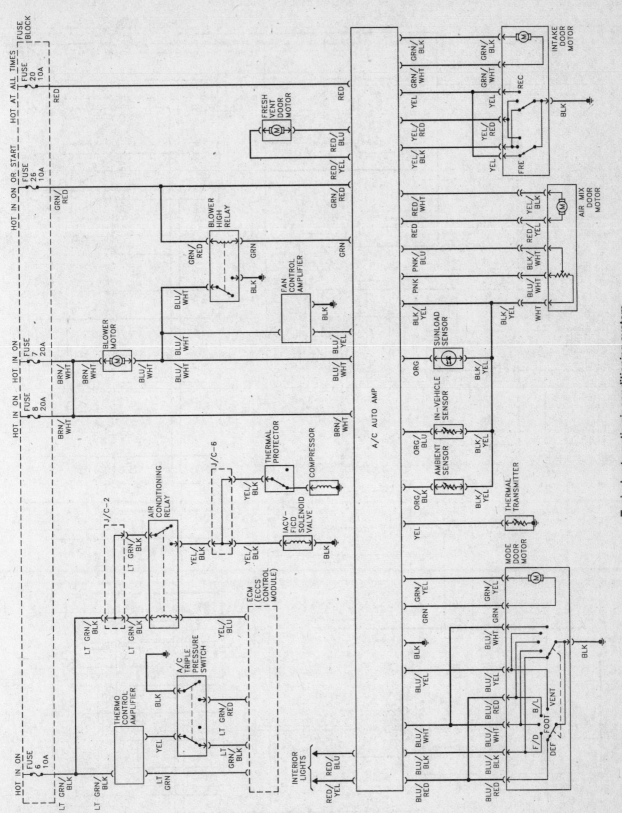

Typical automatic air conditioning system

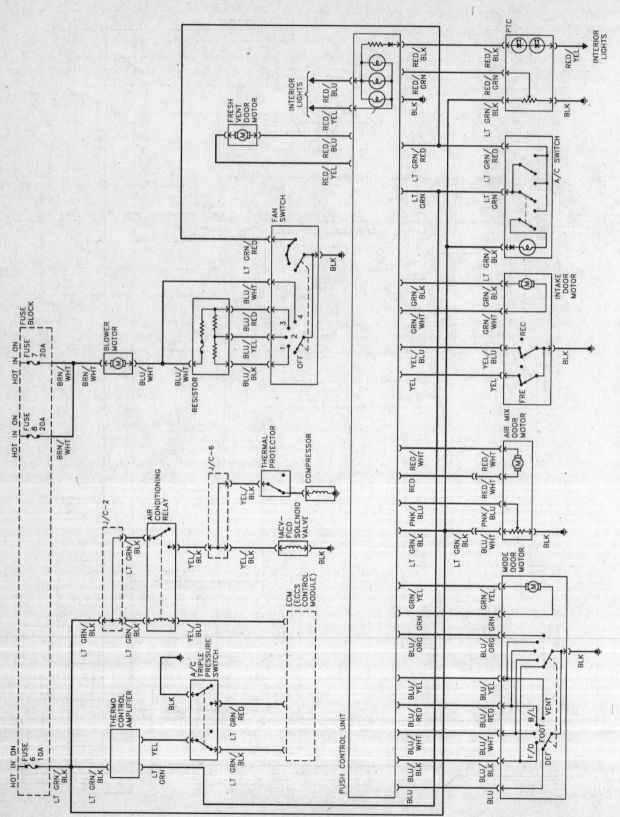

Typical manual air conditioning system

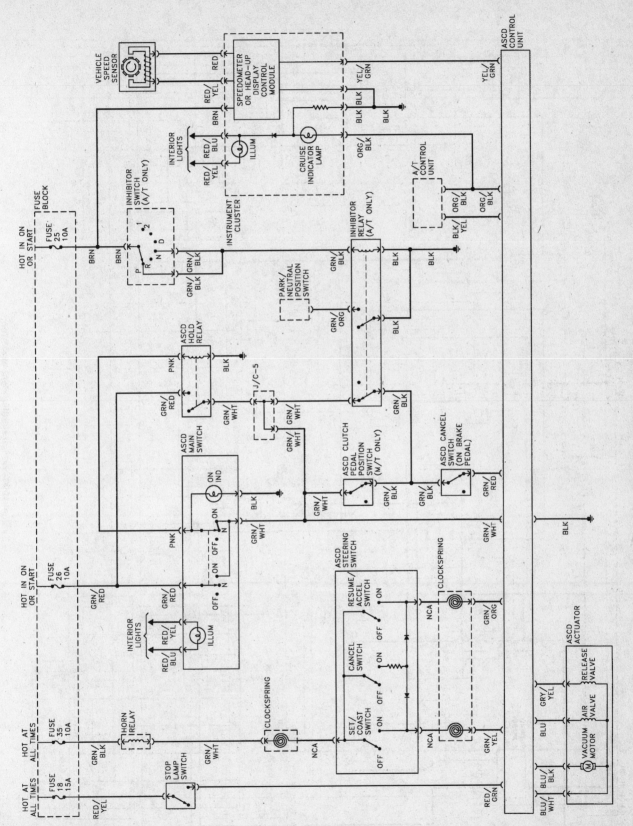

Typical 1995 and earlier cruise control system

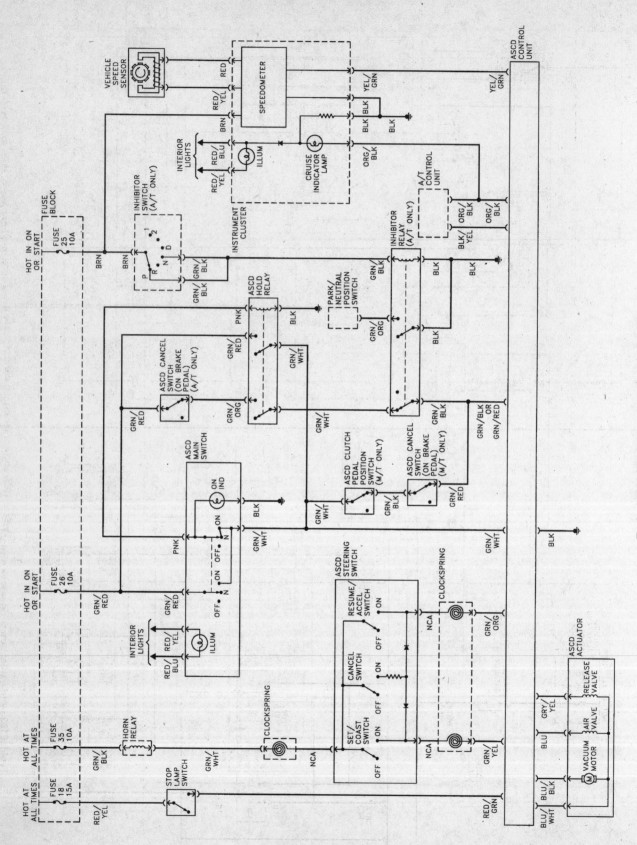

Typical 1996 and later cruise control system

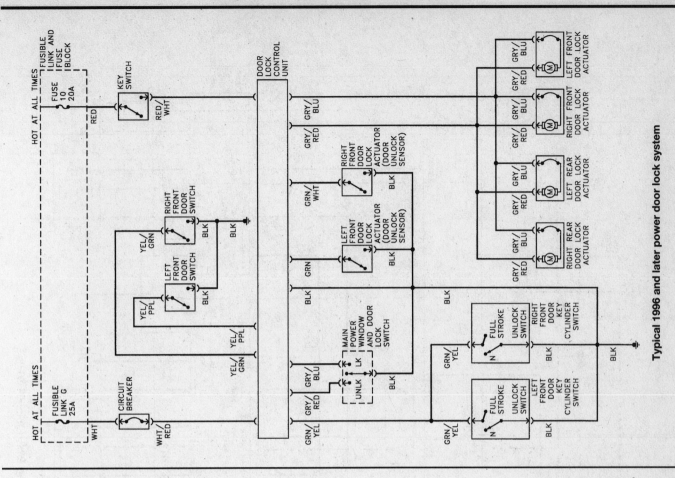

Typical 1996 and later power door lock system

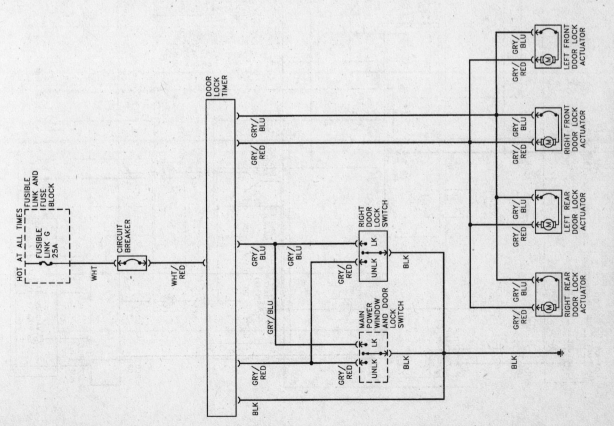

Typical 1995 and earlier power door lock system

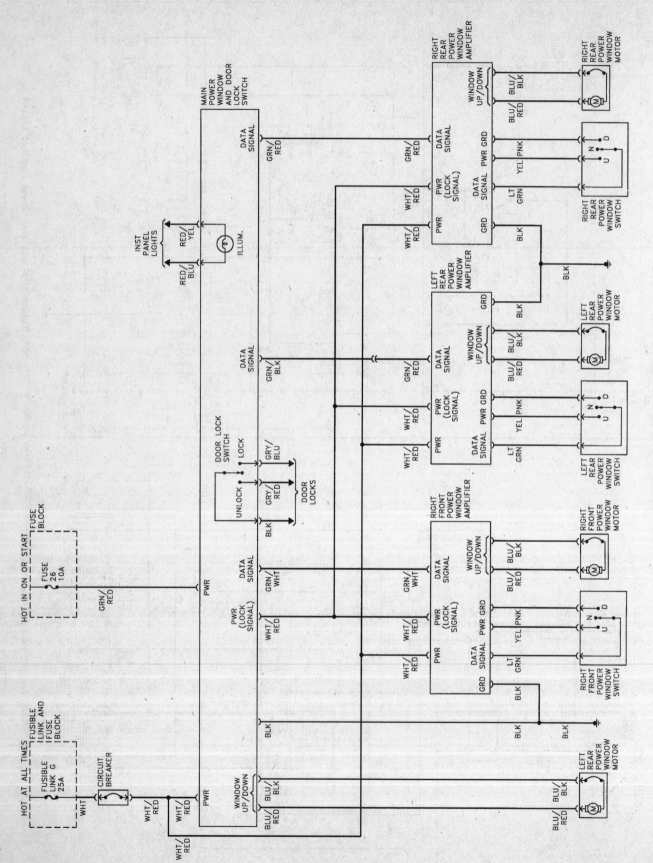

Typical 1994 and earlier power window system

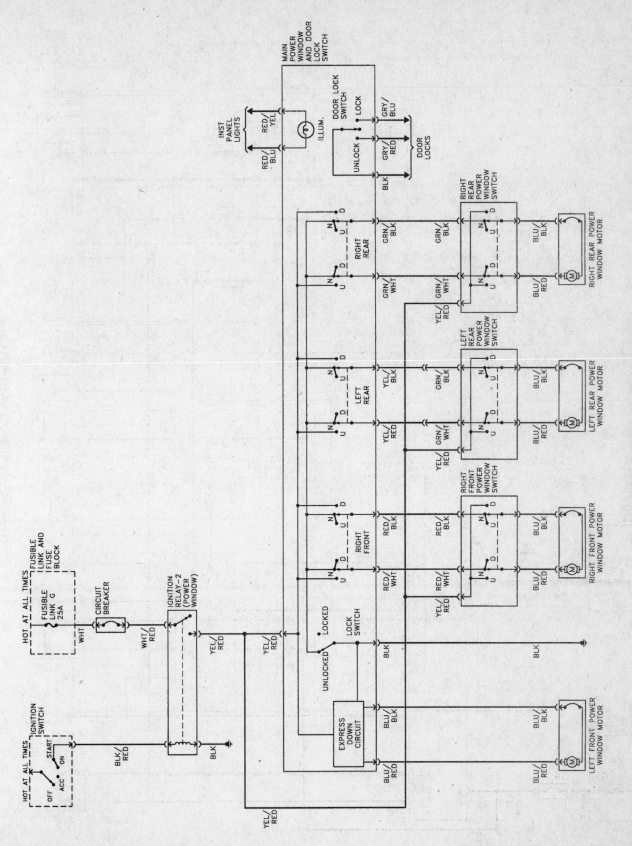

Typical 1995 and later power window system

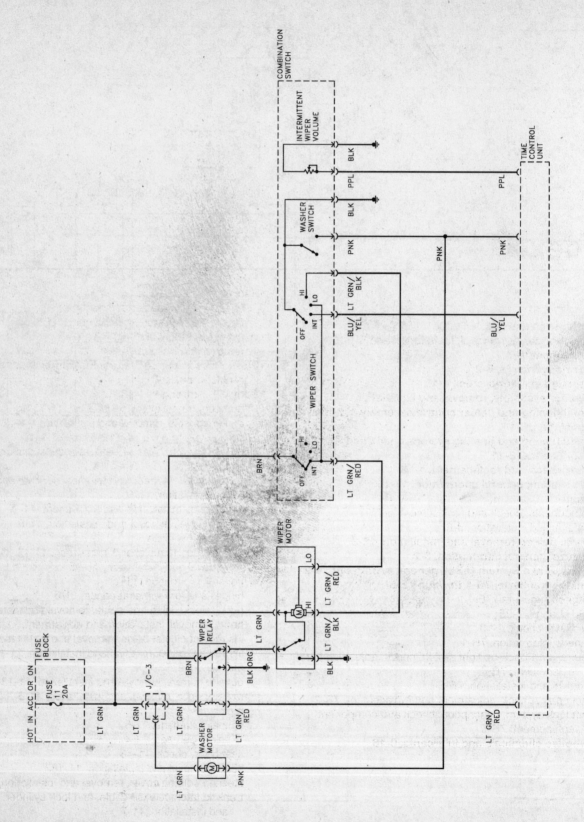

Typical windshield washer and wiper system

Index

Haynes Automotive Manuals

NOTE: New manuals are added to this list on a periodic basis. If you do not see a listing for your vehicle,
consult your local Haynes dealer for the latest product information.

ACURA
*12020 Integra '86 thru '89 & Legend '86 thru '90

AMC
Jeep CJ - see JEEP (50020)
14020 Mid-size models, Concord,
Hornet, Gremlin & Spirit '70 thru '83
14025 (Renault) Alliance & Encore '83 thru '87

AUDI
15020 4000 all models '80 thru '87
15025 5000 all models '77 thru '83
15026 5000 all models '84 thru '88

AUSTIN-HEALEY
Sprite - see MG Midget (66015)

BMW
*18020 3/5 Series not including diesel or
all-wheel drive models '82 thru '92
*18021 3 Series except 325iX models '92 thru '97
18025 320i all 4 cyl models '75 thru '83
18035 528i & 530i all models '75 thru '80
18050 1500 thru 2002 except Turbo '59 thru '77

BUICK
Century (front wheel drive) - see GM (829)
*19020 Buick, Oldsmobile & Pontiac Full-size
(Front wheel drive) all models '85 thru '98
Buick Electra, LeSabre and Park Avenue;
Oldsmobile Delta 88 Royale, Ninety Eight
and Regency; Pontiac Bonneville
19025 Buick Oldsmobile & Pontiac Full-size
(Rear wheel drive)
Buick Estate '70 thru '90, Electra '70 thru '84,
LeSabre '70 thru '85, Limited '74 thru '79
Oldsmobile Custom Cruiser '70 thru '90,
Delta 88 '70 thru '85,Ninety-eight '70 thru '84
Pontiac Bonneville '70 thru '81,
Catalina '70 thru '81, Grandville '70 thru '75,
Parisienne '83 thru '86
19030 Mid-size Regal & Century all rear-drive
models with V6, V8 and Turbo '74 thru '87
Regal - see GENERAL MOTORS (38010)
Riviera - see GENERAL MOTORS (38030)
Roadmaster - see CHEVROLET (24046)
Skyhawk - see GENERAL MOTORS (38015)
Skylark '80 thru '85 - see GM (38020)
Skylark '86 on - see GM (38025)
Somerset - see GENERAL MOTORS (38025)

CADILLAC
*21030 Cadillac Rear Wheel Drive
all gasoline models '70 thru '93
Cimarron - see GENERAL MOTORS (38015)
Eldorado - see GENERAL MOTORS (38030)
Seville '80 thru '85 - see GM (38030)

CHEVROLET
*24010 Astro & GMC Safari Mini-vans '85 thru '93
24015 Camaro V8 all models '70 thru '81
24016 Camaro all models '82 thru '92
Cavalier - see GENERAL MOTORS (38015)
Celebrity - see GENERAL MOTORS (38005)
24017 Camaro & Firebird '93 thru '97
24020 Chevelle, Malibu & El Camino '69 thru '87
24024 Chevette & Pontiac T1000 '76 thru '87
Citation - see GENERAL MOTORS (38020)
*24032 Corsica/Beretta all models '87 thru '96
24040 Corvette all V8 models '68 thru '82
*24041 Corvette all models '84 thru '96
10305 Chevrolet Engine Overhaul Manual
24045 Full-size Sedans Caprice, Impala, Biscayne,
Bel Air & Wagons '69 thru '90
24046 Impala SS & Caprice and
Buick Roadmaster '91 thru '96
Lumina - see GENERAL MOTORS (38010)

24048 Lumina & Monte Carlo '95 thru '98
Lumina APV - see GM (38035)
24050 Luv Pick-up all 2WD & 4WD '72 thru '82
*24055 Monte Carlo all models '70 thru '88
Monte Carlo '95 thru '98 - see LUMINA (24048)
24059 Nova all V8 models '69 thru '79
*24060 Nova and Geo Prizm '85 thru '92
24064 Pick-ups '67 thru '87 - Chevrolet & GMC,
all V8 & in-line 6 cyl, 2WD & 4WD '67 thru '87;
Suburbans, Blazers & Jimmys '67 thru '91
*24065 Pick-ups '88 thru '98 - Chevrolet & GMC,
all full-size pick-ups, '88 thru '98; Blazer &
Jimmy '92 thru '94; Suburban '92 thru '98;
Tahoe & Yukon '98
24070 S-10 & S-15 Pick-ups '82 thru '93,
Blazer & Jimmy '83 thru '94,
*24071 S-10 & S-15 Pick-ups '94 thru '96
Blazer & Jimmy '95 thru '96
*24075 Sprint & Geo Metro '85 thru '94
*24080 Vans - Chevrolet & GMC, V8 & in-line
6 cylinder models '68 thru '96

CHRYSLER
25015 Chrysler Cirrus, Dodge Stratus,
Plymouth Breeze '95 thru '98
25025 Chrysler Concorde, New Yorker & LHS,
Dodge Intrepid, Eagle Vision, '93 thru '97
10310 Chrysler Engine Overhaul Manual
*25020 Full-size Front-Wheel Drive '88 thru '93
K-Cars - see DODGE Aries (30008)
Laser - see DODGE Daytona (30030)
*25030 Chrysler & Plymouth Mid-size
front wheel drive '82 thru '95
Rear-wheel Drive - see Dodge (30050)

DATSUN
28005 200SX all models '80 thru '83
28007 B-210 all models '73 thru '78
28009 210 all models '79 thru '82
28012 240Z, 260Z & 280Z Coupe '70 thru '78
28014 280ZX Coupe & 2+2 '79 thru '83
300ZX - see NISSAN (72010)
28016 310 all models '78 thru '82
28018 510 & PL521 Pick-up '68 thru '73
28020 510 all models '78 thru '81
28022 620 Series Pick-up all models '73 thru '79
720 Series Pick-up - see NISSAN (72030)
28025 810/Maxima all gasoline models, '77 thru '84

DODGE
400 & 600 - see CHRYSLER (25030)
*30008 Aries & Plymouth Reliant '81 thru '89
30010 Caravan & Plymouth Voyager Mini-Vans
all models '84 thru '95
*30011 Caravan & Plymouth Voyager Mini-Vans
all models '96 thru '98
30012 Challenger/Plymouth Saporro '78 thru '83
30016 Colt & Plymouth Champ (front wheel drive)
all models '78 thru '87
*30020 Dakota Pick-ups all models '87 thru '96
30025 Dart, Demon, Plymouth Barracuda,
Duster & Valiant 6 cyl models '67 thru '76
*30030 Daytona & Chrysler Laser '84 thru '89
Intrepid - see CHRYSLER (25025)
*30034 Neon all models '95 thru '97
30035 Omni & Plymouth Horizon '78 thru '90
*30040 Pick-ups all full-size models '74 thru '93
*30041 Pick-ups all full-size models '94 thru '96
*30045 Ram 50/D50 Pick-ups & Raider and
Plymouth Arrow Pick-ups '79 thru '93
30050 Dodge/Plymouth/Chrysler rear wheel
drive '71 thru '89
*30055 Shadow & Plymouth Sundance '87 thru '94
*30060 Spirit & Plymouth Acclaim '89 thru '95
*30065 Vans - Dodge & Plymouth '71 thru '96

EAGLE
Talon - see Mitsubishi Eclipse (68030)
Vision - see CHRYSLER (25025)

FIAT
34010 124 Sport Coupe & Spider '68 thru '78
34025 X1/9 all models '74 thru '80

FORD
10355 Ford Automatic Transmission Overhaul
*36004 Aerostar Mini-vans all models '86 thru '96
*36006 Contour & Mercury Mystique '95 thru '98
36008 Courier Pick-up all models '72 thru '82
36012 Crown Victoria & Mercury Grand
Marquis '88 thru '96
10320 Ford Engine Overhaul Manual
36016 Escort/Mercury Lynx all models '81 thru '90
*36020 Escort/Mercury Tracer '91 thru '96
*36024 Explorer & Mazda Navajo '91 thru '95
36028 Fairmont & Mercury Zephyr '78 thru '83
36030 Festiva & Aspire '88 thru '97
36032 Fiesta all models '77 thru '80
36036 Ford & Mercury Full-size,
Ford LTD & Mercury Marquis ('75 thru '82);
Ford Custom 500,Country Squire, Crown
Victoria & Mercury Colony Park ('75 thru '87);
Ford LTD Crown Victoria &
Mercury Gran Marquis ('83 thru '87)
36040 Granada & Mercury Monarch '75 thru '80
36044 Ford & Mercury Mid-size,
Ford Thunderbird & Mercury
Cougar ('75 thru '82);
Ford LTD & Mercury Marquis ('83 thru '86);
Ford Torino,Gran Torino, Elite, Ranchero
pick-up, LTD II, Mercury Montego, Comet,
XR-7 & Lincoln Versailles ('75 thru '86)
36048 Mustang V8 all models '64-1/2 thru '73
36049 Mustang II 4 cyl, V6 & V8 models '74 thru '78
36050 Mustang & Mercury Capri all models
Mustang, '79 thru '93; Capri, '79 thru '86
*36051 Mustang all models '94 thru '97
36054 Pick-ups & Bronco '73 thru '79
36058 Pick-ups & Bronco '80 thru '96
36059 Pick-ups, Expedition &
Mercury Navigator '97 thru '98
36062 Pinto & Mercury Bobcat '75 thru '80
36066 Probe all models '89 thru '92
36070 Ranger/Bronco II gasoline models '83 thru '92
*36071 Ranger '93 thru '97 &
Mazda Pick-ups '94 thru '97
36074 Taurus & Mercury Sable '86 thru '95
*36075 Taurus & Mercury Sable '96 thru '98
*36078 Tempo & Mercury Topaz '84 thru '94
36082 Thunderbird/Mercury Cougar '83 thru '88
*36086 Thunderbird/Mercury Cougar '89 and '97
36090 Vans all V8 Econoline models '69 thru '91
*36094 Vans full size '92-'95
*36097 Windstar Mini-van '95-'98

GENERAL MOTORS
*10360 GM Automatic Transmission Overhaul
*38005 Buick Century, Chevrolet Celebrity,
Oldsmobile Cutlass Ciera & Pontiac 6000
all models '82 thru '96
*38010 Buick Regal, Chevrolet Lumina,
Oldsmobile Cutlass Supreme &
Pontiac Grand Prix front-wheel drive
models '88 thru '95
*38015 Buick Skyhawk, Cadillac Cimarron,
Chevrolet Cavalier, Oldsmobile Firenza &
Pontiac J-2000 & Sunbird '82 thru '94
*38016 Chevrolet Cavalier &
Pontiac Sunfire '95 thru '98
38020 Buick Skylark, Chevrolet Citation,
Olds Omega, Pontiac Phoenix '80 thru '85
38025 Buick Skylark & Somerset,
Oldsmobile Achieva & Calais and
Pontiac Grand Am all models '85 thru '95
38030 Cadillac Eldorado '71 thru '85,
Seville '80 thru '85,
Oldsmobile Toronado '71 thru '85
& Buick Riviera '79 thru '85
*38035 Chevrolet Lumina APV, Olds Silhouette
& Pontiac Trans Sport all models '90 thru '95
General Motors Full-size
Rear-wheel Drive - see BUICK (19025)

(Continued on other side)

Haynes North America, Inc., 861 Lawrence Drive, Newbury Park, CA 91320-1514 • (805) 498-6703

Haynes Automotive Manuals (continued)

NOTE: New manuals are added to this list on a periodic basis. If you do not see a listing for your vehicle, consult your local Haynes dealer for the latest product information.

GEO

Metro - *see CHEVROLET Sprint (24075)*
Prizm - *'85 thru '92 see CHEVY (24060), '93 thru '96 see TOYOTA Corolla (92036)*
*40030 **Storm** all models '90 thru '94
Tracker - *see SUZUKI Samurai (90010)*

GMC

Safari - *see CHEVROLET ASTRO (24010)*
Vans & Pick-ups - *see CHEVROLET*

HONDA

42010 **Accord CVCC** all models '76 thru '83
42011 **Accord** all models '84 thru '89
42012 **Accord** all models '90 thru '93
42013 **Accord** all models '94 thru '95
42020 **Civic 1200** all models '73 thru '79
42021 **Civic 1300 & 1500 CVCC** '80 thru '83
42022 **Civic 1500 CVCC** all models '75 thru '79
42023 **Civic** all models '84 thru '91
*42024 **Civic & del Sol** '92 thru '95
*42040 **Prelude CVCC** all models '79 thru '89

HYUNDAI

*43015 **Excel** all models '86 thru '94

ISUZU

Hombre - *see CHEVROLET S-10 (24071)*
*47017 **Rodeo** '91 thru '97; **Amigo** '89 thru '94; **Honda Passport** '95 thru '97
*47020 **Trooper & Pick-up**, all gasoline models Pick-up, '81 thru '93; Trooper, '84 thru '91

JAGUAR

*49010 **XJ6** all 6 cyl models '68 thru '86
*49011 **XJ6** all models '88 thru '94
*49015 **XJ12 & XJS** all 12 cyl models '72 thru '85

JEEP

*50010 **Cherokee, Comanche & Wagoneer Limited** all models '84 thru '96
50020 **CJ** all models '49 thru '86
*50025 **Grand Cherokee** all models '93 thru '98
50029 **Grand Wagoneer & Pick-up** '72 thru '91 Grand Wagoneer '84 thru '91, Cherokee & Wagoneer '72 thru '83, Pick-up '72 thru '88
*50030 **Wrangler** all models '87 thru '95

LINCOLN

Navigator - *see FORD Pick-up (36059)*
59010 **Rear Wheel Drive** all models '70 thru '96

MAZDA

61010 **GLC Hatchback** (rear wheel drive) '77 thru '83
61011 **GLC** (front wheel drive) '81 thru '85
*61015 **323 & Protegé** '90 thru '97
*61016 **MX-5 Miata** '90 thru '97
*61020 **MPV** all models '89 thru '94
Navajo - *see Ford Explorer (36024)*
61030 **Pick-ups** '72 thru '93
Pick-ups '94 thru '96 - *see Ford Ranger (36071)*
61035 **RX-7** all models '79 thru '85
*61036 **RX-7** all models '86 thru '91
61040 **626** (rear wheel drive) all models '79 thru '82
*61041 **626/MX-6** (front wheel drive) '83 thru '91

MERCEDES-BENZ

63012 **123 Series** Diesel '76 thru '85
*63015 **190 Series** four-cyl gas models, '84 thru '88
63020 **230/250/280** 6 cyl sohc models '68 thru '72
63025 **280 123 Series** gasoline models '77 thru '81
63030 **350 & 450** all models '71 thru '80

MERCURY

See FORD Listing.

MG

66010 **MGB** Roadster & GT Coupe '62 thru '80
66015 **MG Midget, Austin Healey Sprite** '58 thru '80

MITSUBISHI

*68020 **Cordia, Tredia, Galant, Precis & Mirage** '83 thru '93
*68030 **Eclipse, Eagle Talon & Ply. Laser** '90 thru '94
*68040 **Pick-up** '83 thru '96 & **Montero** '83 thru '93

NISSAN

72010 **300ZX** all models including Turbo '84 thru '89
*72015 **Altima** all models '93 thru '97
*72020 **Maxima** all models '85 thru '91
*72030 **Pick-ups** '80 thru '96 **Pathfinder** '87 thru '95
72040 **Pulsar** all models '83 thru '86
*72050 **Sentra** all models '82 thru '94
*72051 **Sentra & 200SX** all models '95 thru '98
*72060 **Stanza** all models '82 thru '90

OLDSMOBILE

*73015 **Cutlass** V6 & V8 gas models '74 thru '88

For other OLDSMOBILE titles, see BUICK, CHEVROLET or GENERAL MOTORS listing.

PLYMOUTH

For PLYMOUTH titles, see DODGE listing.

PONTIAC

79008 **Fiero** all models '84 thru '88
79018 **Firebird** V8 models except Turbo '70 thru '81
79019 **Firebird** all models '82 thru '92

For other PONTIAC titles, see BUICK, CHEVROLET or GENERAL MOTORS listing.

PORSCHE

*80020 **911** except Turbo & Carrera 4 '65 thru '89
80025 **914** all 4 cyl models '69 thru '76
80030 **924** all models including Turbo '76 thru '82
*80035 **944** all models including Turbo '83 thru '89

RENAULT

Alliance & Encore - *see AMC (14020)*

SAAB

*84010 **900** all models including Turbo '79 thru '88

SATURN

87010 **Saturn** all models '91 thru '96

SUBARU

89002 **1100, 1300, 1400 & 1600** '71 thru '79
*89003 **1600 & 1800** 2WD & 4WD '80 thru '94

SUZUKI

*90010 **Samurai/Sidekick & Geo Tracker** '86 thru '96

TOYOTA

92005 **Camry** all models '83 thru '91
92006 **Camry** all models '92 thru '96
92015 **Celica Rear Wheel Drive** '71 thru '85
*92020 **Celica Front Wheel Drive** '86 thru '93
92025 **Celica Supra** all models '79 thru '92
92030 **Corolla** all models '75 thru '79
92032 **Corolla** all rear wheel drive models '80 thru '87
92035 **Corolla** all front wheel drive models '84 thru '92
*92036 **Corolla & Geo Prizm** '93 thru '97
92040 **Corolla Tercel** all models '80 thru '82
92045 **Corona** all models '74 thru '82
92050 **Cressida** all models '78 thru '82
92055 **Land Cruiser** FJ40, 43, 45, 55 '68 thru '82
92056 **Land Cruiser** FJ60, 62, 80, FZJ80 '80 thru '96
*92065 **MR2** all models '85 thru '87
92070 **Pick-up** all models '69 thru '78
*92075 **Pick-up** all models '79 thru '95
*92076 **Tacoma** '95 thru '98, **4Runner** '96 thru '98, & **T100** '93 thru '98
*92080 **Previa** all models '91 thru '95
92085 **Tercel** all models '87 thru '94

TRIUMPH

94007 **Spitfire** all models '62 thru '81
94010 **TR7** all models '75 thru '81

VW

96008 **Beetle & Karmann Ghia** '54 thru '79
96012 **Dasher** all gasoline models '74 thru '81
*96016 **Rabbit, Jetta, Scirocco, & Pick-up** gas models '74 thru '91 & Convertible '80 thru '92
96017 **Golf & Jetta** all models '93 thru '97
96020 **Rabbit, Jetta & Pick-up** diesel '77 thru '84
96030 **Transporter 1600** all models '68 thru '79
96035 **Transporter 1700, 1800 & 2000** '72 thru '79
96040 **Type 3 1500 & 1600** all models '63 thru '73
96045 **Vanagon** all air-cooled models '80 thru '83

VOLVO

97010 **120, 130 Series & 1800 Sports** '61 thru '73
97015 **140 Series** all models '66 thru '74
*97020 **240 Series** all models '76 thru '93
97025 **260 Series** all models '75 thru '82
*97040 **740 & 760 Series** all models '82 thru '88

TECHBOOK MANUALS

10205 **Automotive Computer Codes**
10210 **Automotive Emissions Control Manual**
10215 **Fuel Injection Manual, 1978 thru 1985**
10220 **Fuel Injection Manual, 1986 thru 1996**
10225 **Holley Carburetor Manual**
10230 **Rochester Carburetor Manual**
10240 **Weber/Zenith/Stromberg/SU Carburetors**
10305 **Chevrolet Engine Overhaul Manual**
10310 **Chrysler Engine Overhaul Manual**
10320 **Ford Engine Overhaul Manual**
10330 **GM and Ford Diesel Engine Repair Manual**
10340 **Small Engine Repair Manual**
10345 **Suspension, Steering & Driveline Manual**
10355 **Ford Automatic Transmission Overhaul**
10360 **GM Automatic Transmission Overhaul**
10405 **Automotive Body Repair & Painting**
10410 **Automotive Brake Manual**
10415 **Automotive Detaiing Manual**
10420 **Automotive Eelectrical Manual**
10425 **Automotive Heating & Air Conditioning**
10430 **Automotive Reference Manual & Dictionary**
10435 **Automotive Tools Manual**
10440 **Used Car Buying Guide**
10445 **Welding Manual**
10450 **ATV Basics**

SPANISH MANUALS

98903 **Reparación de Carrocería & Pintura**
98905 **Códigos Automotrices de la Computadora**
98910 **Frenos Automotriz**
98915 **Inyección de Combustible 1986 al 1994**
99040 **Chevrolet & GMC Camionetas** '67 al '87 Incluye Suburban, Blazer & Jimmy '67 al '91
99041 **Chevrolet & GMC Camionetas** '88 al '95 Incluye Suburban '92 al '95, Blazer & Jimmy '92 al '94, Tahoe y Yukon '95
99042 **Chevrolet & GMC Camionetas Cerradas** '68 al '95
99055 **Dodge Caravan & Plymouth Voyager** '84 al '95
99075 **Ford Camionetas y Bronco** '80 al '94
99077 **Ford Camionetas Cerradas** '69 al '91
99083 **Ford Modelos de Tamaño Grande** '75 al '87
99088 **Ford Modelos de Tamaño Mediano** '75 al '86
99091 **Ford Taurus & Mercury Sable** '86 al '95
99095 **GM Modelos de Tamaño Grande** '70 al '90
99100 **GM Modelos de Tamaño Mediano** '70 al '88
99110 **Nissan Camionetas** '80 al '96, Pathfinder '87 al '95
99118 **Nissan Sentra** '82 al '94
99125 **Toyota Camionetas y 4Runner** '79 al '95

Over 100 Haynes motorcycle manuals also available

5-98

---

* Listings shown with an asterisk (*) indicate model coverage as of this printing. These titles will be periodically updated to include later model years - consult your Haynes dealer for more information.

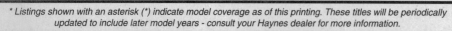

Haynes North America, Inc., 861 Lawrence Drive, Newbury Park, CA 91320-1514 • (805) 498-6703